Boston Studies in the Philosophy and History of Science

Volume 347

The series *Boston Studies in the Philosophy and History of Science* was conceived in the broadest framework of interdisciplinary and international concerns. Natural scientists, mathematicians, social scientists and philosophers have contributed to the series, as have historians and sociologists of science, linguists, psychologists, physicians, and literary critics.

The series has been able to include works by authors from many other countries around the world.

The editors believe that the history and philosophy of science should itself be scientific, self-consciously critical, humane as well as rational, sceptical and undogmatic while also receptive to discussion of first principles. One of the aims of Boston Studies, therefore, is to develop collaboration among scientists, historians and philosophers.

Boston Studies in the Philosophy and History of Science looks into and reflects on interactions between epistemological and historical dimensions in an effort to understand the scientific enterprise from every viewpoint.

Jonathan Y. Tsou • Jamie Shaw • Carla Fehr
Editors

Values, Pluralism, and Pragmatism: Themes from the Work of Matthew J. Brown

Springer

Editors
Jonathan Y. Tsou
Center for Values in Medicine, Science, and Technology
The University of Texas at Dallas
Richardson, TX, USA

Jamie Shaw
Institute of Philosophy
Leibniz University Hannover
Hannover, Germany

Carla Fehr
Department of Philosophy
University of Waterloo
Waterloo, ON, Canada

ISSN 0068-0346 ISSN 2214-7942 (electronic)
Boston Studies in the Philosophy and History of Science
ISBN 978-3-031-92957-1 ISBN 978-3-031-92958-8 (eBook)
https://doi.org/10.1007/978-3-031-92958-8

This Springer imprint is published by the registered company Springer Nature Switzerland AG
The registered company address is: Gewerbestrasse 11, 6330 Cham, Switzerland

To Kathleen Okruhlik (1951–2024): Mentor, teacher, and friend

Introduction: Values, Pluralism, and Pragmatism

This book offers 18 original historical and philosophical essays focused on the topics of values in science, scientific pluralism, and pragmatism. These themes have been central in the work of Matthew J. Brown, and the contributions in this book address these topics often through an engagement with Brown's broad-ranging research on values in science. The themes of this book are integrated and unified in the pragmatic and value-laden ideal of science defended by Brown in his fascinating 2020 book, *Science and Moral Imagination* (Brown 2020). Brown's ideal of moral imagination prescribes that scientists should recognize the contingencies in their work as unforced choices, examine morally salient aspects of these decisions, identify the various interests of relevant stakeholders, explore and construct alternative options, and exercise fair and warranted value judgments to guide those decisions. The interdisciplinary essays in this volume engage with different aspects of Brown's philosophical research on scientific values as well as his historical research on figures such as John Dewey and Paul K. Feyerabend.

The essays in this book include contributions by distinguished and familiar voices from the values in science literature (e.g., Kevin Elliott, Janet Kourany, and Don Howard) as well as newer voices, who articulate and defend novel frameworks and perspectives. The resulting book showcases some of the latest innovative research being done on values, pluralism, and pragmatism, and it will be an important resource for researchers working on values in science and related topics. Specific topics addressed in the book include: moral imagination (Chaps. 10–13), inductive risk (Chaps. 15–17), epistemic priority (Chap. 17), the value-free ideal of science (Chaps. 11, 14–16), values in the contexts of artificial intelligence and medicine (Chaps. 7–9), varieties of pluralism (e.g., epistemic, ethical, and cultural), and the structure of scientific communities (Chaps. 5 and 6). The book also features new historical research on important figures in the traditions of pragmatism, pluralism, and values in science, such as John Dewey, John Stuart Mill, Paul Feyerabend, W. V. Quine, Ludwig Fleck, among others (Chaps. 1–6).

The essays in this book represent a broad range of contemporary methodological perspectives and institutions in philosophy of science, which have influenced Brown's own work. These methodological perspectives include socially relevant philosophy of science (SRPoiSE), history of philosophy of science (HOPOS), feminist philosophy of science, integrated history and philosophy of science (&HPS), and philosophy of science in practice (SPSP). The topics addressed in this book will be of interest to a broad audience of researchers working in history and philosophy of science, philosophy of science, feminist philosophy, history of philosophy of science, philosophy of medicine, and science and technology studies.

Values

Discussions of values in science have been a perennial topic for centuries. In Paul Franco's essay (Chap. 14), Franco carefully traces how many of these discussions evolved in philosophy of science in the early to mid-twentieth century. Considering figures such as C.L. Stevenson, Herbert Feigl, Carl Hempel, and Ernest Nagel's positions, Franco considers arguments for and against the view that values have cognitive content. He further contends that debates about values and science, at least historically, crucially depend on meta-ethical positions. This makes it challenging to defend or attack a position on values in science without taking a meta-ethical stance. More recent discussions on this topic have been framed by Heather Douglas' influential criticism (Douglas 2009) of the "Value-Free Ideal" (VFI). Douglas criticizes the view that science is or should be free of "non-epistemic" values (e.g., moral or political values). Large swaths of literature have ensued since Douglas' pioneering work. While some have sought to defend the VFI (e.g., see Betz 2013; Bright 2018), the larger portion of contributors to these discussions have criticized it from numerous angles. The most influential arguments against the VFI have been the "gap argument" based on underdetermination (Longino 1990), the argument from inductive risk (Douglas 2000), and the argument from thick concepts (Dupré 2007; Alexandrova 2017). Many philosophers take some of these arguments as providing decisive reasons for rejecting the VFI, arguing that we should move on to formulating positive ideals about how values and science should properly interact. Brown's *Science and Moral Imagination* (Brown 2020) is exemplary in this regard.

Section IV contains three contributions that critically engage with Brown's book. P. D. Magnus' contribution (Chap. 12) questions whether Brown's framework can accommodate deeply personal ("existential") choices that scientists make, such as changing career directions. These are decisions that impact others, and so they should fall within the scope of Brown's theory of moral imagination. However, Magnus argues that the requirement of moral imagination to do right by a broad contingency of stakeholders overlooks the intuitive justification that some decisions are deeply private and made by individuals as parts of their self-creation. Magnus further explores how this quandary ties into questions about science, existential freedom, and values in the pragmatist tradition (viz., James, Dewey, and Putnam). Kourany's essay (Chap. 11) considers what kind of eulogy Brown would deliver at the funeral of the VFI. If the VFI is dead, according to philosophers, what should step in and take its place? The VFI—false or not—played an important functional role in mediating between science and society. It remains unclear, Kourany argues, whether following Brown's framework can preserve the *beneficial roles* that the VFI has historically served. Manuela Fernandez-Pinto's essay (Chap. 13) is also critical, arguing that it remains unclear how much the theory of moral imagination can really improve matters in cases of commercial science, which are becoming increasingly prominent. She contends that the ideals of considering stakeholders inclusively and acting accordingly might only be exploited by for-profit science entities, thus creating serious limitations for its practical plausibility.

This volume further contains several essays that are sympathetic to the value-laden dimensions of scientific practice. Section III contains numerous case studies in this regard, with a repeated engagement on issues related to artificial intelligence (AI). Brett Bolander's essay (Chap. 8) engages with the cognitive metaphors frequently used in discourses on machine learning and AI. This essay explores uses and limitations of these metaphors, and it considers multiple frameworks for approaching responsible deployments of such metaphors with regard to existing ethical principles frequently used in AI-related regulatory policies. Emily LaRosa's essay (Chap. 9) focuses on uses of AI in healthcare settings in particular. This analysis reveals the necessity to properly understand what "trust" amounts to and whether AI can be trusted or relied upon, insofar as it is used to replace agents who engage in distinctive fiduciary relationships.

While most, if not all, of the contributors of this volume acknowledge that science and values intersect in interesting and important ways, there are some who reject how this issue is framed in current debates about the VFI. Greg Lusk puts this point bluntly in his contribution (Chap. 15): we should simply stop talking about the VFI. The reason is not because the debate has been settled, but because the VFI on its own is too indeterminate to help us decide what to do. Because it is not sufficiently clear for deriving action-guiding principles, it is difficult to really assess its merits. Rather, Lusk contends that there are other sources of normative ideals—such as from theories of democracy—that carry more potential in this regard. Similarly, Kevin Elliott (Chap. 16) argues that the VFI debate has been framed in an overly rigid fashion. There is neither a coherent "VFI" nor a coherent denial of the VFI that can lay claim to being especially significant. Rather, Elliott investigates the ambiguities present in the terms "values," "freedom," and "ideal" to show that there are many VFIs and, concomitantly, many value-laden ideals that are of greater specificity and can be considered independently from one another. *Pace* Brown (2024), we should shift from "*the* VFI debate" to a variety of more local debates about specific instances of science and values. Kareem Khalifa, Jared Millson, and Mark Risjord (Chap. 17), after clarifying the notion of "epistemic priority" that has been debated by Daniel Steel and Matt Brown, argue for an alternative approach they dub "inquisitive due diligence." Illustrating this position with the case of social scientific research on segregation, they argue that one can accept a hypothesis if all of the relevant pointed questions (i.e., questions whose answers might challenge the hypothesis) have been asked, correctly answered, and do not defeat the hypothesis. While values may play an important role in determining the relevance of some pointed questions, inquisitive due diligence offers a "basic epistemic standard" to hold hypothesis acceptance to. This argument can be interpreted as reconceiving the relationship(s) between epistemic and non-epistemic influences in scientific inquiry challenging the conventional framing implied by the VFI debate.

Pluralism

The idea that science should be pluralistic in some respect has become a truism for contemporary philosophy of science. This was a hard-won philosophical achievement, overcoming powerful and influential arguments attempting to formulate unified methodological or theoretical ideals for scientific inquiry. However, the nature and dynamics of pluralism remain open topics. Many contributions contain applications and affirmations of pluralism. Carla Fehr (Chap. 7) investigates this issue in the cases of probability of errors with facial recognition technologies. She argues that benchmarks in facial recognition over-represented individuals with light skin types, leading to a variety of harmful errors when detecting others. Drawing on Helen Longino's (1990) "contextual constructive empiricism" and Joy Buolamwini's research on race and gender bias in facial recognition systems, Fehr argues that the errors and biases of facial recognition technology reflect insufficient diversity and mechanisms for ensuring diversity. Bennett Knox (Chap. 10) articulates and defends "hermeneutic pluralism" (i.e., a methodological ideal that demands a plurality of incompatible interpretations of psychiatric phenomena) in the context of controversial Spectrum 10K study on the genetics of autism. He argues that engagements with the autistic community provide valuable perspective to the narrower perspective embedded in the Spectrum project. Both Fehr's and Knox's essays make use of the notion that social situatedness and epistemic diversity come together when science is conducted in a socially responsible manner.

Other essays further discussions of pluralism by looking back into history. Kathleen Okruhlik's contribution (Chap. 1) provides a *tour de force* analysis of the influence of John Stuart Mill's political philosophy on the twentieth-century philosophy of science. Okruhlik revisits Mill's understanding of the relationships between science, values, democracy, and truth. She argues that the diverse (and conflicting) uptakes of Mill display varying degrees of faithfulness to Mill's original intentions. Okruhlik's historical analysis is motivated by the judgment that Mill "got some important things right." Her resulting analysis provides an authoritative exegesis of Milll's pluralism, which deepens our understanding and appreciation of Mill's positions. One of the self-identified Millians that Okruhlik discusses, Paul Feyerabend, is focused on in more detail in Jamie Shaw and Jonathan Tsou's essays. Shaw's essay (Chap. 2) concurs with Okruhlik's judgment that Feyerabend misunderstood Mill's position in important ways, which reveals internal conflicts in Feyerabend's commitments to pluralism. Shaw focuses on Mill's and Feyerabend's positions on the ethical limits of pluralism. He argues that Feyerabend was less clear than Mill on this issue, leaving Feyerabend's position ambiguous. He further seeks possible supplementations for Feyerabend's uptake of Mill's ethical pluralism, suggesting resources can be found in existentialist philosophy. Tsou's essay (Chap. 3) explores Feyerabend's changing positions on pluralism by arguing that a key shift in Feyerabend's philosophy is not from realism to relativism, as many Feyerabend scholars have asserted, but from defending pluralism in science to a broader pluralism in society. This shift, which Tsou argues appears in Feyerabend's publications in the late 1960s, owes a great deal to Feyerabend reading Mill and incorporating Mill's arguments for pluralism into his own arguments for "anarchism." While Okruhlik chides Feyerabend's analysis, which was written during Feyerabend's "relativist" period (c. 1975–1978), for misunderstanding Mill's realism about truth, Tsou maintains that Feyerabend endorsed a methodological realist position ("pluralistic realism") during this period.

Pragmatism

While pragmatism is a recurring theme in many of the contributions of this volume, two essays investigate this topic historically. Don Howard's essay (Chap. 4) tackles the thorny question of whether W. V. Quine was a pragmatist and, if so, what kind of a pragmatism he endorsed. Howard argues that Quine can be read as Dewey's successor for the bulk of the second half of the twentieth century, making him deeply pragmatist in several respects. Reconstructing Quine's views on a variety of topics, including naturalism, holism, ontological commitment, behavioral semantics, truth, realism, and foundationalism demonstrates, according to Howard, how deeply Quine's commitment to pragmatism runs. Paul Howatt's contribution (Chap. 5) focuses specifically on Dewey. Specifically, he explores Dewey's conception of democratic inquiry that formed a crucial pillar of his vision of a "Great Community." Reconstructing Dewey's rethinking of what it is to be a citizen in a democratic society, Howatt engages Dewey's attempt to close the gap between average citizens and experts. This provides a way for thinking about democratically legitimate theories of scientific inquiry that addresses the concerns raised about public ignorance famously enunciated by Walter Lippmann. This problem addressed in Dewey's social approach to inquiry is also addressed in Victoria Wang's contribution (Chap. 6), which aims to revitalize Ludwik Fleck's views on science communication. Fleck's account of "thinking collectives," or communities who talk to each other in intellectual ways, Wang suggests, provides a valuable way for theorizing the "expert-lay" divide that often plagues democratic accounts of inquiry like Dewey's. Maintaining that these boundaries are and should be porous, providing crucial spaces for science journalists or K-12 educators to maintain or expand thinking collectives, allows one to acknowledge epistemic asymmetries between different members of society while still seeing them as co-contributors in inquiry. While not explicitly engaging with pragmatist thinkers, Wang's Fleckian approach takes seriously many pragmatic principles, such as the view that inquiry is a collective practice, best pursued by diversely situated people aiming to solve specific problems.

Acknowledgments

Many of the essays of this book originated from the 2023 Values in Medicine, Science, and Technology Conference (VMST-11), which was hosted by the Center for Values in Medicine, Science, and Technology (CVMST) at the University of Texas at Dallas in May 2023. The VMST-11 conference was a unique VMST conference that celebrated Matthew Brown's contributions to the CVMST as its outgoing director (2011–2022).

The CVMST was established in 2009 by Dennis M. Kratz, who was then Dean of the School of Arts and Humanities at the University of Texas at Dallas. In 2009, Magdalena G. Grohmann was appointed as the Associate Director of the CVMST (2009-2022). In 2011, Matthew J. Brown was appointed as the first Director of the CVMST (2011–2021). Over the next decade, Matt and Magda established an international reputation for the CVMST as a top-rank interdisciplinary venue for research in history and philosophy of science and science and technology studies, especially in the fields of philosophy of science and philosophy of medicine. The signature event of the CVMST is its annual VMST conference. Under the stewardship of Matt and Magda, the first VMST conference ("Human Enhancement Symposium") was held in 2011 and the tenth VMST conference was held in 2020. In 2022, Dr. Brown and Dr. Grohmann left their positions at the University of Texas at Dallas to pursue new opportunities. Dr. Brown was hired at Southern Illinois University as Professor of Philosophy, the Jo Ann and Donald N. Boydston Chair of American Philosophy, and the Director of the Center for Dewey Studies. Dr. Grohmann was hired at the University of North Texas as Clinical Associate Professor in Design. In 2022, Jonathan Tsou was hired at the University of Texas at Dallas as Professor of Philosophy and the Marvin and Kathleen Stone Distinguished Professor of Humanities in Medicine and Science. In light of Matt and Magda's departures, Dr. Tsou was appointed as Director of the CVMST in 2023. The present volume is intended as a show of gratitude to Matt and Magda for all of their efforts and successes in establishing the CVMST as an internationally recognized center for research in history and philosophy of science.

Jonathan Tsou is grateful to many people who helped organize the 2023 VMST-11 conference. Thanks to members of the VMST-11 program committee: Carla Fehr, Magda Grohman, Kareem Khalifa, Joyce Havstad, and Jamie Shaw. Thanks to the Bass School in Arts, Humanities, and Technology (BAHT) at the University of Texas at Dallas for financial support. Special thanks to Pale Almissouri, Magda Grohman, Matt Brown, Holly Miori, Lisa Lyles, and Nils Roemer for their invaluable guidance and assistance.

In addition to chapters that originated from the 2023 VMST-11 conference, several chapters were solicited by the editors. Janet Kourany's and Manuela Fernandez-Pinto's chapters were originally presented as part of an author-meets-critics symposium focused on Brown's *Science and Moral Imagination*, which was sponsored by the Philosophy of Science Association and held at the American Philosophical Association's Eastern Division meeting in New York City in January

2024. Kathleen Okruhlik's contribution was composed over a decade prior to this volume and was shared with Jamie Shaw during her supervision of his dissertation which engaged with similar themes. A variant of this paper, with a greater focus on Feyerabend's use of Mill, was presented at the conference entitled "Feyerabend 2018: Commemorating the 40th Anniversary of Science in a Free Society" held in conjunction with VMST-8 and organized by Matt Brown and Ian James Kidd. We are grateful to Kathleen's husband, James Brown, for his aid in securing the rights for Kathleen's essay after her passing in 2024.

Jonathan Y. Tsou
Jamie Shaw
Carla Fehr

Bibliography

Alexandrova, Anna 2017. *A Philosophy for the Science of Well-Being*. New York: Oxford University Press.

Betz, Gregor. 2013. In Defence of the Value Free Ideal. *European Journal for Philosophy of Science* 3:207–220. https://doi.org/10.1007/s13194-012-0062-x

Bright, Liam Kofi. 2018. Du Bois' Democratic Defence of the Value Free Ideal. *Synthese* 195:2227–2245. https://doi.org/10.1007/s11229-017-1333-z

Brown, Matthew J. 2020. *Science and Moral Imagination: A New Ideal for Science and Values*. Pittsburgh: University of Pittsburgh Press.

Brown, Matthew J. 2024. For Values in Science: Assessing Recent Arguments for the Ideal of Value-Free Science. *Synthese* 204:112. https://doi-org.libproxy.utdallas.edu/10.1007/s11229-024-04762-1

Douglas, Heather 2000. Inductive Risk and Values in Science. *Philosophy of Science* 67 (4): 559–579.

Douglas, Heather E. 2009. *Science, Policy, and the Value-Free Ideal*. Pittsburgh: University of Pittsburgh Press.

Dupré, John 2007. Fact and Value. In *Value-Free Science? Ideals and Illusions*, ed. Harold Kincaid, John Dupré, and Alison Wylie, 21–41. Oxford: Oxford University Press.

Longino, Helen. 1990. *Science as Social Knowledge: Values and Objectivity in Scientific Inquiry*. Princeton: Princeton University Press.

Contents

Contributors

Brett Bolander is a PhD Candidate in the Department of Philosophy at the University of Pennsylvania.

Matthew J. Brown is a Professor of Philosophy, the Jo Ann and Donald N. Boydston Chair of American Philosophy, and the Director of the Center for Dewey Studies at Southern Illinois University.

Kevin C. Elliott is a Red Cedar Distinguished Professor of Philosophy at Michigan State University, with joint appointments in Lyman Briggs College, the Department of Fisheries and Wildlife, and the Department of Philosophy.

Carla Fehr is a Professor of Philosophy and the Wolfe Chair in Scientific and Technological Literacy at the University of Waterloo.

Paul L. Franco is an Associate Teaching Professor in the Department of Philosophy at the University of Washington.

Don Howard is a Professor in the Department of Philosophy and the former Director of the John J. Reilly Center for Science, Technology, and Values at the University of Notre Dame.

Paul Howatt is a PhD Candidate in the Department of Philosophy at Indiana University, Bloomington.

Kareem Khalifa is a Professor in the Department of Philosophy at the University of California, Los Angeles.

Bennett Knox is a PhD Candidate in the Department of Philosophy at the University of Utah.

Janet A. Kourany is a Professor in the Department of Philosophy and Gender Studies Program at the University of Notre Dame.

Emily LaRosa is a PhD Candidate in the Department of Philosophy at Michigan State University.

Greg Lusk is an Assistant Professor at the Department of Philosophy at Durham University.

P. D. Magnus is a Professor of Philosophy at the University at Albany, State University of New York.

Jared Millson is an Assistant Professor in the Department of Philosophy at Rhodes College.

Kathleen Okruhlik was a Professor Emerita in the Department of Philosophy at the University of Western Ontario.

Manuela Fernández Pinto is an Associate Professor in the Philosophy Department and Center of Applied Ethics at the Universidad de los Andes.

Mark Risjord is a Professor in the Department of Philosophy and the Director for Institute for the Liberal Arts at Emory University.

Jamie Shaw is a Postdoctoral Fellow at the Institute for Philosophy at Leibniz University Hannover.

Jonathan Y. Tsou is a Professor of Philosophy, the Marvin and Kathleen Stone Distinguished Professor of Humanities in Medicine and Science, and the Director of the Center for Values in Medicine, Science, and Technology in the Bass School of Arts, Humanities, and Technology at the University of Texas at Dallas.

Victoria Min-Yi Wang is a PhD Candidate in the School of Philosophy at the University of Edinburgh.

Part I
Mill, Feyerabend, and Pluralism

Chapter 1
John Stuart Mill, the Marketplace, and the Democratization of Science

Kathleen Okruhlik

Abstract The purpose of this chapter is to demonstrate that John Stuart Mill's chief influence on twenith and twenty-first century philosophy of science was through *On Liberty*. This includes discussions about uptakes of Mill's ideas in Feyerabend's relativist and his notion of a 'Mutual Killing Society', Kitcher's 'well-ordered science', and Longino's contextual empiricism. After clarifying that Mill would not have endorsed the ideology underpinning the 'marketplace of ideas' metaphor often used to describe Mill's views on democracy, science, and truth, it is shown the various uptakes and misunderstandings of Mill's political epistemology in late twentieth century and early twenty-first century philosophy of science. Here, it is shown that Feyerabend originally wrongfully interprets Mill as someone who endorse a 'Mutual Killing Society', demonstrating a misunderstanding of Mill's understandings of utility and freedom. Eventually Feyerabend comes to recognize their differences, where Mill was a realist about truth while Feyerabend was a relativist. However, Feyerabend did enact Mill's call to proliferate marginal viewpoints with gusto. It is finally shown that Longino has enacted Mill's call for pluralism both by criticizing dominate androcentric hypotheses and promoting feminist science.

Keywords John Stuart Mill · Marketplace of ideas · Pluralism · Democracy · Feminism · Well-ordered science

---

Kathleen Okruhlik was deceased at the time of publication.

K. Okruhlik (Deceased) (✉)
Department of Philosophy, University of Western Ontario, London, ON, Canada
e-mail: okruhlik@uwo.ca

J. Y. Tsou et al. (eds.), *Values, Pluralism, and Pragmatism: Themes from the Work of Matthew J. Brown*, Boston Studies in the Philosophy and History of Science 347, https://doi.org/10.1007/978-3-031-92958-8_1

1.1 Introduction

Although John Stuart Mill's *System of Logic* and his *Principles of Political Economy* were extraordinarily influential texts in Victorian philosophy of science, they had little impact on the discipline during the twentieth century (except that the *Logic* was occasionally invoked as a foil for rival views or even an object of ridicule). Until at least the 1970s, people who bothered to think about the question at all might have said that Mill's chief positive importance during the long period dominated by logical empiricism was as intermediary between Hume and Russell in the tradition of British empiricism. This, however, is a mistake. I shall argue that Mill's influence on philosophy of science during the closing years of the twentieth century was considerable and that it stemmed not from the *System of Logic* or the *Principles*, but from Mill's political philosophy—specifically, *On Liberty*. I shall argue further that some recent invocations of Mill by philosophers of science have misrepresented his views in significant ways. Because questions about truth and democracy, science and values are not likely to go away any time soon, it is important to get Mill right. Doing so will not solve our twenty-first century problems, but it may be a step in the right direction.

Mill is typically invoked as the heroic champion of a "marketplace of ideas" with the attendant belief that truth will emerge from unfettered competition among rival hypotheses or systems of belief. This is perhaps the most common and the most egregiously mistaken of interpretations; it is also the one that will be considered first in the discussion below.

Other, more specific, employments of Mill will be discussed subsequently. Paul Feyerabend famously invoked Mill in support of epistemological anarchism. More recently, Philip Kitcher has cited him as an ally in his defense of "well-ordered science." Despite the rather striking contrast between anarchy on the one hand and order on the other, both Feyerabend and Kitcher manage to misrepresent Mill's position in various ways. Finally, the class of arguments sometimes described as "feminist empiricism" is said to have its roots in the liberal tradition. Helen Longino's "contextual empiricism" is the best-developed work in this genre. Longino makes explicit reference to Mill and acknowledges similarities between their positions. I shall suggest, however, that there are resources in Mill that might be more thoroughly exploited.

The question of Millian resources for twenty-first-century philosophy of science is raised more generally in the concluding section of this paper. The focus there is on how our situation resembles Mill's and how it differs.

1.2 Mill and the Marketplace of Ideas

The main problem with arguments that cite "Mill's marketplace of ideas" is that Mill never used that figure of speech and would not have endorsed the concept that lies behind it. The metaphor was introduced long after Mill was dead by U.S. Supreme Court Justice Oliver Wendell Holmes in his dissenting opinion in the case of *Abrams v. United States* (1919). The defendant in this case had distributed leaflets during World War I urging workers to undermine domestic war industries through general strikes as a way of showing solidarity with Russian revolutionaries. He was sentenced to twenty years' imprisonment for promoting subversive activities. In his dissenting opinion, Holmes argued that publication of the leaflets should be protected as free speech under the First Amendment of the U.S. Constitution:

> [W]hen men have realized that time has upset many fighting faiths, they may come to believe even more than they believe the very foundation of their own conduct that the ultimate good desired is better reached by free trade in ideas – *that the best test of truth is the power of the thought to get itself accepted in the competition of the market…* (Abrams v. U.S. 616 1919, emphasis added)

Even here we do not find the phrase "marketplace of ideas", but the base metaphor is clearly present, and Holmes returns to it in later writings. There are certainly similarities with Mill's position, which is not surprising since Holmes seems to have been inspired by *On Liberty* as well as by Milton's 1644 "Aereopagitica." The differences between the two positions are, however, large and the importance of honoring these differences increases as twenty-first century debates about truth, liberty, democracy, and the marketplace heat up. Some disciplines may have attended to the differences between Mill and the marketplace; philosophy of science has paid too little attention.

The argument here proceeds in three parts. First, we review Mill's position. Then we contrast it with the "marketplace of ideas" introduced by Holmes as a metaphor, and finally, we consider how that the marketplace has become literal (rather than metaphorical) in the present context of much scientific research.

1.2.1 Mill on Liberty of Thought and Discussion

The emphasis in Justice Holmes's dissenting opinion is on free trade in ideas and on a judicial policy of non-intervention. It is clear that diversity must be tolerated, but there is nothing in Holmes's argument to suggest that it should be actively generated or promoted. And although the justification offered for protecting free speech is largely epistemic ("the best test of truth is the power of the thought to get itself accepted in the competition of the market"), there is no real defense of the thesis that a *laissez-faire* approach will lead to truth rather than falsehood (in any independently defined sense). The underlying assumption that most people will choose the truest opinion if exposed to multiple competitors is built into the free trade metaphor, perhaps by definition.

Mill, on the other hand, was motivated to write "Of the Liberty of Thought and Discussion" (chapter two of *On Liberty*), precisely because he had no confidence whatsoever that the majority of people would embrace the best ideas or that truth would triumph in the competition for acceptance:

> …the dictum that truth always triumphs over persecution, is one of those pleasant falsehoods which men repeat after one another till they pass into commonplaces, but which all experience refutes. History teems with instances of truth put down by persecution. If not suppressed forever, it may be thrown back for centuries…. It is a piece of idle sentimentality that truth, merely as truth, has any inherent power denied to error, of prevailing against the dungeon and the stake. (Mill [1859] 1977, p. 238)

Mill famously offers "four distinct grounds" for protecting freedom of expression *and* promoting diversity of opinion: First, the silenced opinion may be true. Second, even if it is an error, the silenced opinion may contain some portion of the truth, which will be supplied only through collision with adverse opinions. Third, even if the received opinion is the *whole* truth, if it is not vigorously contested, it will be held only as a prejudice "with little comprehension or feeling of its rational grounds." Finally, if it becomes an uncontested dogma, the meaning of the doctrine may be lost or "deprived of its vital effect on the character or conduct" (p. 258). Even if only one person holds an opinion, it must not be silenced. The injury that results from repressing an unpopular opinion is not primarily a personal injury. If it were, then it would make some difference whether the injury was inflicted on many or just a few. Instead, the harm that is done constitutes a *public* injury because it robs the human race of the benefits listed above (p. 229).

It is especially important for our purposes to note that Mill is advocating not simply a policy of non-interference, but an interventionist program for the promotion of diversity. The less popular opinion is the one with the greater claim to be heard:

> …[I]f either of the two opinions has a better claim than the other, not merely to be tolerated, but to be encouraged and countenanced, it is the one which happens at the particular time and place to be in a minority. That is the opinion which, for the time being, represents the neglected interests, the side of human well-being which is in danger of obtaining less than its share. (p. 254)

Tolerance is not enough; instead, we require a sort of affirmative action program for unpopular ideas and underrepresented points of view. Even if threats of persecution and prosecution are removed, says Mill, people still hesitate to dissent from majority opinion. In England, it is the threat of social stigma that keeps people quiet—even those financially secure enough to risk that stigma without risking their livelihood. Such a society "cannot send forth the open, fearless characters, and logical, consistent intellects who once adorned the thinking world" (p. 242).

One very striking aspect of Mill's argument against this kind of tyranny is his insistence that when there are no rivals to established opinion, we must *contrive* to create a system of meaningful challenges. Among the contrivances he considers is that of the "devil's advocate." Mill, who rarely has anything positive to say about the Catholic Church, thinks this a splendid device, worthy of emulation: "The most intolerant of churches, the Roman Catholic Church, even at the canonization of a saint, admits, and listens patiently to, a "devil's advocate." The holiest of men, it appears, cannot be admitted to posthumous honours, until all that the devil could say against him is known and weighed" (p. 232).[1]

> A similar practice is central to the best science:
>
> If even the Newtonian philosophy were not permitted to be questioned, mankind could not feel as complete assurance of its truth as they now do. The beliefs which we have most warrant for, have no safeguard to rest on, but a standing invitation to the whole world to prove them unfounded. (ibid)

The allusion here to "Newtonian philosophy" is important in the context of philosophy of science, because it confirms that Mill did not limit his concern to social, political, and ethical theories. While admitting the greater role played by empirical evidence in the sciences, he explicitly includes them in his arguments:

> Even in natural philosophy, there is always some other explanation possible of the same facts; some geocentric theory instead of heliocentric, some phlogiston instead of oxygen; and it has to be shown why that other theory cannot be the true one: and until this is shown and until we know how it is shown, we do not understand the grounds of our opinion. (p. 244)

It is important to note that pluralism is not valued for its own sake but as a means to the discovery of truth. It must also be stressed that for Mill, truth is not created by our debates or by the choices we make. Truth is antecedent to and independent of all human activity; there is no guarantee that we will get it right. In the course of history, however, societies can progress in their social arrangements, and this in turn can facilitate progress in the discovery of truth. But how is truth related to liberty? And how is liberty related to utility?

There is a natural tendency to think of liberty and utility as two principles that will often be in conflict—that liberty of individuals must frequently be curtailed in order to promote the general welfare. This, however, was not Mill's way of conceptualizing the relationship, and recent work on Mill has significantly altered our understanding of many aspects of his thought. The "libertarian" Mill, the "negative liberties" Mill has been replaced with a much more complex and interesting character who values liberty *because* it promotes utility, understood "in the largest sense, grounded on the permanent interests of man as a progressive being" (p. 224). John Skorupski has stressed that "these 'permanent interests' arise from human beings' potential for free self-development… a potential realized not only in the course of individual lives but progressively in human history itself… [The human good is] dynamic, developmental and individual" (Skorupski 2006, p. 25). Liberty promotes the sort of self-development that is associated with the highest forms of happiness;

[1] Mill would have been saddened to learn that John Paul II eliminated the role of Devil's Advocate in 1983 as part of a complete overhaul of the canonization process.

so liberty and utility are not generally at cross-purposes for Mill. Instead, they usually go hand-in-hand.

The relationship between liberty and democracy is, however, less intimate. Democracy is good because, once societies have reached a certain historical stage, it seems to be the form of government that best facilitates self-development. Mill says that "the test of what is right in politics is not the *will* of the people, but the *good* of the people…" Mill [1832] 2006, p. 175). Democracy has merit, *not* because it gives the people what they want, but because it is better than other forms of government at promoting the good of the people, understood as their potential for free self-development. But democracy has disadvantages, too—most famously, the "tyranny of the majority", Mill's main target in *On Liberty*.

Perhaps the most important thing to recognize is that Mill's model for competition among ideas, whether in science or elsewhere, is unlike two other nineteenth-century models that employ competition as a way of selecting the best. It differs from both the Darwinian struggle for existence and *laissez-faire* capitalism in that it does not *define* the "best" as the contender that vanquishes its competitors and emerges victorious from the struggle. For Mill, truth is defined independently of victory, independently of acceptance. Justice Holmes's "marketplace of ideas", on the other hand, seems to have more in common with capitalist and Darwinian models of competition.[2]

1.2.2 The Market Metaphor

Jill Gordon has argued forcefully for the importance of drawing a sharp distinction between Mill and Holmes on the issue of free speech (Gordon 1997). Her emphasis is not on Holmes's judicial tolerance versus Mill's epistemic activism but on the market metaphor itself. The model embraced by Mill is, she says, very different from that implied by Holmes's metaphor: "the 'best'—whether product or opinion—is defined after the fact in the market as that product or idea demanded or adopted by the most powerful or most numerous" (Gordon 1997, p. 241). The ideas that would dominate in a marketplace of ideas are popular ideas (exactly what Mill feared) rather than those with epistemic merit. "In the marketplace of ideas, the ideas that survive or prevail will be those espoused either by the most powerful or the most numerous in the society. The metaphor therefore implies an outcome that runs counter to Mill's entire conception of freedom of expression and is conducive to the very tyranny he wishes to prevent" (pp. 239–240).

What is at stake here is not simply an issue concerning use of the phrase "Mill's marketplace of ideas", which still occurs in scholarly publications but is perhaps

[2] NB: We should not, however, automatically accept the common conflation of Darwinian selection with "competition" in the (colloquial, non-technical) sense. See Evelyn Fox Keller's essay, "Language and Ideology in Evolutionary Theory: Reading Cultural Norms into Natural Law" (Fox Keller [1991] 1996).

more common in casual conversation. What matters is the extent to which market analogies have come to dominate our thinking in all areas of life.

One is left to wonder, in the final analysis, what the appeal of the market metaphor is, despite its inconsistency with Mill's text. What makes the marketplace an appealing metaphor for one of our dearest freedoms? It would seem that the marketplace has become the paradigm for all freedoms we value, but this paradigm, I want to suggest, stems from an ideological framework that Mill did not share (p. 245).

Here, there is a shift in the focus of Gordon's argument. Having argued that the marketplace metaphor is not one that Mill would have accepted (in any event), she also wants to stress that Mill's own view of the market is very different from the neo-classical economic theory that permeates our own culture. The neo-classical framework, in fact, "seems to be founded implicitly on a bastard form of Darwinian theory" that makes little room for discussion of the common good (ibid). The moral considerations that informed classical economic theory, including Mill's own theory, have no place in neo-classical economic theories, which are built on a conception of humans as purely self-interested beings. The situation is further exacerbated by neo-classical theorists' adoption of formal models of instrumental rationality that leave no room at all for rational discussion of the desirability of various social ends.

Mill's *science* of political economy has no room for such discussions either. But much of the argument in his *Principles* concerns the *art* of political economy, which is required to supplement the science. The system of wealth distribution a state chooses will depend upon the sort of society it wishes to create. So the political economist must engage in rational discussion about intellectual, social, and moral ends and about how different economic systems might bring about the desired ends. Laws of production are natural laws, but the laws of distribution are dependent on social structure and the ends that a society wishes to achieve.

Mill is, of course, rightly associated with a form of *laissez-faire* economics. Although he was strongly attracted to some aspects of socialism, he thought that in his time and place (i.e., Victorian England) capitalism would probably create a greater abundance of wealth. This wealth could then be partially redistributed to promote the good of all. As Gordon says: "[h]is endorsement of capitalism and competition was to serve the ultimate and higher end of humanity's improvement; competition, and ultimately capital growth, could ease working conditions, increase leisure time, improve educational opportunities, increase overall wealth, and generally decrease disparities of wealth and well-being" (ibid). Antagonistic individualism has no place in Mill's liberalism. Competition is a form of cooperation, a social mechanism to produce the best theories *and* the greatest abundance of goods for all.

Jonathan Riley (1998, p. 317) makes the important point that we must not conflate *laissez-faire* with liberty. Trade is a social act, says Mill (Mill [1848] 2006, p. 700); so in principle the vendor's conduct comes within the jurisdiction of society. As a result the doctrine of free trade (*laissez-faire*) "rests on grounds different from the principle of liberty."

Riley goes on to say that according to Mill's account of *laissez-faire*, the place where we ought to draw the line between free competition and government intervention is justified on grounds of expedience for the achievement of abundance. It must also be noted that the place where this line should be drawn may vary across societies. The situation, however, is completely different for liberty. There is just 'one very simple principle' of liberty, which is the same for all societies. It is grounded on considerations of justice and security, and it is inseparable from the individual self-development associated with the highest forms of happiness.

Although Riley doesn't say this, capitalism seems to be (more-or-less) on the same plane with democracy in that neither capitalism nor democracy is best at all time in all places. Each is justified only insofar as (in certain historico-social circumstances) it promotes liberty and utility.

1.2.3 The Literal Marketplace

By the late twentieth century, it had become clear that Justice Holmes's "marketplace of ideas" metaphor had ceased in many respects to function as a metaphor at all. Instead, the increasing "commodification of knowledge" had led to something like a *literal* marketplace of ideas. This is a development that has had sometimes catastrophic effects on scientific research as well as on individuals who would have benefitted from an older conception of scientific knowledge as a public resource.

To get some sense of how quickly (and how disastrously) things have changed, we need only remind ourselves of an incident involving Jonas Salk in the 1950's, following his discovery of a vaccine against polio. When he was asked in a television interview with Edward R. Murrow about patenting the vaccine, he expressed incredulity and suggested that doing so would be like patenting the sun. The vaccine, like the sun, belongs to everyone.[3] Salk's answer reflects adherence to one of Robert Merton's norms of science in a democratic order: "The substantive findings of science are a product of social collaboration and are assigned to the community" (Merton [1942] 1973, pp. 273–275). (This principle was originally referred to as "communism" and later, in a changed political climate. as "communalism").

Things have changed dramatically in the years since Jonas Salk's discovery, as corporations have privatized a bigger and bigger portion of scientific research, including pharmaceutical research. The ethical problems are obvious, since the chief goal of corporations is to maximize profits rather than enhance human well-being. The epistemic problems are also devastating because basic norms of scientific research are violated when adherence to those norms might cut into the profit margin. Critical scrutiny of the sort that Mill advocates is ruled out, and secrecy is the order of the day. Democratically elected governments, for the most part, do not protest on behalf of their citizens but instead defend the right of corporations to literally *own* knowledge and to deny access to others. The conflation of democratic

[3] It should be acknowledged here that some commentators have suggested that Salk's response may have been more calculated than sincere.

rights to freedom of thought and expression with the principles of neoclassical economics is an important part of the mindset that allows people to think that this is how things must be in a "free society."

In discussing another norm of science, one that he called "disinterestedness", Robert Merton remarked on the "virtual absence of fraud in the annals of science" compared to other areas of human endeavor (p. 275). He attributed this not to any exceptionally high degree of moral virtue in individual scientists but to the institution itself: "the activities of scientists are subjected to rigorous policing, to a degree perhaps unparalleled in any other field of activity" (p. 277). It is no longer possible to make Merton's claim about the virtual absence of fraud in science. When policing is absent because of secrecy, private ownership of knowledge, and the profit motive, fraud abounds. Unfavorable results are suppressed, corners are cut in the development and implementation of experimental designs, prominent scientists sign off on research in which they had no active role, and journal editors cannot find referees who do not themselves have significant conflicts of interest.[4]

These problems are not limited to pharmaceutical research but are common in many areas of science. Sometimes, as in environmental research, private ownership of knowledge is not the issue but rather the dependence of future profits on getting the "right" theory accepted. This is an especially severe problem in areas of scientific research that have direct implications for public policy (McGarity and Wagner 2008). Unfortunately, these issues have received less attention than they deserve in mainstream philosophy of science because "it's just not what we do." These are not abstract problems that can be tackled using purely formal methods that yield elegantly simple solutions. They are instead complex problems, the solution to which will require reference to ethics, politics, and economics. In treating these issues as outside our purview, we show that our view of philosophy of science is pinched, narrow, and impoverished compared to Mill's view. This is a point to which we shall return later.

1.3 Mill in Late Twentieth-Century Philosophy of Science

Despite our distance from Mill and despite his relative neglect during most of the twentieth century, Mill was occasionally invoked in the period following the demise of logical empiricism by philosophers of science who wanted to kick start a new kind of discussion about science and society. This section provides a brief examination of the three best-known cases.

---

[4] See, for example: Angell (2004).

1.3.1 Paul Feyerabend on Epistemological Anarchy

Paul Feyerabend's most famous book is *Against Method,* in which he argues that there is no identifiable set of rules that jointly constitute "the scientific method" (Feyerabend 1975). Feyerabend claims that so-called "rationalist" methodologies, if applied rigorously, would have blocked the greatest episodes in the history of science; Galileo's use of the telescope provides the central case study. The only epistemology that lets science be science is epistemological anarchism.[5] And science itself (having no method that gives it special access to truth or a monopoly on reason) should be regarded as just one tradition among others. In a free society, all traditions should enjoy equal rights and equal access to power. Feyerabend argued that science is a threat to democracy and that it must be placed under democratic control. Scientific "experts" should be consulted only if lay juries exercise democratic control over the process.

In an earlier essay, also called "Against Method", Feyerabend had expressed his great admiration for Mill's pluralism, and seemed to derive a striking consequence from it.

> The possibilities of Mill's liberalism can be seen from the fact that it provides room for any human desire, and for any human vice. There are no general principles apart from the principle of minimal interference with the lives of individuals, or groups of individuals who have decided to pursue a common aim. For example, *there is no attempt to make the sanctity of human life a principle that would be binding for all.* Those among us who can realize themselves only by killing their fellow human beings and who feel fully alive only when in mortal danger are permitted to form a subsociety of their own where human targets are selected for the hunt, and are hunted down mercilessly, either by a single individual or by specially trained groups… So whoever wants to live a dangerous life, whoever wants to taste human blood, will be permitted to do so within the domain of his own subsociety. (Feyerabend 1981a, p. 69)

There are a number of problems with Feyerabend's use of Mill. We have already seen that his liberalism cannot be reduced to a principle of non-interference. Instead it is intimately bound up with Mill's realist, strongly normative accounts of truth and utility. Liberty facilitates both the search for true beliefs and the process of development of the self by the self, which is at the heart of his conception of utility.

Can the choice to join a Mutual Killing Society be consistent with liberty understood in this way? Suicide might seem to provide a rough analogue to the case at hand, but Mill never deals directly with that issue. In any event, mutual killing has the added complication of being an other-regarding act (even if no one outside the Society is harmed by a member's participation). It is perhaps Mill's discussion of selling oneself into slavery that is most relevant to the case at hand, and it provides a fairly strong indication that he would not endorse Feyerabend's Mutual Killing

[5] Feyerabend frequently pointed out later that he had also described epistemological anarchism as a "medicine" that would help to cure what was ailing science and society. There might be times in the future during which a good dose of methodological rationalism would be just what the doctor ordered.

Society. The basic argument is that such an act would be inconsistent with Liberty because it would preclude (or would almost certainly preclude) the ability to make future choices: "[t]he principle of freedom cannot require that he should be free not to be free. It is not freedom, to be allowed to alienate his freedom" (Mill [1859] 1977, p. 300).

Feyerabend also calls on Mill for support in a slightly different context, suggesting that pluralism is desirable for its own sake and that consensus is a sign of failure. He says that, as opposed to those who believe in an "approach to the truth",

> a thinker following Mill... regards any prolonged stability, either of ideas and impressions or of background knowledge which one is not willing to give up (realism; separation of subject and object; commensurability of concepts), as an indication *of failure*, pure and simple. (Feyerabend 1981a, p. 77)

But this is a claim that Mill rejects at considerable length and with some passion:

> But what! (it may be asked) Is the absence of unanimity an indispensable condition of true knowledge? Is it necessary that some part of mankind should persist in error, to enable any to realize the truth?... I affirm no such thing. As mankind improve, the number of doctrines which are no longer disputed or doubted will be constantly on the increase: and the well-being of mankind may almost be measured by the number and gravity of the truths which have reached the point of being uncontested. (Mill [1859] 1977, p. 250)

The disagreement here is pretty basic. Mill is a realist about truth, and when he wrote the passage cited above, Feyerabend was a relativist. By the time he wrote the Introduction to his collected papers, he had recognized and acknowledged this difference:

> There is much to be said in favor of a pluralistic realism... John Stuart Mill has explained the arguments in his immortal essay *On Liberty* which is still the best modern exposition and defence of a critical philosophy.... *But the drawbacks are considerable*... there may be excellent reasons for resisting... universal realism. (Feyerabend 1981b, pp. xi–xii)

Feyerabend suggests that according to his earlier realist account of pluralism, criticism leads to the proliferation of theories, which in turn leads to realism regarding the theory that emerges victorious from the fray. On his later relativist account of pluralism, however, we choose a form of life and in doing so reject any universal standard for criticism and any sort of universal realism.

The point here is that Mill's pluralism is not a form of relativism, and the diversity that he promotes is not diversity for its own sake. Pluralism serves as a means to moral, social, political, and epistemic improvement—understood objectively. Mill is in fact one of "those who believes in an 'approach to the truth'." He simply recognizes that there are no guarantees of epistemic progress. Truth cannot speak for itself; wrong turns and backsliding are ever-present dangers.

In an essay called "Feyerabend, Mill, and Pluralism", Elisabeth Lloyd (1997) argues that:

> (19) Many of Feyerabend's more unusual stances are best understood as enactments of Mill's call for defenders of unpopular minority opinions.

(20) "…Feyerabend's appeals to Mill's views were, in fact, genuine endorsements: Feyerabend never distanced himself from Mill's position…" (Lloyd 1997, p. S397 italics in original)

I disagree with Lloyd's second claim for reasons sketched above. While continuing to admire Mill, Feyerabend did come to recognize the strong form of realism that underlies the argument in *On Liberty* and to reject that realism in favour of his own quite different philosophical project.

On the other hand, Lloyd's suggestion that we think of Feyerabend as *enacting* Mill's call for defenders of unpopular minority opinions is both plausible and useful. This interpretation not only makes sense of much Feyerabendian behavior, it also highlights Feyerabend's unusual appreciation of Mill's call for *action* in the development and promotion of views that challenge popular opinion. There is no question here of standing back while some quasi-Darwinian or quasi-capitalist competition produces winners that are the best simply in virtue of their survival or their success in the marketplace of ideas. If there are no existing rivals, we must create them. Feyerabend took on the work of devil's advocate with unusual gusto. He and Mill, in their different ways, recognized the necessity of actively opposing the tyranny of the majority: Mill because it hampered the search for truth and hence opportunities for improvement of the self; Feyerabend because it imposed a single form of life on everyone (and produced, among other bad outcomes, some really incompetent philosophers of science).

1.3.2 Philip Kitcher on "Well-Ordered Science"

Whereas Feyerabend enlisted Mill's support in defense of epistemological anarchism, Philip Kitcher's *Science, Truth, and Democracy* argues that Millian principles can be adduced to support new constraints on free scientific inquiry (Kitcher 2001). Kitcher argues for a notion of "well-ordered science", in which research agendas and the applications of research findings would be subject to public control informed by democratic deliberation. The resulting science would be well-ordered if the deliberations of actual citizens yielded the same results that would be arrived at by ideal agents deliberating under ideal conditions.

The plot thickens when Kitcher suggests that ideal deliberation among ideal agents may sometimes lead to the conclusion that certain sorts of research ought not be pursued. (It must be noted that this is a moral 'ought'; it does not imply that such research should be legally proscribed). The suggestion would be uncontroversial if Kitcher were considering just the usual sorts of example, such as the need for limits on human experimentation. His specific examples, however, involve the types of research that explore various aspects of alleged inferiority in women and in members of some ethnic groups. In certain (clearly specified) epistemic and political situations, ideal deliberators could decide that such research would almost inevitably be interpreted in ways that would limit the well-being of members of the groups in question. In such cases, they might decide that the duty to seek truth must be subordinated to "the duty to care for those whose lives go less well and to protect them against foreseeable occurrences that would further decrease their well-being" (p. 103).

Kitcher is suggesting, therefore, that a proper understanding of Mill's conception of freedom as "the freedom to define and purse our own vision of the good" forces us to recognize that unconstrained freedom to pursue scientific truth may sometimes interfere with freedom of the most fundamental sort for some members of society. So, he says, Mill's position can be understood as supporting (under specified conditions) moral proscriptions against studies of women's suitability for leadership as well as research on the intellectual abilities of minority group members. In other words, "we can envisage a Millian argument *against* freedom of inquiry, one that proceeds by trying to show that certain types of research would be likely to undermine a more fundamental freedom" (p. 95).

Speculating about what a person long dead would (or would not) say if confronted by a situation we face today is, of course, a dangerous game. Nevertheless I shall argue that Kitcher's conscription of Mill in this instance is unconvincing. Some interpretive points are pretty straightforward. Kitcher seems to suggest that Mill holds the "rosy" view that "truth will out" (p. 96). We have already seen that Mill explicitly repudiates such a position and argues that it is precisely because falsehood often triumphs over truth that we must be vigilant in not simply tolerating but encouraging dissent.

Other complaints cannot be formulated quite so sharply. The language of "ideal deliberation by ideal agents", for example, seems so alien to Mill's project that speculating about what would count as "Millian" under those circumstances requires that we abstract from many of the defining characteristics of Mill's empiricist epistemology. The inquiry that concerns him is carried out by real people, whose epistemic abilities and shortcomings are central to his understanding of how decisions are made, how we deliberate, how we sometimes make good decisions and sometimes go badly off the rails. The social and epistemic situations of real (rather than ideal) agents are also central to Mill's account of the relationships among liberty, truth, and utility.

Even if we set aside any discomfort with the notion of ideal deliberators, we can still doubt that the straightforward trade-off between liberty and utility (described in Kitcher's terms as freedom to pursue truth vs. freedom to define and pursue our own vision of the good) is something that finds a comfortable home in Mill's thought. Although he argues that it is wrong to force superior forms of self-development on someone who prefers lower pleasures,[6] it is much harder to contemplate instances in which Mill might morally proscribe the investigation of truth claims.

[6] Mill ([1859] 1977, p. 272) notes that the spirit of liberty and the spirit of progress or improvement, although both the enemy of Custom, may sometimes clash: "The spirit of improvement is not always a spirit of liberty, for it may aim at forcing improvements on an unwilling people; and the spirit of liberty, in so far as it resists such attempts, may ally itself locally and temporarily with the opponents of improvement; but the only unfailing and permanent source of improvement is liberty..."

Finally, there is something just plain weird about enlisting Mill in a project to "democratize science", since his discussion of liberty of thought and expression is motivated by *fear* about what happens to truth within a democratic framework. Liberty and truth rank outrank democracy in Mill's hierarchy of allegiances. It is not enough to reply that in Kitcher's well-ordered science, method (as employed in the "context of justification") is protected from democratic oversight, if certain research agendas are proscribed at the outset. This claim seems to rely on views about the difference between "discovery" and "justification" that have been pretty thoroughly discredited in recent years.

Perhaps, though, it is irresponsible to engage in too much counterfactual speculation about what Mill *would say* were he here today. Another strategy is to look at what he actually *did* say about the sorts of questions that Kitcher makes the centerpiece of his argument. Luckily for us, Mill had a great deal to say about the allegedly innate inferiority of women, the Irish, and various other ethnic groups. He didn't hesitate for a moment to acknowledge that the achievements of these groups were in many ways inferior to those of white men. *On the Subjection of Women* provides his most sustained discussion of this type of issue (Mill [1869] 2006). Instead of denying that the achievements of women are inferior, he argues that this is because they have not been given the same opportunities for liberty and self-development as men. His point is that the experiment has not been done. Until some significant period of time has passed during which women have the same liberties and opportunities as men, we have no meaningful evidence. Mill is not backing away from empirical investigation of the abilities of women (or the Irish or people of color). He is saying that such investigation can yield meaningful results only under conditions of political and educational equality.

The sociobiological cases examined by Kitcher are precisely the sort in which Mill calls for *more* research, more experiments, more hypotheses that challenge the status quo. Racist and sexist views are so deeply entrenched in our culture that it will take decades of oppositional research to locate the effects of such bias in existing science, propose and test alternatives, implement change, and test again. Charges of "political correctness" do nothing to alter the fact of widespread social bias against women and some ethnic minorities, and pretending that regressive or "politically incorrect" views represent genuine challenges to the scientific status quo is laughable.

Mill would be looking for "contrivances" to develop and promote antiracist and anti-sexist theories with testable hypotheses. In our society, such contrivances include (but are not limited to) differential research funding, changes to publication policies, and an increase in the diversity of the scientific community. At this point, one might be tempted to say that the gap between Kitcher and Mill is not very great, since neither wishes to actively promote research based on assumptions of innate inferiority in some groups of people. Nonetheless the difference in approach is significant: moral constraints on sociobiological research vs. active support for egalitarian hypotheses, *including support for the kinds of social change that would make good experiments possible.*

There is a further point that must be stressed as well. Although Mill is very confident that most differences in ability between social groups can be accounted for by nurture rather than nature, it is possible that evidence to the contrary could be produced even under egalitarian social conditions. If there *are* innate differences in ability between social groups (as well as among individuals), Mill would point out that nothing follows from that about how society should order itself or distribute its goods. (Recall earlier comments about the laws of production vs. distribution in political economy). Considerations of utility may lead us to use the results of science to create conditions under which those with innate disadvantages receive additional support in their efforts at self-development.

1.3.3 Helen Longino and Feminist Critiques of Science

Earlier we granted considerable credence to Elisabeth's Lloyd claim that Feyerabend is best understood as *enacting* Mill's call for defenders of unpopular minority opinions. I shall suggest here that recent feminist critics of science have provided an even more faithful, sustained, and productive enactment of that call. Perhaps the best example is Helen Longino, who for more than 25 years has not only criticized androcentric science but also provided alternative accounts that throw into sharp relief the weaknesses of more traditional approaches. Longino's "contextual empiricism" is bound up with her view that science is a form of social knowledge, and she has cited Mill (as well as Charles Peirce and Karl Popper) as a forerunner of present interest in the social dimensions of knowledge. While acknowledging that she and Mill differ on many issues, including the forms of empiricism they embrace, Longino says: "[n]evertheless, on the necessity of critical interaction for the integrity of knowledge Mill's view and the one I defend do coincide" (Longino 2002, p. 4).

We can distinguish two different ways in which Longino's approach has been Millian in character. (1) She has *enacted* Mill's call for the formulation and advancement of ideas that challenge cherished beliefs by advocating specific hypotheses, background assumptions, values, and epistemic virtues that are framed in conscious opposition to the scientific status quo. (2) As theoretician, she has offered a general account of science as social knowledge that reflects Millian themes.

In *Science as Social Knowledge*, Longino argued that scientific communities will be objective to the degree that they satisfy four criteria necessary for achieving the transformative dimension of critical discourse: (1) there must be recognized avenues for the criticism of evidence, of methods, and of assumptions and reasoning; (2) there must exist shared standards that critics can invoke; (3) the community as a whole must be responsive to such criticism; (4) intellectual authority must be shared equally among qualified practitioners (Longino 1990, p. 76).

These criteria certainly have a Millian flavor, and Longino explicitly highlights the importance of diversity: "the greater the number of different points of view included in a scientific community, the more likely it is that its scientific practice will be objective" (p. 80). What seems to be missing, however, is the activism that

Mill calls for in *creating* pluralism where it does not exist in a sufficiently robust form.[7] These aspects of the Millian program have greater visibility in Longino's more explicitly feminist work.

Just three years earlier, in a 1987 *Hypatia* article called "Can There Be A Feminist Science?", Longino had argued "for the deliberate and active choice of an interpretive model and for the legitimacy of basing that choice on political considerations" (Longino 1987, p. 61).[8] She described such deliberate and active choices as "feminist interventions" (p. 62). These interventions are meant to be local (in the sense of being specific to a particular area of research), non-exclusive (in the sense that different feminist perspectives may be represented), and in some way continuous with existing scientific work. Nevertheless they have revolutionary potential: "The accretion of such interventions, of science done by feminists as feminists, and by members of other disenfranchised groups, has the potential… ultimately to transform the character of scientific discourse" (ibid).

The language of "deliberate and active choice" and "feminist interventions" is certainly more reminiscent of Mill's "contrivances" than is the language found in *Science and Social Knowledge*. This is not to suggest that there is a contradiction between these two expressions of Longino's position—only that she is perhaps more Millian in the *Hypatia* article than in the book that was published three years later.

Some of the present analysis may, however, seem too quick. We have seen that Mill believed that the ability to arrive at true beliefs depends in part on social conditions and that the search for truth is best served by promoting diversity of opinion. We know that he places a high premium on individuality and even idiosyncrasy. But is there any reason to think that Mill advocates diversity of *social groups* as an advantage in the search for knowledge? This is important if the argument for a better scientific community is to be made not just in terms of individual diversity but also diversity of the social groups that are included in that community.

In chapter three of *On Liberty*, we do in fact find substantial evidence for this claim. Mill quotes Wilhelm von Humboldt not once, but twice (Mill [1859] 1977, p. 260 and p. 273), to the effect that the two requirements in order that human beings achieve their end (the individuality of power and development) are "freedom, and variety of situations." We have discussed freedom in the context of chapter two, but *variety of situations* adds a new dimension to our discussion. We saw earlier that

[7] Nonetheless there is something vaguely disappointing about the list insofar as it does not seem to require action to *promote* or *create* diversity—either social diversity in the scientific community or diversity of opinion. In this respect, feminist standpoint theory may seem to have a significant advantage over Longino's contextual empiricism. Standpoint theory is associated with the political theories of Hegel and Marx rather than with the liberal tradition. Marx provides a source of justification for positive action in favor of the disadvantaged. According his theory, important sorts of epistemic privilege attach to the standpoints of the disadvantaged (in this case the proletariat as opposed to the bourgeoisie). So there is a pretty straightforward argument for seeking out and differentially attending to underrepresented groups and their beliefs about the world.

[8] The extent to which such considerations are legitimate will vary, depending in part on the nature of the science.

Mill worried that the very progress of truth, though good in itself, would lead to less diversity of opinion, a negative effect. Here we see Mill worried that the social, economic, and cultural homogenization of society is having a similar effect. He attributes Europe's status as "an improving, instead of a stationary portion of mankind" not to any superior excellence in Europeans, but to "their remarkable diversity of character and culture." His concern is that this diversity is rapidly disappearing:

> The circumstances which surround different classes and individuals, and characters, are daily becoming more assimilated. Formerly, different ranks, different neighborhoods, different trades and professions, lived in what might be called different worlds; at present to a great degree in the same. Comparatively speaking, they now read the same things, go to the same places, have their hopes and fears directed to the same objects, have the same rights and liberties, and the same means of asserting them. Great as are the differences of position which remain, they are nothing to those which have ceased. (p. 273)

If we conjoin this emphasis on diversity of social situation[9] with the activist character of Mill's support for underrepresented opinions, it would seem that Longino has all the resources she needs to add a fifth criterion that scientific communities must satisfy if they are to be objective and achieve the transformative dimension of critical discourse: In at least some cases, a scientific community should actively promote intellectual diversity by recruiting members of underrepresented groups.

1.4 Conclusions

What conclusions, if any, can we draw about Mill's theory and its relevance to issues in twenty-first-century philosophy of science? I would suggest the following:

1. Mill got some important things right.
2. We should get Mill right.
3. We should also emulate Mill in seeing both science and philosophy of science as inseparably bound up with other dimensions of life—including ethics, economics, politics, and culture. We must also recognize, as he did, the historical and social dimensions of the pursuit of knowledge.

None of this is to say that Mill is right about everything or that he anticipated all our challenges. On the one hand, we have just seen that he was already sensitive to the homogenizing effects of "globalization." On the other hand, he could not have foreseen the commodification and privatization of knowledge that has come with the growth of large multinational corporations. He was worried about the power of the *state* as a threat to individual liberty of thought and expression. Now, however, many citizens worry that their democratically elected governments will not or

[9]Two qualifications may be in order: First, Mill is interested in diversity of social groups largely because these contribute to diversity of *individuals*. (Nonetheless the fact remains that social diversity promotes diversity of opinion, which promotes truth). Second, although Mill mentions diversity of class and culture, he does not explicitly mention race or gender in this context.

cannot rein in the power of private interests to shape and control systems of belief. They want the state to preserve freedom of expression by taking a *more* active role in these matters.

Mill lived in an era when science was still small and pursued (in some important sense) by individuals rather than by teams and networks dependent on governments or corporations for the big funding that big science requires. One of the major concerns people have today regarding truth, free inquiry, and the role of criticism is often summed up with the adage "He who pays the piper calls the tune." This was not one of Mill's problems.

It is worth pausing for a moment to reflect briefly on some surprising overlaps between the concerns of Feyerabend, the anarchist, and Kitcher, the advocate of good order. Both of them argue (albeit in very different ways) for *more* democratic control over science, which sounds decidedly unMillian, despite Kitcher's arguments to the contrary. What are we to make of this somewhat bizarre turn of events? The best answer perhaps is that the status of science in society has changed since Mill's time. It is now perceived by some as having too much authority, too much power, too few effective critics, and too few competitors when it comes to telling us what we ought to believe and how we ought to live. In attempting to address the altered relationships among science, democracy, and the marketplace in the twenty-first century, we might learn a few lessons from two of the best Victorian philosophers of science.

Laura Snyder, in her 2006 book, *Reforming Philosophy: A Victorian Debate on Science and Society* argues that for all their differences, John Stuart Mill and William Whewell agreed that the reform of society and the reform of the sciences must proceed hand-in-hand (Snyder 2006). This is reflected in the ways they tried to reform philosophy of science as well. The need for such an approach is at least as urgent today as it was in Victorian times—not just because social reform depends upon the intelligent use of scientific knowledge but also because good science depends upon a social order that promotes both the self-development of citizens and rational debate about ends.

References

Abrams v. United States, 250 U.S. 616 (1919).

Angell, M. 2004. *The Truth about the Drug Companies: How They Deceive us and What to do About it*. New York: Random House.

Feyerabend, P. 1975. *Against Method: Outline of an Anarchistic Theory of Knowledge*. London: New Left Books.

Feyerabend, P. 1981a. Two Models of Epistemic Change: Mill and Hegel. In *Philosophical Papers, Problems of Empiricism*, ed. P. Feyerabend, vol. 2, 65–79. Cambridge: Cambridge University Press.

Feyerabend, P. 1981b. Introduction to Volumes 1 and 2. In *Philosophical Papers, Rationalism, Realism and Scientific Method*, ed. P. Feyerabend, vol. 1, ix–xiv. Cambridge: Cambridge University Press.

Fox Keller, E. 1991/1996. Language and Ideology in Evolutionary Theory: Reading Cultural Norms into Natural Law. In *Feminism and Science: Oxford Readings in Feminism*, ed. E. Fox Keller and H. Longino, 154–172. Oxford: Oxford University Press.

Gordon, J. 1942/1997. John Stuart Mill and the "Marketplace of Ideas.". *Social Theory and Practice* 23 (2): 235–249.

Kitcher, P. 2001. *Science, Truth, and Democracy*. Oxford: Oxford University Press.

Lloyd, E. 1997. *Feyerabend, mill, and pluralism. Philosophy of Science, 64*(S4), S396–S407.

Longino, H. 1987. Can There be a Feminist Science? *Hypatia* 2 (3): 51–64.

Longino, H. 1990. *Science as Social Knowledge: Values and Objectivity in Scientific Inquiry*. Princeton: Princeton University Press.

Longino, H. 2002. *The Fate of Knowledge*. Princeton: Princeton University Press.

McGarity, T., and W. Wagner. 2008. *Bending Science: How Special Interests Corrupt Public Health Research*. Cambridge: Harvard University Press.

Merton, R. 1973. The Normative Structure of Science. In *The Sociology of Science: Theoretical and Empirical Investigations*. Chicago: University of Chicago Press.

Mill, J. S. 1832/2006. Pledges. In *Collected Works of John Stuart Mill*, Newspaper Writings Part II, ed. A. Robson and J. Robson, vol. XXIII, 162–177. Toronto: University of Toronto Press.

Mill, J. S. 1859/1977. On Liberty. In *Collected Works of John Stuart Mill*, Essays on Politics and Society Part I, ed. J. Robson, vol. XVIII, 230–358. Toronto: University of Toronto Press.

Mill, J. S. 1869/2006. On the Subjection of Women. In *Collected Works of John Stuart Mill*, Essays on Equality, Law, and Education, ed. J. Robson, vol. XXI, 259–371. Toronto: University of Toronto Press.

Mill, J. S. 1848/2006. *The collected works of John Stuart Mill, volume III - Principles of political economy part I.I.* University of Toronto Press.

Riley, J. 1998. Mill's Political Economy: Ricardian Science and Liberal Utilitarian Art. In *Cambridge Companion to Mill*, ed. J. Skorupski, 293–337. Cambridge: Cambridge University Press.

Skorupski, J. 2006. *Why Read Mill Today?* New York: Routledge.

Snyder, L. 2006. *Reforming Philosophy: A Victorian Debate on Science*. Chicago: University of Chicago Press.

Chapter 2
The Limits of Ethical Pluralism: Mill, Feyerabend, and Experiments in Living

Jamie Shaw

Abstract In this chapter, I explore the idea of limits on ethical pluralism in the philosophy of Paul Feyerabend. This centers around Feyerabend's appropriation of John Stuart Mill's notion of 'experiments in living', although Feyerabend takes Mill's ethical pluralism much further than Mill did. Specifically, I show that Feyerabend was conflicted between embracing an ethical pluralism without limits and a pluralism that does not allow undue suffering to be caused by an entirely unhindered ethical pluralism. I consider Jurgen Habermas' discourse ethics as a potential remedy for Feyerabend's conundrum but find it wanting as it stands. I consider a more fruitful path through the notion of an 'appeal' in existentialist philosophy.

Keywords John Stuart Mill · Paul Feyerabend · Experiments in living · Pluralism · Discourse ethics · Existentialism

2.1 Introduction

> I favor any *skepsis* to which I may reply 'Let us try it!' But I no longer wish to hear anything of those things and questions that do not permit any experiment. This is the limit of my 'truthfulness'; for there courage has lost its right (Nietzsche [1882] 1974, §51).[1]

> Man's capacities have never been measured; nor are we to judge of what he can do by any precedents, so little has been tried. Whatever have been thy failures hitherto, "be not afflicted, my child, for who shall assign to thee what thou hast left undone?" (Thoreau [1854] 2012, p. 6).

[1] See Bamford (2016) for an interesting discussion of Nietzsche's experimentalism and Bearn (1986) for related discussion while also connecting insights from Nietzsche to Feyerabend.

J. Shaw (✉)
Institute of Philosophy, Leibniz University Hannover, Hannover, Germany
e-mail: jamieco.shaw@utoronto.ca

J. Y. Tsou et al. (eds.), *Values, Pluralism, and Pragmatism: Themes from the Work of Matthew J. Brown*, Boston Studies in the Philosophy and History of Science 347, https://doi.org/10.1007/978-3-031-92958-8_2

"Ethics", Feyerabend claims, is "the basis of everything else" (Feyerabend 1965a, p. 219 fn. 5). This is one of the few ideas that persists in his corpus from the early 1960s, at least,[2] until his final writings (see Brown 2016). Despite this bold stance, a detailed ethical (or value) theory is conspicuously absent from his otherwise wide-ranging corpus. For a philosopher of science who bases his entire outlook on ethics, it is strange in itself that he never provides a sustained treatment on the subject. Instead, we find scattered, sporadic remarks on ethical topics and, in his more theoretical discussions, he defers to his hero, John Stuart Mill, whose *On Liberty* had a profound impact on Feyerabend's thinking (Tsou 2025). Specifically, Feyerabend refers to Mill's argument for a pluralism of *experiments in living* (henceforth, 'pluralism'), and states that the "reasons [for pluralism] were explained by Mill in his immortal essay *On Liberty*. It is not possible to improve upon his arguments" (Feyerabend 1978, p. 86). Given such unqualified comments, we might read Feyerabend as nothing but a Millian in ethics. However, as we shall see, some of Feyerabend's contentions are at odds with Mill's pluralism. Specifically, I show that Feyerabend was conflicted about the very idea of firm limits to pluralism. Mill, on the other hand, was not. Because of this, we should not interpret Feyerabend's adoption of Mill's pluralism at face value. Rather, incorporating Mill's pluralism into Feyerabend's broader philosophy reveals hidden tensions and lacunae. However, given the brevity of Feyerabend's discussions on ethics and his interpretation of Mill's pluralism, there remain several questions about what a Mill-Feyerabend hybrid pluralism might look like. My primary goal, in this paper, is to fill this gap and sketch an ethical theory that a Feyerabendian could endorse.

I begin by reconstructing Mill's view with a focus on the 'Harm Principle' and its role in delimiting pluralism—something Feyerabend skips over in a footnote once (Feyerabend 1981, p. 69 fn. 7). I then formulate two criticisms of Mill's pluralism that are applications of aspects of Feyerabend's philosophy, yet he didn't notice that they are criticisms against Mill's pluralism. However, archival and published material suggests that Feyerabend was conflicted as he both wants to delimit pluralism and leave it unconstrained. To solve this dilemma, I consider Jürgen Habermas' discourse ethics and show that it cannot be appealing to a Feyerabendian at face value. Finally, I suggest that the notion of an appeal from existentialist philosophy, specifically in the works of Jean-Paul Sartre and Simone de Beauvoir, provides a more potentially fruitful solution.

2.2 Experiments in Living and the Harm Principle in Mill

Despite the voluminous literature on Mill's 'experiments in living', the term only appears in one section of *On Liberty*. He writes "that there should be different experiments in living; that free scope should be given to varieties of character, short

[2] While Feyerabend's methodological views on ethics and knowledge are unclear in his earlier philosophy, he was almost certainly thoughtful about the matter given that his PhD supervisor, Viktor Kraft, wrote extensively on the subject (see Kuby 2021, especially pp. 3289–3992).

of injury to others; and that the worth of different modes of life should be proved practically, when anyone thinks fit to try them" (Mill [1859] 1991, p. 63). Experiments in living are tests of hypotheses that can support counterfactuals concerning how one would appraise their life if they engaged in specific activities.[3] These hypotheses can concern abstract values, including 'dignity' or 'respect', or the activities that are constitutive of the good life for individuals (e.g., basketball, poetry, sky-diving). Experiments in living, while primarily for the benefit of individuals, can be conducted by groups and benefit civilization as a whole (Muldoon 2015; Pearce 2020, Chap. 6). The context the notion is brought up in is the cultivation of individuality—namely, what the good life entails for individuals or the instrumental conditions for happiness.

Given that 'experiments in living' is not a technical, well-defined term in Mill's writings, it is unclear what is required for an action (or set of actions) to constitute an experiment of living. Two interpretive questions, in particular, stand out:

1. Is *living* the experiment a necessary condition for learning the lessons that experiments in living are meant to teach?
2. Must experiments in living be *intentional* actions? That is, must the agent think of themselves as conducting an experiment of living?

I think that there is textual support for a 'yes' answer for (1) and a 'no' answer for (2). For (1), Mill is fairly clear-cut: "the worth of different modes of life should be proved practically." Mill is emphatic that, for the purposes of cultivating individuality, one can never be confident that the experiments in living conducted by *others* will show similar results for *them* (Mill [1859] 1991, p. 65). Moreover, in Mill's discussion of freedom of thought and expression, he repeatedly emphasizes that pluralism only genuinely works when it is a *practice* rather than a mere intellectual possibility. For instance, Mill writes:

> when [a belief] has come to be an hereditary creed, and to be received passively, not actively – when the mind is no longer compelled, in the same degree as at first, to exercise its vital powers on the questions which its belief presents to it, there is a progressive tendency to forget all of the belief except the formularies, or to give it a dull and torpid assent, as if accepting it on trust dispensed with the necessity of realizing it in consciousness, or testing it by personal experience, until it almost ceases to connect itself at all with the inner life of the human being. (p. 49)

To legitimately believe something, one must continuously exercise their mind just like one must exercise other muscles to prevent atrophy. "The mental and moral, like the muscular, powers are improved only by being used" (p. 65). Mill is not being a puritanical Victorian here. Practical pluralism, Mill thinks, is essential for the growth of knowledge as well. Unless one participates in pluralistic practices, where people of different opinions genuinely believe what they are arguing for, one is liable to make epistemic errors. The lessons, therefore, of experiments in living require living life in certain ways.

---

[3] In this function, experiments in living are an interesting premonition of 'behavioral experiments' used in various forms of therapy.

For (2), not only does Mill never claim that experiments in living must be consciously conducted, but it seems like an unnecessary requirement.[4] Surely individuals and societies can learn from experience even if one was not intentionally testing a hypothesis. If I were to fall and break my foot, although not intentional, the ensuing mobility impairment can be a learning experience. Thus 'experiments in living' can be *any experience that can evaluate moral hypotheses and personal preferences, insofar as they are reflected upon to some degree* (even if retrospectively and by different agents).[5]

When so conceived, it becomes clear that experiments in living play a fundamental role in Mill's moral philosophy despite the limited use of the phrase. Elizabeth Anderson (1991) describes how experiments in living are an essential part of Mill's argument for 'higher pleasures' (e.g., philosophy) that are superior than 'lower pleasures' (e.g., the taste of cherry blasters). But experiments in living go beyond this particular distinction as can be seen all over *Utilitarianism*. For instance, in discussing the foundations of duty, Mill writes

> The internal sanction of duty, whatever our standard of duty may be, is one and the same - a feeling in our own mind; a pain, more or less intense, attendant on violation of duty, which in properly cultivated moral natures rises, in the more serious cases, into shrinking from it as an impossibility. This feeling, when disinterested, and connecting itself with the pure idea of duty, and not with some particular form of it, or with any of the merely accessory circumstances, is the essence of Conscience; though in that complex phenomenon as it actually exists, the simple fact is in general all encrusted over with collateral associations, derived from sympathy, from love, and still more from fear; from all the forms of religious feeling; from the recollections of childhood and of all our past life; from self-esteem, desire of the esteem of others, and occasionally even self-abasement. This extreme complication is, I apprehend, the origin of the sort of mystical character which, by a tendency of the human mind of which there are many other examples, is apt to be attributed to the idea of moral obligation, and which leads people to believe that the idea cannot possibly attach itself to any other objects than those which, by a supposed mysterious law, are found in our present experience to excite it. Its binding force, however, consists in the existence of a mass of feeling which must be broken through in order to do what violates our standard of right, and which, if we do nevertheless violate that standard, will probably have to be encountered afterwards in the form of remorse. Whatever theory we have of the nature or origin of conscience, this is what essentially constitutes it. (Mill [1863] 1991, p. 161)

The feeling of duty and obligation and our moral understanding doesn't merely arise from a single experience, which wouldn't support the counterfactuals necessary for moral evaluation, but as a lesson from a whole host of experiences and social practices. Its existence is strengthened by violations of duty—for then we feel remorse, thus reassuring us of the existence and nature of duty. Similarly, Mill says of the "Principle of Utility" that "the sole evidence it is possible to produce that

[4] Mill does say that "[h]e who lets the world, or his own portion of it, choose his plan of life for him has no need of any other faculty than the ape-like one of imitation" (Mill [1859] 1991, p. 65). But this only means that one who is an intellectual slave to custom or habit is living sub-optimally, rather than saying that experiments in living require conscious choices to *be* experiments in living.

[5] This seems assumed in Anderson's (1991) treatment of Mill's conception of experiments in living.

anything is desirable, is that people do actually desire it" (p. 168).[6] Moreover, this principle is continuously proven by the advancement of civilizations:

> this firm foundation is that of the social feelings of mankind… which is already a powerful principle in human nature, and happily one of those which tend to become stronger, even without express inculcation, from the influences of advancing civilisation… this association is riveted more and more, as mankind are further removed from the state of savage independence. (p. 164)

Again, we see experiments in living (in the broad sense defined here) playing essential roles in justifying moral claims. This, especially in conjunction with Mill's rejection of intuitions as a source for moral knowledge (Mill [1859] 1991, p. 10; see also Snyder 2006, 4), suggests that that experiments in living are not relegated to this or that part of Mill's moral methodology, but foundational or all (or at least most) of Mill's major moral claims. Therefore, despite the immediate context in which the term is used, the concept plays a vital and ineliminable role in Mill's moral philosophy.

Mill makes it clear that experiments in living should be practiced in a pluralistic way: "unity of opinion, unless resulting from the fullest and freest comparison of opposite opinions, and diversity is not desirable, and diversity not an evil, but a good… As it is useful that while mankind are imperfect there should be different opinions, so it is that there should be different experiments in living" (p. 63). This suggests that there should be sufficient freedom to engage in experiments in living of various sorts. While this argument may sound close to Mill's arguments for pluralism and freedom of *thought and expression*, the parallel between speech and action is not absolute: "No one pretends that actions should be as free as opinions" (p. 62). Rather, experiments in living should be constrained by the 'Harm Principle' (henceforth, HP).[7] "As soon as any part of a person's conduct affects prejudicially the interests of others, society has jurisdiction over it, and the question of whether the general welfare will or will not be promote by interfering with it, becomes open to discussion" (pp. 83–84). In some cases, HP covers violations of human rights. Here, no deliberation, however democratic, can justify violations of human rights. In other cases, the HP concerns actions that do not violate rights but involve unwanted encroachments on others. Here, decisions on the legitimacy of such behaviors are left to the court of public opinion. This leaves us with hard and soft restrictions of pluralism. While the

[6] One prominent criticism of 'Mill's proof' is that Mill "moves from the premise that each person desires his own happiness to the conclusion that each person's happiness is desirable for him" (Millgram 2000, p. 283). Following Railton (1986), we might think that what is desirable is not necessarily what is in fact desired (Millgram reads Mill in this fashion) but what a beneficent being with perfect understanding of the laws of nature would suggest. Mill would likely be sensitive to this issue, given his experiences with depression. Mill seems open to the idea that what is desired by individuals is not what is desirable for them but might be better sought by consulting people of *experience* (people who have conducted relevant experiments in living), and so Mill does not seem to make the inferential leap he is accused of. But this exegetical task is complex, and I will not attempt to resolve it satisfactorily here.

[7] It's not clear how the disconnect between freedom of speech and action should be interpreted, since Mill clearly thought that the HP could restrict the expression of opinions. For relevant discussions, see Bell (2021) and Tunick (2022).

notion of 'harm' has been subject to much scrutiny (see Turner 2014), Mill takes the justification of HP to be straightforward: "every one who receives the protection of society owes a return for the benefit, and the fact of living in society renders it indispensable that each should be bound to observe a certain line of conduct towards the rest" (Mill [1859] 1991, p. 82). Similar to Hobbes, a stateless society permitting any and all experiments in living would be a horrible place to live. Insofar as stateless societies have existed, they should be regarded as failed experiments.[8]

2.3 Feyerabend and the Limits of Pluralism

Despite Feyerabend's claim that it is "not possible to improve upon" Mill's pluralism, there are two Feyerabendian criticisms relevant for the idea of delimiting pluralism. The first criticism runs like this. Some practices do not share the tacit assumptions in Mill's conception of experiments in living. The idea that we must *test* moral convictions itself is contentious—some lifestyles do not allow for this possibility, and so it would be arbitrary to restrict their practice on Millian grounds. Dewey gives the example of religion which he claims serves to "magnify doctrines that are not subject to critical inquiry and test" (Dewey 1963, p. 151).[9] Further examples can be proliferated *ad nauseum*: people who prefer a lazy life over an informed life;[10] people who choose ignorance to remain happy (ignorance is bliss!);[11] lifestyles that seek detachment from worldly things;[12] and many others[13]

[8] More modern evaluations of stateless societies should be more nuanced, as since the Kibbutz movement or the stateless BaMbuti peoples suggest more mixed results than the traditional portrayals of the state of nature.

[9] For a more nuanced assessment of religious reasoning, see Murphy (1993). Dewey sees these practices as a threat to democracy.

[10] Feyerabend writes to Lakatos: "I want to live a quiet, peaceful, lazy life" (in Motterlini 1999, p. 281). Or, consider Rousseau's experience (in Dent's words) where "it was only when he left Paris in 1756 to live at the Hermitage that he began to be able to experience true happiness, to take delight in the simple pleasures of existence and to enjoy his own lazy and dreamy nature" (Dent 2005, p. 183). See also Scollon and King (2004) and the massive sympathetic literature on Epicureanism.

[11] Mill rejects this view in *Utilitarianism*, claiming that no experienced judge would choose ignorance and happiness over intelligence and misery (Mill [1863] 1991, p. 140). Mill is softer in *On Liberty* when he says that the "superior worth of simplicity of life" is "never entirely absent from cultivated minds" (Mill [1859] 1991, p. 53). For a fascinating related case study in terms of Chinese philosophy, see Matthyssen (2021).

[12] The Catholic mystic, Jeanne Guyon, anticipating aspects of Nietzsche's *amor fati*, for example, writes: "When the soul is docile, and leaves itself to be purified, and emptied of all that which it has of its own, opposite to the will of God, it finds itself by little and little, detached from every emotion of its own, and placed in a holy indifference, wishing nothing but what God does and wills" (Guyon [1791] 1995, p. 36). Similarly, in the Pali Canon of Buddhism, Upeksha ("state of entire indifference") is the fourth and final stage of Dhyana (which is what is ultimately desirable) which was influential on Schopenhauer. For related discussion, see Dumoulin (1981).

[13] Feyerabend uses the example of the BaMbuti peoples (using the antiquated and ambiguous term 'Pygmies'), but only cites research showing that they don't wish to engage with experiments in

cannot be criticized by the Millian without begging the question. While these practices, insofar as they do not violate HP, would be permissible for Mill, they would still involve a violation of their positive duties to promote happiness. Feyerabend, as we shall see, actively embraces such experiments in living that Mill seems committed to condemning.

While Feyerabend was often keen to showcase circularities, he goes further than this. Specifically, his criticism of 'rationalism', defended adamantly in Popper's *The Open Society and its Enemies*, points us towards an *embrace* of such 'arational' or 'irrational' lifestyles rather than mere toleration. Popper envisages a society in which we make collective decisions through exchanging reasons. In cases of peoples who do not live life in this way, Popper allows for "some form of imperialism" (Popper [1946] 1962, p. 181). While Mill was subtler than Popper on this point (see Tunick 2006), it remains a commitment for a Millian who values happiness more fundamentally than liberty.[14] Not only was Feyerabend personally repulsed by Popper's suggestion, but he thinks that this social philosophy "removes ideas and blocks actions we might want to retain" (Feyerabend 1987, p. 15). The reason is that many achievements prized by Popper (and Mill), like modern science, require 'irrationality' as a part of their discovery. More acutely, the development of modern science required flaunting rules of 'rationality' without context-specific reasons to justify their behavior (the reasons that "justify" their behavior are the *result* of their activities, rather than the *cause* of them). This is the central point of *Against Method* and repeated throughout his later writings. Galileo had (little to) no legitimate grounds to develop the Copernican hypothesis as he did—the fact that his work was essential for progress is only a retrospective judgment vindicated by the eventual unforeseen epistemic consequences of his research. While one might object that this shifts the target from moral decision-making to science, the change of terrain would not be worrisome for Feyerabend. Following Kierkegaard, he writes "[s]cientific method is a part of a general theory of man. It receives its rules from this theory and is built up in accordance with our ideas of a worthwhile human existence" (Feyerabend 1981, p. 67). Feyerabend argues for pluralism in science and ethics,

---

living outside of the Ituri forest rather than avoiding experiments in living altogether (Feyerabend 1987, p. 62 fn. 34).

[14] Mill is unequivocal on this point: "I regard utility as the ultimate appeal on all ethical questions; but it must be utility in the largest sense, grounded on the permanent interests of man as a progressive being. Those interests, I contend, authorize the subjection of individual spontaneity to external control only in respect to those actions of each which concern the interest of other people. If anyone does an act hurtful to others, there is a *prima facie* case for punishing him by law or, where legal penalties are not safely applicable, by general disapprobation. There are also many positive acts for the benefit of others which he may rightfully be compelled to perform, such as to give evidence in a court of justice, to bear his fair share in the common defense or in any other joint work necessary to the interest of the society of which he enjoys the protection, and to perform certain acts of individual beneficence, such as saving a fellow creature's life or interposing to protect the defenseless against ill usage—things which whenever it is obviously a man's duty to do he may rightfully be made responsible to society for not doing" (Mill [1859] 1991, p. 15). Feyerabend, therefore, entirely misreads Mill when he says that "[m]aximum happiness plays no role in *On Liberty*" (Feyerabend 1981, p. 70 fn. 10). For further discussion, see Okruhlik (2025).

seeing the two as (ideally) a continuous practice, from *Against Method* onwards. Feyerabend's criticism, therefore, does not merely point out the circularity inherit in some applications of a Millian ethos but adds that, even if this criticism was valid, it would erase the conditions under which science thrives:

> And this is not an exception: it is the normal case: theories become clear and 'reasonable' only after incoherent parts of them have been used for a long time. Such unreasonable, nonsensical, unmethodical foreplay thus turns out to be an unavoidable precondition of clarity and of empirical success.[15] (Feyerabend [1975] 1993, pp. 17–18)

A defender of Mill might respond that we can reconstruct these practices in such a way that they are rational after all. Although historical agents may not have been consciously following rational rules in the moment, their 'irrationality' could be construed as a means to progress and thereby can be permitted as 'rational' means to worthwhile ends. Feyerabend's response was representative of the practice-centered approach that characterizes his thought from (at least) *Against Method* onwards:

> Occasionally [philosophers] succeed in finding a machinery that might work in certain ideal conditions, but they never inquire, or even find it worth inquiring, whether the conditions are satisfied in this real world of ours. Such an inquiry, on the other hand, will have to explore the way in which scientists actually deal with their surroundings, it will have to examine the actual shape of their product, viz. 'knowledge', and the way in which this product changes as a result of decisions and actions in complex social and material conditions. In a word, such an inquiry will have to be anthropological. (pp. 196–197)

Rational reconstructions are only helpful if they can be de-idealized to concrete situations. How would we know that particular instances of 'irrational' behavior will lead to progress, like what has happened historically? To know this, we'd have to be able to predict their epistemic consequences, which Feyerabend thought was impossible (at least for the long-term) (see Shaw 2022a).[16] This criticism, moreover, would severely impact the viability of a Peircian retreat[17] whereby Mill's pluralism is restricted to specific audiences (thus conceding to Feyerabend's 'relativism'): even those who highly prize modern science must embrace the irrationality of

---

[15] At various points, Feyerabend makes the same point for science and morality in the same breath (e.g., "[p]rogress has always been achieved by probing well-entrenched and well-founded forms of life with unpopular and unfounded values. This is how man gradually freed himself from fear and from the tyranny of unexamined system" (Feyerabend 1970, pp. 209–210). Feyerabend's case study of the Galileo affair might fruitfully be read as a case study for the history of science and values since Galileo violated the best scientific and moral arguments of his day.

[16] Laurie Paul's argument that 'transformative experiences', by their very nature, are impossible to predict as to whether they would be rational to engage in (Paul 2014) is analogous to Feyerabend's argumentative structure: transformative experiences change one's value priorities in unpredictable ways, and so our reasons for or against engaging in transformative experiences will differ pre- and post-experience. For related discussion about unpredictability and long-term moral decision making, see Tarsney (2023).

[17] This is in reference to Peirce's strategy in the "Fixation of Belief" (Peirce 1877/1955) where he restricts his conception of the 'scientific method' to people who are willing to flea the security of cherished beliefs and learn from experience.

human behavior, where 'experiments' can be conducted even in the face of seemingly rational constraints. This would mean that, ironically, Mill himself would not be the appropriate audience for Mill's pluralism.

A second criticism of Mill's pluralism is methodological. The Millian view rests on a notion of moral progress where we gradually eliminate some kinds of experiments in living as legitimate practices. The limits of pluralism are discovered *a posteriori*. On this point, Gerald Gaus writes,

> But a claim to liberty to explore these discredited ways of living cannot be based on their manifestation of core values on individuality, reason, and progress. The point of experimentation is to learn, and to learn is to discard discredited claims. It is simply incoherent to adopt the progress-through-experimentation approach and yet argue that all options remain as viable and worthy alternatives. (Gaus 2021, p. 120)

The HP must be established as inviolable by experiments in living (if we can violate the HP, then the HP does not genuinely constrain pluralism). This means that, once upon a time, there were no legitimate limits on pluralism but, through experimentation, we discovered the HP. Mill's tone suggests that he accepted this position, at least for some of his moral claims. For instance, in his discussion of whether the higher pleasures are properly so-called, he writes that there can "be no appeal" and that the conclusion "must be admitted as final" (Mill [1863] 1991, p. 141). More to the point, Mill explicitly adopts this view of moral progress. In defending himself from the criticism that his view entails that "the fruits of conquest [moral discoveries] perish by the very completeness of the victory", he writes:

> I affirm no such thing. As mankind improve, the number of doctrines which are no longer disputed or doubted will be constantly on the increase; and the well-being of mankind may almost be measured by the number and gravity of the truths which have reached the point of being uncontested. The cessation, on one question after another, of serious controversy is one of the necessary incidents of the consolidation of opinion – a consolidation as salutary in the case of true opinions as it is dangerous and noxious when the opinions are erroneous. But though this gradual narrowing of the bounds of diversity of opinion is necessary in both senses of the term, being at once inevitable and indispensable, we are not therefore obliged to conclude that all its consequences must be beneficial. The loss of so important an aid to the intelligent and living apprehension of a truth as is afforded by the necessity of explaining it to, or defending it against, opponents, though not sufficient to outweigh, is no trifling drawback from the benefit of its universal recognition. Where this advantage can no longer be had, I confess I should like to see the teachers of mankind endeavoring to provide a substitute for it – some contrivance for making the difficulties of the question as present to the learner's consciousness as if they were pressed upon him by a dissentient champion, eager for his conversion. (Mill [1859] 1991, p. 49)

If this were applied to the HP, Mill would be committed to the view that we might engage in academic exercises defending it from criticisms and alternatives, but we cannot seriously engage in *practices* that violate its imperatives. That is, the HP is not beyond criticism but it is beyond reproach in terms of experiments in living. Thus Mill's assurances of the finality of particular discoveries are far from rhetorical flourishes: they are constitutive of his view.

Feyerabend did not accept this view of progress. As Luca Tambolo (2015) masterfully demonstrates, for Feyerabend, progress consists in "an ever-increasing ocean of alternatives each of them forcing the others into greater articulations, all of them contributing, via this process of competition, to development of our mental faculties" (Feyerabend [1965b] 1981, p. 107). From early in Feyerabend's career, he argues that there is no point at which we can say that a particular theory is definitely eliminated (Feyerabend 1960, p. 233). Moreover, he admonishes us to keep 'falsified' research programs alive to be vital parts of the pluralistic competition of ideas (da Silva Oliveira 2021).[18] This view (the 'principle of tenacity') is originally justified by the fact that seemingly hopeless research programs might make comebacks; as Ryan Muldoon puts it, "perspectives we have abandoned can, in different circumstances, become useful anew" (Muldoon 2015, p. 183). Later in his writings, this is further buttressed by the Millian reasons that keeping them alive is necessary for ensuring that we understand our own views better and why we have rejected them. This is constitutive of maximizing pluralism. By 'keeping alive', Feyerabend means retaining 'falsified' practices *as practices*—not merely talking points. Quoting Daniel Cohn-Bendit with approval, Feyerabend writes "only if the plurality of ideas is allowed *to express itself in social practice* does this idea have any meaning" (Feyerabend 1981, p. 66 fn. 4, emphasis in original[19]). Because of this, pluralism only *grows* with progress rather than shrinks. Therefore, Feyerabendian progress does not allow us to establish the HP, or any limit for pluralism, such that those limits are beyond testing. Feyerabend's pluralism runs much deeper than Mill's and his refusal to accept any claim as final, even the most well-confirmed *a posteriori* suppositions, permits the continued exploration of pluralism without limits. As Nietzsche put it: "Whether we immoralists do any harm to virtue? - Just as little as anarchists do to princes. It is only because they have been shot at that they once again sit securely on their thrones. Moral: we must shoot at morals" (Nietzsche [1888] 1990, §36). Interestingly enough, Mill adopts the same methodological view:

> In the case of any person whose judgment is really deserving of confidence, how has it become so? Because he has kept his mind open to criticism of his opinions and conduct. Because it has been his practice to listen to all that could be said against him; to profit by as much of it as was just, and to expound to himself, and upon occasion to others, the fallacy of what was fallacious. Because he has felt that the only way in which a human being can make some approach to knowing the whole of a subject is by hearing what can be said about it by persons of every variety of opinion, and studying all modes in which it can be looked at by every character of mind. No wise man ever acquired his wisdom in any mode but this; nor is it in the nature of human intellect to become wise in any other manner. The steady habit of correcting and completing his own opinion by collating it with those of others, so

[18] This claim puts Feyerabend at odds with more moderate thinkers like Milton, who claimed that some views should be allowed to be pursued in public space but "if they be found mischievous and libelous, the fire and the executioner will be the timeliest and the most effectual remedy that man's prevention can use" (Milton [1957] 2003, p. 749). There is no period at which this retrospective judgment becomes definitive, for a Feyerabendian.

[19] Similarly, in an early (currently unpublished draft of *Against Method*, Feyerabend writes "That is why it is necessary not only to think through alternative forms of life (Buddhism; magical thinking; schizophrenia) that one wants to try out, but to experience them concretely" (translation, Matteo Collodel). He quotes Mill's concept of experiments of living with approval a couple of pages later.

> far from causing doubt and hesitation in carrying it into practice, is the only stable foundation for a just reliance on it; for, being cognizant of all that can, at least obviously, be said against him, and having taken up his position against all gainsayers – knowing that he has sought for objections and difficulties instead of avoiding them, and has shut out no light which can be thrown upon the subject from any quarter – he has a right to think his judgment better than that of any person, or any multitude, who have not gone through a similar process. (Mill [1859] 1991, p. 25)[20]

Matt Brown has argued, convincingly I think, that Feyerabend is the more consistent of the two since Mill's own arguments for pluralism should apply to the HP as well.[21] While Mill tries to distinguish freedom of speech from freedom of experiments in living, this would not work for the Millian because the two are practically interconnected. In his fourfold defense for freedom of speech,[22] Mill constantly asserts that diversity of opinions must be a *practical* reality rather than mere academic exercises or reminders about possible opinions that might conflict with one's own (p. 51). Mill, seemingly arbitrarily, has removed the HP from his own arguments for pluralism. In other words, a consistent Millian should be a Feyerabendian: alternatives to the HP should be considered *and practiced* as well.

We should, however, not pretend as if the stakes are not much higher when discussing moral progress rather than scientific progress. The worst thing that would happen if we were allowed to explore science without epistemic constraints would be that we wasted time and resources. While these are important considerations, morally open-ended experimentation can ruin and end lives and societies. Intellectual arguments, no matter how powerful or cogent they may seem, falter when confronted with some concrete miseries. "Suffering", as Gandhi rightly put it, "appeals not to the intellect, it pierces the heart" (Gandhi [1939] 1999, p. 196).[23] In Feyerabend's words: "[a]rguments do have power—I admit this—but they effect only a small minority and they affect their brains not their heart unless we find a way to combine reason and emotion" (Feyerabend 1991, p. 512). The same is true for happiness, whose glow is hard to deny once felt or witnessed with due empathy. Feyerabend was well-aware that he wasn't playing an intellectual game without consequences. Because of this, it should be unsurprising that Feyerabend was not entirely welcoming of an open-ended pluralism in ethics. This can be seen in his correspondence with Ivan Jakšić.[24] Jakšić writes

> Before being able to saw off the branch one's sitting on, it is first necessary to get up on to it. A very pedagogical approach, I grant, but useful. It is more powerful to hear you say that models for living are far more necessary than theoretical discussions, after seeing the many battles that you have fought against theoretical rigidity. (August 1981)

[20] Feyerabend approvingly quotes a part of this passage (Feyerabend 1981, p. 68).

[21] This argument was made in the Q&A of Kathleen Okruhlik's talk on Mill and Feyerabend at VSMT in 2018 (the corresponding paper is published in this volume, Okruhlik 2025). It has taken me five years to appreciate the insightfulness of Brown's comment.

[22] See Lloyd (1997) and Okruhlik (2025) for discussions of these arguments as they appear in Mill and Feyerabend.

[23] Or, in Arne Næss' words, "man has a heart, not only a brain" (Næss 1984, p. 266).

[24] The exchange can be found in the Feyerabend archive at Konstanz University (PF 2–9-2).

The metaphor of sawing off the branch one is sitting on concerns the central idea behind the notion of an experiment in living: the 'test' comes from *lived experience*. Jakšić expects Feyerabend to be warm to this idea, given Feyerabend's frequent favoring of the richness of particulars experiences over abstract principles. This theme repeats itself in *Farewell to Reason* (Feyerabend 1987, pp. 294–297) and is central to *Conquest of Abundance* (Feyerabend 1999). To Jakšić's surprise, Feyerabend rejects this claim in his responding letter:

> This point of view contains ingredients that I feel rather uncomfortable with. For example, you say 'before being able to saw off the branch one's sitting on, it is first necessary to get up on to it.' I wholeheartedly disagree. For example, would you apply this to Nazism which is a branch of sorts? All I need to do is look at it and know that I neither want nor need to get up there.

There are two claims here. First, lived experience is not necessary, though it might be sufficient, for not engaging in certain experiments in living (cf. Feyerabend 1987, p. 314). This is a departure from most of Feyerabend's published views,[25] although one that seems reasonable. Experiments in living require that the 'test' is *public*. Otherwise, we cannot learn from second-hand testimonies about experiences and the would-be pluralist is committed to the claim that every individual, to know that torture is not for them, would need to experience torture for themselves (over and over again, moreover, in case they forget the experience). For pluralism to be at its best, experiments in living should not be construed as necessary for testing ethical theories—although they may be helpful epistemic exploration or direct tests. A more modest view, then, would be that experiments in living are more powerful and epistemically secure teachers of moral lessons but they are not, strictly speaking, necessary for testing moral claims. The second point is that Feyerabend claims that there are limits to experiments in living. This point is repeated and made more substantive in a later interview, where he says

> There may still be some people who say we cannot interfere because this is a culture of its own and we must not touch any culture which has gone according to different history and so on and to this I say "radically no." I mean where there is misery, there must be interference and sometimes even warlike interference. So my limit is when people start suffering.

Similarly, when discussing freedom of thought, Feyerabend makes it abundantly clear that he supports restrictions on pluralism:

> Writers love to point out that the pen is mightier than the sword. Well if they are right, then it is also more dangerous. Just imagine the following case: a society on the verge of civil war; a writer writing a book that might start it. As a responsible ruler I would order the book burned and the writer imprisoned if he did not give his solemn promise to wait for less dangerous times; as far as I'm concerned human lives are much more important than words pretending to represent ideas. (Feyerabend 1991, p. 500)

[25] This is Feyerabend's stance in his more careful moments: "[i]t makes no sense to me to condemn and even attack a movement, a culture, an idea from afar, without having tried to live with it, *or without detailed reports from people on the spot*" (Feyerabend 1991, emphasis added, p. 511).

However, in each case, Feyerabend uses the first person ('*I* feel rather uncomfortable…', '*my* limit is…', 'As far as *I'm* concerned…'). This is not an accident:

> "How will you decide the question [between right and wrong]?" means "Whom will you ask?", or "What method will you use?" or "To which principles will you refer?". Now there are of course people who do recognize outside authorities. I am not one of them. My only authority in such difficult matters is my very limited experience and my love of particular people… (p. 520)

I will return to this point in 2.5.

So Feyerabend was conflicted about delimiting pluralism and, possibly, the necessity of experiments in living in moral learning. How can this tension within Feyerabend's thought be reconciled? Much work in the liberal tradition that values pluralism takes its cue from, roughly, the HP and so showing how Feyerabend's criticisms can be addressed without allowing for an unlimited pluralism would be significant. Given that Mill and Feyerabend reject *a priori* approaches to ethics, and Feyerabend rejects *a posteriori* constraints on pluralism, we must look elsewhere to resolve Feyerabend's conflict. A distinct place to start would be a Kantian approach. One possible candidate for this approach can be found in the works of Jürgen Habermas.

2.4 Habermas and Deliberation

Habermas' approach shares many similarities with Mill and Feyerabend, with an emphasis on discourse. His discourse ethics starts with idea that moral questions, to be settled intersubjectively, are answered deliberatively with deliberation that satisfies the following two principles:

> Discourse Principle: "only those norms can claim to be valid that meet (or could meet) with the approval of all affected in their capacity as participants in a practical discourse." (Habermas 1990, p. 66)

> Universal Discourse Principle: "All affected can accept the consequences and the side effects its general observance can be anticipated to have for the satisfaction of everyone's interests (and these consequences are preferred to those of known alternative possibilities for regulation)." (Habermas 2004, p. 146)

These principles are rich in content, deriving much from Habermas' conception of 'universal pragmatics.' While I cannot go over all the details here, sufficed to say that they imply concrete norms binding on all such as abstention from violence, avoiding contradictions, or only asserting what one believes. As a result, the foundations of ethics reside in the norms governing the conditions under which discourse is possible. Violating these norms would be tantamount to violating the conditions under which ethical judgments can be made in the first place.

This approach is neo-Kantian in nature. It is Kantian because it employs a transcendental approach, in this case it grounds moral norms on the conditions of the possibility of deliberation, rather than *a priori* stipulations or inductive generalizations from raw facts. It is 'neo' Kantian because these conditions are not purely formal [as they are in Apel (1972)], but also require empirical facts about the processes of deliberations.[26] To illustrate the basic idea, considering the following from Matt Brown:

> Biological values come *first* in two respects. They are historically or chronologically the first kinds of value that come into existence with very basic organisms. (Or, if you are uncomfortable attributing values to nonhuman organisms or some subset thereof, with the very first humans). They also come first in that they are basic preconditions for other types of value. It is difficult if not impossible to fight for justice when one lacks food to eat or sufficient resources to nurture one's own children. (Brown 2020, p. 119)

Analogously, it seems impossible to deliberate when one is dead, interlocutors refuse to listen to each other, or severely malnourished. At times, similar arguments are used to prevent some kinds of deliberations; namely, those that practically arrest other deliberations. Catherine MacKinnon, for example, contends that some kinds of speech or expressions (e.g., sexist or racist arguments within sexist or racist societies, pornography) not only violate HP by causing harm (MacKinnon 2007, p. 950), but they also do not *add* to diversity of opinions but subtract from it by arresting, marginalizing, or downgrading arguments or testimonies from some peoples. The contingent needs of beings like us to deliberate must be met to discover or evaluate particular moral claims. Since norms for behavior depend on deliberation, arresting deliberation is akin to assuming infallibility—something Mill and Feyerabend rejected full stop.

One might think that the circularity worry applies here as well. Why deliberate in the first place? Many will see deliberation as pointless, at best, and so there is no need to respect the derived norms of behavior.[27] The response Habermas (and Apel) offer is that refuting discourse ethics involves a pragmatic contradiction. Specifically, defenders of non-deliberative worldviews must assume that their views should be respected. This is a moral claim, that is supposed to be binding on all parties. To justify this, they would need to rely on the results of deliberation, thus forcing them by their own lights into the deliberative worldview. A naturalistic approach like Mill's struggles to be objective for those who don't take its assumptions on board, but a neo-Kantian approach, according to Apel and Habermas, is binding for all (Apel 1980, p. 277; Habermas 2004).[28]

Prima facie, this approach might be conciliatory to a Feyerabendian whose pluralism requires ongoing dialogues. However, Feyerabend's pluralism makes the transcendental approach more worrisome than it might first appear. In *Science in a*

[26] I am unaware of any place where Mill outlines a general view on deliberative procedures, but says they should be determined "by the circumstances of the individual case" (Mill [1859] 1991, p. 61). He does, however, unequivocally condemn "want of candour, or malignity, bigotry, or intolerance" (ibid).

[27] "Anyone who posits inspiration must consistently regard every critical deliberation-whether for or against- as something dubious, a kind of temptation" (Kierkegaard [1846] 1992, p. 26).

[28] It's not clear, to me at least, how Habermas' view would be appealing to nihilists.

Free Society, Feyerabend wishes to extend equality to *all existing* traditions. He writes,

> A free society is a society in which all traditions are given equal rights, equal access to education and other positions of power... choosing one tradition as a basis of a free society is an arbitrary act that can be justified only by resort to power. A free society thus cannot be based on any particular creed... The basic structure of a free society is a protective structure, not an ideology, it functions like an iron railing and not like a conviction. (Feyerabend 1978, p. 30)

This epistemic egalitarianism has implications for the proper carrying out of deliberative procedures:

> in the first case some or all participants adopt a well specified tradition and only those responses that correspond to its standards. If one part has not yet become a participant of the chosen tradition he will be badgered, persuaded, 'educated' until he does – and then exchange begins... A rational debate is a special case of a guided exchange. If the participants are rationalists then all is well and the debate can start right away. If only some participants are rationalists [then]... a society based on rationality is not entirely free; one has to play the game of the intellectuals. An open exchange, on the other hand, is guided by a pragmatic philosophy. The tradition adopted by the parties is unspecified in the beginning and develops as the exchange goes along. The participants get immersed into each others' way of thinking, feeling, perceiving to such an extent that their ideas, perceptions, world views may be entirely changed – they become different people participating in a new and different tradition. An open exchange respects the partner whether he is an individual, or an entire culture while a rational exchange promises respect only within the framework of a rational debate. (pp. 29–30)

To illustrate, imagine a society with a significant amount of people who belong to a tradition wherein it is believed that we can speak to the dead sufficiently well enough for the purposes at hand [see e.g., Pieters (2019)]. In this society, we have a deep disagreement about whether death arrests deliberations and, therefore, the conditions for the possibility of deliberation. (Or, if one prefers a less extravagant example, those who deeply distrust fellow humans to enter into good-willed deliberation will deny that legitimate deliberations are possible). Notice that the death-speaker response does not deny the value of deliberation, and so they can agree abstractly with the neo-Kantian, but disagree about substantive procedural norms. While some might take this response to be the last refuge of a desperate mind, it is difficult to refute without overcoming the circularity argument mentioned above. To refuse to respect beliefs because of how irrational they may seem, moreover, requires the assumption that future deliberations would not reveal their truth. Thus there is a paradox for the neo-Kantian who wishes to discard death-speakers' arguments: they must presume what deliberation will reveal while presuming that deliberation has stable set of preconditions.[29] As a result, Feyerabend paralyzes transcendental arguments for restrictions on pluralism as they only work in societies

[29] One might respond by defending an iterative approach, like that suggested by Muldoon (2015). Feyerabend's rejection of the iterative approach is grounded in the view that pluralism is often not the *result* of problems with established views (as in Kuhn) but the *cause* of those problems. *First*, we allow death-speakers to develop their practices, thus forcing us to take their concerns seriously

where the relevant kinds of pluralism don't exist. Moreover, even in situations where we have sufficient agreement to deliberate, Feyerabend's principle of proliferation, where we should "invent and elaborate theories which are inconsistent with the accepted point of view, even if the latter should happen to be highly confirmed" (Feyerabend [1965b] 1981, p. 105), admonishes us to create such traditions thus inviting difficulties into deliberative procedures where none existed. He writes,

> There is always the possibility that new forms of thought will arrange matters in a different way and will lead to a transformation even of the most immediate impressions we received from the world. Considering this possibility, we may say that the long-lasting success of our categories and the omnipresence of a certain point of view is not a sign of excellence or an indication that the truth or part of the truth has at last been found. It is, rather, the indication of a *failure of reason* to find suitable alternatives which might be used to transcend an accidental intermediate stage of our knowledge. This remark leads to an entirely new attitude towards success and stability. (Feyerabend 1981, p. 72)

Because of this, even in societies where we have sufficiently agreed upon standards for deliberation, we still wouldn't have grounds to constrain pluralism. Thus even in the safest possible scenarios, the Feyerabendian cannot avail themselves to a neo-Kantian approach to the limits of pluralism.

2.5 Appeals in Moral Philosophy

Given that *a priori*, *a posteriori*, and neo-Kantian approaches to delimiting pluralism seem untenable, it looks like the Feyerabendian must abandon some central commitment or accept a pluralism without limits. We should not, I think, fall into the pits of despair this quickly. Rather, I would like to try one more route for fleshing out a Feyerabendian pluralism through the notion of *appeals* found in existentialist philosophy.

The notion of 'appeals' appears throughout the writings of several existentialist thinkers, especially Jean-Paul Sartre, in his *Notebooks for an Ethics* (Sartre 1992), and Simone de Beauvoir in, amongst other places, *Pyrrhus et Cinèas* (de Beauvoir [1944] 2020). Sartre and de Beauvoir see the existentialist as trapped in a predicament. On the one hand, our freedom to decide how to live is absolute—the death of God permits everything, making each individual the author of their own ethos. Existence precedes essence. On the other hand, their experiences with the Nazi invasion of France compelled them to formulate stronger approaches to ethics. If the Nazis cannot be condemned, then an existentialist ethics would hardly be worth the paper it is printed on. But how can this be done, without running into the problems mentioned earlier?

No deep knowledge of existentialism is necessary to see that the existentialist is captured in a similar dilemma as Feyerabend: between an open-ended ethical

(in a Feyerabendian free society), and *then* we figure out what merit their views might have (which, as mentioned earlier, can never be established definitively).

philosophy on the one hand, and a critically impotent one on the other. Sartre and de Beauvoir address this situation through the notion of an *appeal*. Appeals are honest and sincere expressions aiming to convince interlocutors. Appeals might rely on reasons, emotions, artwork, or rhetoric and they may not be verbal—screeches in pain or beaming smiles might be appeals as well (cf. Feyerabend 1987, p. 81; Herzog and Serrano-Zamora 2024).[30, 31] Appeals are not equivalent to intuitions, although appeals are more likely succeed if there are shared intuitions. The goal of appeals is to get interlocutors—similarly given free reign to choose whichever morals they wish—to join one in holding certain values, or trying to jointly realize certain goals. While de Beauvoir, and Levinas (1969) in more detail, stressed the phenomenological effectiveness of this strategy in face-to-face situations, appeals might also be made in an indefinite number of forum. The quality of an appeal will depend on contingent circumstances: "Differences of class, of nationality, of form limits the appeal in its nature" (Sartre 1992, p. 285). Because of this, as they say in the arts, one must know their audience. While we might try to construct situations in which appeals are 'best' made, in keeping with the ambitions of Habermas, Feyerabend, and Mill, appeals alone do not necessarily carry this baggage. To illustrate, a victim of assault may scream for mercy and while this would be a horrendous and ineffectual situation to make an appeal, the existentialist acknowledges that this might be the best (or only) way forward given the contingencies of the situation. While Feyerabend appears unaware of this aspect of existentialist thought, he was no stranger to the concept. His experiences with Berthold Brecht and the theatre revealed the poverty of reasoning when compared to the dramatic impact that plays can have on their audiences (see Couvalis 1987).[32] Appeals aim to secure common bases for ethical reasoning that are not purely rational. The 'objectivity' of ensuing discourses, therefore, is rooted in manufactured consensus. Appeals have no guarantees to be successful, but they can bridge gaps that ethical or practical reasoning cannot cover.

Sartre never completed his project to base an ethics on appeals, although de Beauvoir did. Similar to Habermas, de Beauvoir takes a neo-Kantian approach where appeals are only possible when one can speak and others listen, and builds a phenomenology of discourse about how this process must proceed. For the reasons given in the previous section, this approach is unavailable to a Feyerabendian. Still, appeals might help Feyerabend establish limits to his pluralism. Rationally speaking, there is nothing that makes appeals intrinsically more desirable than exercises of power. However, what drives de Beauvoir, Sartre, Mill, Habermas, and

[30] Sartre and de Beauvoir emphasise authenticity, and so appeals cannot be propagandistic. Feyerabend, on the other hand, is explicitly open to propaganda. However, he says nothing about paternalistic approaches to ethics (where propaganda has an advantage over honesty) so this issue must be left unresolved.

[31] For related discussion on use of artistic expressions (e.g., dance, theatre) in deliberative democratic forum see Herzog and Serrano-Zamora (2024).

[32] Relatedly, recent developments in cognitive psychology have emphasized the importance of *images* over *thoughts* (Blackwell 2021).

Feyerabend[33] is the desire to avoid coercion and violence wherever possible (see e.g., Feyerabend 1981, p. 70 fn. 7). This desire cannot be construed as an ethical principle, otherwise we'd be back to square one, but it might be desirable for enough of us just by virtue of how well-entrenched this desire is across history and cultures. Human beings typically don't like violence, especially when it is done to them. However, proliferation must be tempered to make this solution viable. While proliferation makes common bases intrinsically unstable, it will become drastically muted depending on how effective appeals have been. Without a whiff of conservatism,[34] where the burden is on proliferators to convince others to suspend or abandon accepted ethical norms, the notion of an appeal will not help delimit pluralism. If we accept this approach to Feyerabend's conundrum, then specific claims of Mill's or Habermas' might be saved within a Feyerabendian ethics *given their overlapping desires* without worrying about potential conflicts of foundational principles.

As a brief aside, it is interesting to consider whether Habermas' view is ultimately conciliatory with the existentialist view. Habermas explicitly, though unclearly, distinguishes his view from Feyerabend's in that he sees a central place for reason in philosophy (Habermas 1990, p. 9).[35] When considering the objection that non-linguistic, non-rational animals do not count as moral stakeholders (e.g., torturing animals for fun might be a legitimate conclusion of a discourse), he writes:

> What moral theory can do and should be trusted to do is to clarify the universal core of our moral intuitions and thereby to refute value skepticism. What it cannot do is make any kind of substantive contribution. By singling out a procedure of decision making, it seeks to make room for those involved, who must then find answers on their own to the moral-practical issues that come at them, or are imposed upon them, with objective historical force. Moral philosophy does not have privileged access to particular moral truths. (p. 211)

This, and other similarly confusing passages, suggest that 'substantive morality' is not given by philosophical analysis or moral theories, but substance (e.g., don't torture animals) is only given by historically situated deliberators expressing their values and allowing them to be 'discursively tested.' This appears, however, to lead to the normative bankruptcy of discourse ethics unless these norms are interpreted as *appeals* rather than a "proper appreciation" of language, reason, and philosophy.

If we accept this solution to Feyerabend's conundrum, we are left with the question: what function would appeals have in a Feyerabendian ethics? One approach might be particular: we make appeals to specific individuals in specific situations to

---

[33] The same is true of Popper, whose distaste of violence is fundamental to his philosophy and unprovable: "I frankly confess that I choose rationalism because I hate violence, and I do not deceive myself into believing that this hatred has any rational grounds... I do not see that we can do beyond this" (Popper 1963, p. 357).

[34] "[S]ocial experiments should be made carefully, and in a piecemeal manner… Hence ideas with little support should not be introduced in an aggressive manner, they should be checked for their consequences and strengthened only when these are acceptable and better arguments become available" (Feyerabend 1987, p. 178; see also Muldoon 2016, pp. 33–35). This passage, however, should be taken with a grain of salt since Feyerabend is defending St. Bellarmino from Popper which is precisely the context in which Feyerabend's contrarian instincts kick in.

[35] For careful discussion and criticism of Habermas' views on rationality, see Hesse (1995).

engage in specific experiments in living. If appeals can occupy any nook and cranny of Feyerabend's ethics, then his pluralism would dissolve into a series individual appeals. Moreover, the resulting state of affairs would leave ethical decision-making entirely in the hands of local deliberators, which might be hopelessly naïve.[36] Another method of using appeals restricts them to *generic* pleas to practice pluralism. On this view, we might only make appeals for some very basic norms—e.g., anti-violence, pro-happiness, etc.—, build institutions and community norms around them, assuming the appeals are successful, that constrain more specific, 'lower level' ethical decisions. These appeals may still be revisable but they should stand the test of time fairly well if the appeals are sufficiently strong given their audiences.

Of course, generality is a matter of degree. The strength of the appeal will depend partly on the audience, but also partly on the purported strengths of the recommended lifestyles. Clearly Mill and Feyerabend think that an extremely, but not entirely, permissive pluralism has much fruits to offer for many audiences. The appeal, therefore, while not entirely 'rational', is not beyond reason altogether. Rather than thinking of the (purported) fruits of pluralism as something one is rationally compelled to embrace, the pluralist would simply gather fodder for their appeals. The audience for the Feyerabendian pluralist appeal is incredibly broad, at least according to Feyerabend. His thesis, in *Against Method*, is that his pluralism "helps to achieve progress in any one of the senses one cares to choose" (Feyerabend [1975] 1993, p. 18). However, this view cannot be coherently held (why progress?) without some appeal. In parts of Feyerabend's corpus where he makes calls for particular reforms (e.g., in education, foreign policy), this could be fruitfully read as appeals stemming from Feyerabend's *personal* convictions. From this, we can revisit Feyerabend's letter to Jakšić: he is making an *appeal* that Nazism serves as an example of particular experiments in living that are out-of-bounds. Even when confronting Nazis, Feyerabend's appeal has a chance if he is given more room to make substance for the basis of his appeal. (Successes at converting Nazis testify to the promise of appeals which might provide comfort for those who are worried about the dangerous possibilities opened up by grounding ethical deliberation in particular appeals).[37] While this appears to end with relativism defined by conditions under which appeals are successful, the 'boundaries' can be changed as more appeals or experiments in living are performed. Closer investigations of what this would practically look like depend on careful analyses of cultural and individual moral changes and the role(s) appeals play in them.

[36] Even most ethical particularists nowadays recognize that moral decision making cannot be entirely local (see Bakhurst 2007).

[37] See, for example, the testimonies of Daryl Davies for exemplars.

2.6 Concluding Remarks

As one can glean from this, Feyerabend's pluralism touches on deep questions concerning the nature and foundations of moral and political philosophy. As a result, I cannot pretend to have either provided a thoroughgoing picture of a Feyerabendian ethics nor can I say that, even when supplemented with the notion of an appeal, it should be adopted. Questions about its coherence and plausibility must be raised and answered carefully. Since Feyerabend scarcely engaged with research in ethics or political philosophy except around the edges (see Shaw 2022b) and so much has been written since then, we have a long-road ahead of ourselves if we were to devote ourselves to this task. However, I hope to have taken at least the first steps towards understanding and evaluating Feyerabend's ethical philosophy. The renaissance of Feyerabend's thought has multiplied over the past few decades, and we should first have an understanding of his views before discerning their merits. Feyerabendian pluralism, like all views, deserves to have its day in court.

Acknowledgements I'd like to thank Jon Tsou for the generous invitation to present and write this paper and his careful feedback, as well as the helpful comments and suggestions from the 2023 VMST meeting in Dallas, Texas. I'm also grateful for Ryan Muldoon's helpful suggestions and the thoughtful input from Dietmar Hübner although I was not able to follow up on all of their comments. I'm also honored to be a part of a volume in celebration of Matt Brown, who has been nothing short of wonderful to me since day one.

References

Anderson, E. 1991. John Stuart Mill and Experiments in living. *Ethics* 102 (1): 4–26.

Apel, K. O. 1972. The A Priori of Communication and the Foundation of the Humanities. *Man and World* 5 (1): 3–37.

Apel, K. O. 1980. *Towards a Transformation of Philosophy*. New York: Routledge.

Bakhurst, D. 2007. Pragmatism and Ethical Particularism. In *New Pragmatists*, ed. C. Misak, 122–141. Oxford: Oxford University Press.

Bamford, R. 2016. The Ethos of Inquiry: Nietzsche on Experience, Naturalism, and Experimentalism. *Journal of Nietzsche Studies* 47 (1): 9–29.

Bearn, G. 1986. Nietzsche, Feyerabend, and the Voices of Relativism. *Metaphilosophy* 17 (2/3): 135–152.

Bell, M. 2021. John Stuart Mill's Harm Principle and Free Speech: Expanding the Notion of Harm. *Utilitas* 33 (2): 162–179.

Blackwell, S. 2021. Mental Imagery in the Science and Practice of Cognitive Behaviour Therapy: Past, Present, and Future Perspectives. *International Journal of Cognitive Therapy* 14 (1): 160–181.

Brown, M. 2016. The Abundant World: Paul Feyerabend's Metaphysics of Science. *Studies in History and Philosophy of Science Part A* 57:142–154.

Brown, M. 2020. *Science and Moral Imagination: A New Ideal for Values in Science*. Pittsburgh: University of Pittsburgh Press.

Couvalis, S. 1987. Feyerabend's Epistemology and Brecht's Theory of the Drama. *Philosophy and Literature* 11 (1): 117–123.

De Beauvoir, S. 1944/2020. *Pyrrhus et Cinèas*. Paris: Gallimard.

Dent, N. 2005. *Rousseau*. New York: Routledge.
Dewey, J. 1963. *Freedom and Culture*. New York: Capricorn Books.
Dumoulin, H. 1981. Buddhism and Nineteenth-Century German Philosophy. *Journal of the History of Ideas* 42 (3): 457–470.
Feyerabend, P. 1960. Professor Bohm's Philosophy of Nature. *The British Journal for the Philosophy of Science* 10 (40): 321–338.
Feyerabend, P. 1965a. Problems of Empiricism. In *Beyond the Edge of Certainty: Essays in Contemporary Science and Philosophy*, University of Pittsburgh Series in the Philosophy of Science, ed. R. Colodny, vol. 2, 145–260. Englewood Cliffs: Prentice-Hall.
Feyerabend, P. 1965b. Reply to Criticism: Comments on Smart, Sellars, and Putnam. In *Realism, Rationalism and Scientific Method: Philosophical Papers, Volume 1*, ed. P. Feyerabend, 104–131. Cambridge: Cambridge University Press.
Feyerabend, P. 1970. Consolations for the Specialist. In *Criticism and the Growth of Knowledge*, ed. I. Lakatos and A. Musgrave, 197–231. Cambridge: Cambridge University Press.
Feyerabend, P. 1975/1993. *Against Method*. 3rd ed. London: Verso Books.
Feyerabend, P. 1978. *Science in a Free Society*. London: Verso Books.
Feyerabend, P. 1981. Two Models of Epistemic Change. In *Problems of Empiricism: Philosophical Papers, Volume 2*, ed. P. Feyerabend, 65–80. Cambridge: Cambridge University Press.
Feyerabend, P. 1987. *Farewell to Reason*. London: Verso Books.
Feyerabend, P. 1991. Concluding Unphilosophical Conversation. In *Beyond Reason: Essays on the Philosophy of Paul K. Feyerabend*, ed. G. Munèvar, 487–527. London: Boston Studies in the History and Philosophy of Science.
Feyerabend, P. 1999. *Conquest of Abundance: A Tale of Abstraction Versus the Richness of Being*. Chicago: University of Chicago Press.
Gandhi, M. 1939/1999. Requisite Qualifications. In *The Collected Works of Mahatma Gandhi*, ed. Navajivan Trust, vol. 75, 195–196. New Delhi: Government of India.
Gaus, G. 2021. *The Open Society and its Complexities*. Oxford: Oxford University Press.
Guyon, J. 1791/1995. *Autobiography of Madame Guyon*. Chicago: Moody Press.
Habermas, J. 1990. *Moral Consciousness and Communicative Action*. Translated by C. Lenhardt and S. Nicholsen. Cambridge: MIT Press.
Habermas, J. 2004. Discourse Ethics. In *Ethics: Contemporary Readings*, ed. H. Gensler, E. Spurgin, and J. Swindal, 146–153. London: Routledge.
Herzog, L., and J. Serrano-Zamora, J. 2024. The Democratic Public and the Practices of the Oppressed. *The British Journal of Politics and International Relations*, 13691481241287183.
Hesse, M. 1995. Habermas and the Force of Dialectical Argument. *History of European Ideas* 21 (3): 367–378.
Kierkegaard, S. 1846/1992. *Concluding Unscientific Postscript to Philosophical Fragments*. Edited and Translated by H. Hong and E. Hong. Princeton: Princeton University Press.
Kuby, D. 2021. Decision-Based Epistemology: Sketching a Systematic Framework of Feyerabend's Metaphilosophy. *Synthese* 199 (1–2): 3271–3299.
Levinas, E. 1969. *Totality and Infinity: An Essay on Exteriority*. Translated by A. Lingis. Pittsburgh: Duquesne University Press.
Lloyd, E. 1997. Feyerabend, Mill, and Pluralism. *Philosophy of Science* 64 (S4): S396–S407.
MacKinnon, C. 2007. Only Words. In *Law and Morality: Readings in Legal Philosophy*, ed. D. Dysenhaus, S. Moreau, and A. Ripstein, 3rd ed., 948–960. Toronto: University of Toronto Press.
Matthyssen, M. 2021. *Ignorance is Bliss: The Chinese Art of Not Knowing*. Springer Nature.
Mill, J. S. 1859/1991. On Liberty. In *On Liberty and Other Essays*, ed. J. Gray, 5–128. Oxford: Oxford University Press.
Mill, J. S. 1863/1991. Utilitarianism. In *On Liberty and Other Essays*, ed. J. Gray, 131–201. Oxford: Oxford University Press.
Millgram, E. 2000. Mill's Proof of the Principle of Utility. *Ethics* 110 (2): 282–310.

Milton, J. 1957/2003. In *The Complete Poems and Major Prose*, ed. M. Hughes. Indianapolis: Hackett Publishing Company, Ltd.

Motterlini, M., ed. 1999. *For and Against Method, Including Lakatos's Lectures on Scientific Method, and the Lakatos-Feyerabend Correspondence*. Chicago: University of Chicago Press.

Muldoon, R. 2015. Expanding the Justificatory Framework of Mill's Experiments in living. *Utilitas* 27 (2): 179–194.

Muldoon, R. 2016. *Social Contract Theory for a Diverse World: Beyond Tolerance*. New York: Routledge.

Murphy, N. 1993. *Theology in the Age of Scientific Reasoning*. Ithaca: Cornell University Press.

Næss, A. 1984. A Defence of the Deep Ecology Movement. *Environmental Ethics* 6 (3): 265–270.

Nietzsche, F. 1882/1974. *The Gay Science*. New York City: Vintage.

Nietzsche, F. 1888/1990. *Twilight of the Idols*. New York City: Vintage.

Okruhlik, K. 2025. John Stuart Mill, the Marketplace, and the Democratization of Science. In *Values, Pluralism, and Pragmatism: Themes from the Work of Matthew J. Brown*, Boston Studies in the History and Philosophy of Science, ed. J. Tsou, J. Shaw, and C. Fehr. Cham: Springer.

Paul, L. 2014. *Transformative Experience*. Oxford: Oxford University Press.

Pearce, T. 2020. *Pragmatism's Evolution: Organism and Environment in American Philosophy*. Chicago: University of Chicago Press.

Peirce, C. S. 1877/1955. In *The Philosophical Writings of Peirce*, ed. J. Buchler. New York: Dover Publications.

Pieters, J. 2019. *Speaking with the Dead: Explorations in Literature and History*. Edinburgh: Edinburgh University Press.

Popper, K. 1946/1962. *The Open Society and its Enemies*. London: Routledge and Keegan Paul.

Popper, K. 1963. *Conjectures and Refutations: The Growth of Scientific Knowledge*. London: Routledge & Kegan Paul.

Railton, P. 1986. Moral Realism. *The Philosophical Review* 95 (2): 163–207.

Sartre, J. -P. 1992. *Notebooks for an Ethics*. Translated by D. Pallauer. Chicago: University of Chicago Press.

Scollon, C., and L. King. 2004. Is the Good Life the Easy Life? *Social Indicators Research* 68:127–162.

Shaw, J. 2022a. On the Very Idea of Pursuitworthiness. *Studies in the History and Philosophy of Science* 19:103–112.

Shaw, J. 2022b. Feyerabend's Anarchism: A Political and Epistemological Analysis. *Zilsel* 2:191–204.

da Silva Oliveira, D. G. 2021. The Cosmological Divergent Proliferation in Feyerabend's Pluralism. *Principia: An International Journal of Epistemology* 25 (3): 2.

Snyder, L. 2006. *Reforming Philosophy: A Victorian Debate on Science*. Chicago: University of Chicago Press.

Tambolo, L. 2015. A Tale of Three Theories: Feyerabend and Popper on Progress and the Aim of Science. *Studies in History and Philosophy of Science Part A* 51:33–41.

Tarsney, C. 2023. The Epistemic Challenge to Longtermism. *Synthese* 201 (6): 195.

Thoreau, H. D. 1854/2012. *Walden*. Empire Books.

Tsou, J. 2025. Feyerabend's Realism and Expansion of Pluralism in the 1970s. In *Values, Pluralism, and Pragmatism: Themes from the Work of Matthew J. Brown*, Boston Studies in the History and Philosophy of Science, ed. J. Tsou, J. Shaw, and C. Fehr. Cham: Springer.

Tunick, M. 2006. Tolerant Imperialism: John Stuart Mill's Defense of British Rule in India. *The Review of Politics* 68 (4): 586–611.

Tunick, M. 2022. John Stuart Mill's Passage on Pimps and the Limits on Free Speech. *Utilitas* 34 (4): 392–408.

Turner, P. 2014. "Harm" and Mill's Harm Principle. *Ethics* 124 (2): 299–326.

Chapter 3
Feyerabend's Realism and Expansion of Pluralism in the 1970s

Jonathan Y. Tsou

Abstract This chapter aims to clarify the nature of a shift in Feyerabend's philosophical thinking in the 1970s, focusing on issues of realism, relativism, and pluralism. Against John Preston, who argues that Feyerabend shifts from a realist position to a relativist position in publications after *Against Method*, I argue that realism-relativism is a misleading variable for characterizing Feyerabend's shift. Rather, I characterize this shift as Feyerabend's expansion of pluralism and suggest that this shift appears in Feyerabend's publications starting in the late-1960s. Adopting the terminology of Matthew J. Brown and Ian James Kidd, Feyerabend's shift is characterized by a move from arguments for *pluralism within science* to broader arguments for *pluralism in society*. I argue that this shift was largely due to: (1) Feyerabend's rejection of an empiricist demarcation criterion for science in the mid-1960s, and (2) Feyerabend's incorporation of John Stuart Mill's arguments for pluralism into his own publications starting in 1970. This shift can be characterized as a broadening or expansion of the scope of Feyerabend's arguments for pluralism from contexts of scientific knowledge to the broader context of social or cultural (including ethical) knowledge.

Keywords Methodological pluralism · Scientific realism · Cultural relativism · Demarcation problem · Paul Feyerabend · John Stuart Mill · Karl Popper

J. Y. Tsou (✉)
Center for Values in Medicine, Science, and Technology, University of Texas at Dallas, Richardson, TX, USA
e-mail: Jonathan.Tsou@utdallas.edu

J. Y. Tsou et al. (eds.), *Values, Pluralism, and Pragmatism: Themes from the Work of Matthew J. Brown*, Boston Studies in the Philosophy and History of Science 347, https://doi.org/10.1007/978-3-031-92958-8_3

3.1 Introduction

This chapter addresses some outstanding historical issues concerning Paul Feyerabend's positions on realism and relativism in the 1970s. In the Feyerabend literature, there is general agreement that there is a discernible shift in Feyerabend's philosophical thinking sometime in the mid-1970s (Preston 1997a; Tsou 2003; Brown and Kidd 2016). John Preston (1997b) characterizes this shift as Feyerabend's 'retreat from realism to relativism': after the publication of the first edition of *Against Method*, Feyerabend (1975a) shifts away from realism towards a relativist position, as exemplified in *Science in a Free Society* (1978). Eric Oberheim (2006) contests Preston's presentation, suggesting that there's no discontinuity in Feyerabend's views on the issue of realism. On Oberheim's reading, Feyerabend consistently defended a pluralist metaphilosophical view (see Kuby 2021a), but this metaphilosophy did not commit Feyerabend to any specific philosophical positions, such as scientific realism. Hasok Chang (2021) also rejects Preston's narrative, arguing that Feyerabend never abandoned realism in his career. For Chang, Feyerabend began with a narrow view of realism ('pluralist realism') and broadened it over time.

My aim in this chapter is to clarify the nature of the shift in Feyerabend's philosophical thinking in the 1970s, focusing on issues of realism, relativism, and pluralism. Contra-Preston, I argue that realism-relativism is a misleading variable for characterizing Feyerabend's shift in the 1970s. Rather, I characterize this shift as Feyerabend's *expansion of pluralism* and suggest that this shift appears in Feyerabend's publications starting in the late-1960s (e.g., Feyerabend 1968b, 1969b, 1970a, 1970c). Adopting the terminology of Brown and Kidd (2016), this shift is characterized by a move from arguments for *pluralism within science* to broader arguments for *pluralism in society*. I argue that this shift was largely due to: (1) Feyerabend's rejection of an empiricist demarcation criterion for science in the mid-1960s, and (2) Feyerabend's incorporation of Mill's arguments for pluralism into his own publications starting in 1970. This shift can be characterized as a *broadening or expanding the scope of Feyerabend's arguments for pluralism* from contexts of scientific knowledge to the broader context of social or cultural (including ethical) knowledge. I articulate this argument with reference to the framework for classifying Feyerabend's works formulated by Brown and Kidd (2016). Against Brown and Kidd (and Preston), I argue that realism-relativism is a red-herring as a variable for distinguishing Feyerabend's early and middle works. Brown and Kidd's distinction between pluralism within science and pluralism in society is a more useful variable for characterizing Feyerabend's shift in thinking in the 1970s. Whereas Preston (1997) and Brown and Kidd (2016) identify the *beginning* of Feyerabend's pluralism in society phase in 1978 with the publication of *Science in a Free Society* (Feyerabend 1978), I suggest that the relevant shift occurs much earlier, and it appears in Feyerabend's publications starting in 1970 (Feyerabend 1970a, 1970c). On this understanding, Feyerabend's philosophical shift (c. 1970) should be viewed as an expansion (or broadening) of the scope of application of his pluralist arguments, inspired by Mill's ethical arguments for pluralism in society (Mill [1859] 1977).

With respect to Feyerabend's apparent shift to relativism in works after *Against Method* (Feyerabend 1975a), I present Feyerabend's path to relativism in terms of his changing position on the superiority of empirically testable scientific methods compared to the methods adopted in non-scientific theories (e.g., myths, metaphysical theories, religious theories). In works prior to 1966, Feyerabend assumed some Popperian demarcation criterion (scientific theories are distinguished by their empirical testability), and he held that non-scientific theories were useful resources for criticizing accepted scientific theories, but they themselves did not possess meaningful (cognitive) content. In works after 1966, Feyerabend rejects the idea of a demarcation criterion for science, and he regards non-scientific theories *as full-fledged epistemic alternatives to science* that possess their own specific forms of cognitive content. This shift marks Feyerabend's abandonment of empiricism insofar as he no longer views the empiricist methodology characteristic of modern science as epistemically superior to alternative 'non-scientific' methods. Feyerabend's retreat from empiricism in the mid-1960s forms the philosophical basis for his expansion of pluralism and eventual defense of 'anarchism' in the 1970s (Feyerabend 1970a, 1970d, 1975a). On the basis of these considerations, I date the beginning of Feyerabend's 'middle works' in 1970.

3.2 A Framework for Classifying Feyerabend's Works

Matthew J. Brown and Ian James Kidd (2016) propose a promising theoretical framework for distinguishing different phases of Feyerabend's published work.[1] In their taxonomy, Feyerabend's work is divided into early, middle, and later phases:

- *Early Work* (c.1951–1975): Here Feyerabend defends a type of semantic realism that is opposed to phenomenalism or verificationism, argues for pluralism within science, and his main goal is to defend science against philosophies of science that would damage it.

[1] For convenience, I follow Brown and Kidd's periodization of Feyerabend's career in terms of his *published works*, which simplifies historical issues regarding exactly when Feyerabend arrived at various views. However, this methodology creates some complications given that the first edition of *Against Method* (Feyerabend 1975a) was largely compiled from Feyerabend's 1960s publications (see Feyerabend 1995, pp. 139–140) that were published between 1963–1970 (Feyerabend 1963, 1965b, 1970a, 1970d). Regardless of the chronology of Feyerabend's published work, there are good reasons for following Brown and Kidd's proposal that *Against Method* is a 'transitional work' that features substantial aspects of his early work, but also touch on some themes from his middle work (Brown and Kidd 2016, p. 3). For a more comprehensive discussion of methodological issues involved in interpreting Feyerabend, see Collodel (2016, pp. 35–36) and Shaw (2020, p. 110). The usual exegetical disclaimers regarding interpreting Feyerabend apply in this chapter: Feyerabend is difficult to interpret given: (1) his frequently changing views and claim that he has "no position" (Feyerabend, quoted in Horgan 1996, p. 50), and (2) his tendency to joke, posture, and exaggerate (Feyerabend 1995, p. 142).

- *Middle Work* (c.1978–1987): Here, Feyerabend argues for pluralism in society, with science understood as one social tradition amongst others, and calls the view "relativism." His main goal is to defend society from science, in the sense of its having undue authority over society.
- *Later Work* (c.1989–1994): Feyerabend returns to an interest in realism and what he calls 'the problem of reality', explicitly denies relativism, and takes up metaphysics in earnest. The period is marked by a broader agenda of engaging with classics and the arts in addition to science and society. (Brown and Kidd 2016, p. 3)

This framework is useful for categorizing different phases of Feyerabend's work. Brown and Kidd identify two main variables, which distinguish Feyerabend's early works from middle works: (1) realism versus relativism, and (2) pluralism within science versus pluralism in society. Like Preston, Brown and Kidd identify a significant shift in Feyerabend's works sometime between *Against Method* (Feyerabend 1975a) and *Science and a Free Society* (Feyerabend 1978).

Brown and Kidd (2016) argue that their framework can reconcile Preston's and Oberheim's contrasting positions on Feyerabend's alleged shift in publications after *Against Method* (Feyerabend 1975a). In particular, they agree with Oberheim that there is significant continuity throughout Feyerabend's career, but they agree with Preston that there are noteworthy shifts of *emphasis* and *strategy* in Feyerabend's works. Regarding the question of whether there is a significant shift in Feyerabend's philosophical thought after 1975, Brown and Kidd argue that Feyerabend shifts towards relativism (and away from realism) and towards questions about how to defend society from science. In this regard, they largely endorse Preston's argument that Feyerabend makes a shift from realism to relativism in the 1970s and the beginning of this shift is marked by Feyerabend's publication of *Science in a Free Society* (Feyerabend 1978).

3.3 Problems with Brown and Kidd's Framework

In broad strokes, Brown and Kidd's framework offers an illuminating picture of different phases of Feyerabend's career arc and central topics that motivated his projects. While their framework is useful in identifying variables for distinguishing Feyerabend's works, I want to quibble with the details of their periodization of Feyerabend's early and middle phases.

With respect to Brown and Kidd's periodization of Feyerabend's early works (c. 1951–1975), I argue that Feyerabend is committed to a deflationary methodological ideal of realism ("pluralistic realism") that appears in his published works as late as 1981. Pluralistic realism is an unorthodox and deflationary *methodological ideal of realism*, which recommends that *scientists* should interpret scientific theories realistically rather than instrumentally. In his early arguments for pluralistic realism, Feyerabend (1958, 1960, 1962, 1964, 1965a, 1965c) defends a fallibilist and methodological ideal of scientific realism intended to *facilitate criticism in science*.

Feyerabend's methodological realism falls in the tradition of Popper's conjectural realism (Preston 1997a) insofar as it is formulated to promote *criticism* in science and actualize the self-critical ideal of science prescribed by Popper's method of conjectures and refutations (Popper 1963). Feyerabend's defense of realism is summarized by the methodological recommendation that scientific theories should be interpreted realistically (i.e., as descriptions of the world), rather than instrumentally (i.e., as useful instruments for summarizing observations or making predictions). Evidence that Feyerabend was a pluralistic realist as late as 1981 stems from Feyerabend's introduction to volumes 1 and 2 of his *Philosophical Papers* (Feyerabend 1981b). Therein, Feyerabend presents his earlier arguments for realism under the banner of pluralistic realism with no indication that he no longer accepts this form of realism. Hence, if scientific realism is a relevant variable for distinguishing different phases of Feyerabend's published work, there are compelling reasons to date the end of his 'early works' later than Brown and Kidd's analysis presents (e.g., 1981 rather than 1975). In this chapter, I argue that realism-relativism is a red-herring for distinguishing different phases of Feyerabend's corpus.

With respect to Brown and Kidd's periodization of Feyerabend's middle works (c. 1978–1987), Feyerabend presents arguments for pluralism in society in published works as early as 1968 (e.g., Feyerabend, 1968b, 1969b, 1970a, 1970c). In this regard, it is important to note that Feyerabend's arguments for pluralism in society coincide with when he read John Stuart Mill's *On Liberty* (Mill [1859] 1977) and started incorporating Mill's arguments for pluralism in his own. Based on materials from Feyerabend's correspondence (see Collodel 2016), Feyerabend read and started incorporating Mill's arguments for pluralism sometime around 1968 or 1969. In published work, Feyerabend's first discussion of Mill's *On Liberty* appears in his (book-length) chapter "Against method" (Feyerabend 1970a), which was published in the *Minnesota Studies in the Philosophy of Science* book series. Hence, if Feyerabend's shift from arguments for pluralism within science to arguments for pluralism in society is a relevant variable for distinguishing phases of Feyerabend's work, then there are compelling reasons to date the beginning of his middle works much earlier than Brown and Kidd's analysis presents (e.g., 1968 rather than 1978).

The discrepancies between Brown and Kidd's framework for classifying Feyerabend's works and Feyerabend's published works highlighted above indicate that Brown and Kidd's variables of (1) realism-pluralism within science and (2) relativism-pluralism in society do not coincide as neatly as their framework suggests. In separating these variables, I argue that the more salient variable for distinguishing Feyerabend's early and middle works is the variable of pluralism within science versus pluralism in society. In what follows, I explicate Feyerabend's pluralistic realism, and I argue that realism vs. relativism is a red-herring for distinguishing different phases of Feyerabend's work. By contrast, Feyerabend's shift from arguments for pluralism within science to pluralism in society does represent a major shift of emphasis in his philosophical thinking. I call this shift Feyerabend's *expansion of pluralism* and offer historical reasons for dating the start of this shift in 1970.

3.4 Feyerabend's Pluralistic Realism (c. 1958–1981)

In published works from 1958 to 1981, Feyerabend defends a fallibilistic and pluralistic *methodological ideal* of scientific realism, which is formulated against instrumentalist and logical empiricist views on the interpretation of scientific theories. Retrospectively, Feyerabend (1981b) calls this methodological realist ideal "pluralistic realism" (p. xi), which is the first and only time that he uses this phrase.[2] Pluralistic realism holds that scientific development (and epistemological development more generally) requires the proliferation of competing theories of reality ('scientific theories,' 'forms of thought,' 'forms of life'), which function to *criticize accepted scientific theories and promote the formulation of better theories of reality*. Within this broadly realist (and Popperian) picture of evolutionary epistemology, Feyerabend (1981b) presents the epistemological benefit of alternative theories (e.g., Copernicus's cosmology, quantum theory) in terms of their capacity to criticize or overturn firmly accepted scientific theories:

> The procedure is always the same: attempts to retain well-entrenched conceptions are criticized by pointing out that the excellence of a view can be asserted only after alternatives have been given a chance, that the process of knowledge acquisition and knowledge improvement must be kept in motion and that even the most familiar practices and the most evident forms of thought are not strong enough to deflect it from its path. The cosmologies and forms of life that are used as alternatives need not be newly invented; they may be parts of older traditions that were pushed aside by overly eager inventors of New Things. The whole history [of ideas] is mobilized in probing what is plausible, well established and generally accepted. (p. xi)

In pluralistic realism, the proliferation of *alternative theories of reality* provides a means for criticizing currently accepted scientific theories, while a realistic understanding ('interpretation') of theories encourages *genuine criticism* and *metaphysical engagement* among alternative theories (Feyerabend 1958, 1960, 1962, 1964). With respect to the latter, the motivation to *resolve conflicting (metaphysical) accounts of reality,* for Feyerabend, is a crucial element of scientific criticism and the articulation of *better theories of reality* (see Feyerabend 1963, 1965a). By contrast, instrumentalists judge the merits of competing theories exclusively in terms of predictive accuracy or some other empiricist ideal (e.g., 'empirical accuracy,' 'saving the phenomena'), which has the detrimental consequence of bypassing (metaphysical) questions concerning the specific ways in which competing theories agree and disagree (i.e., as alternative descriptions of reality). For Feyerabend, the instrumentalist interpretation of theories is detrimental *for scientific practice* insofar as it discourages engagement with conflicting theories and the potential resolution (or synthesis) of conflicting theories. In this precise sense, Feyerabend argues that

[2] For convenience, I adopt Feyerabend's terminology in this chapter to describe a distinctively realist (and metaphysical) aspect of Feyerabend's arguments for pluralism in the 1960s and 1970s. The only other publication that Feyerabend adopts the terminology of 'pluralistic realism' is his review of Popper's *Objective Knowledge* (Popper 1972), wherein Feyerabend (1974) presents Popper as defending a pluralistic account of realism or 'ontological pluralism' (p. 476).

realism promotes scientific criticism and progress, whereas anti-realism (e.g., instrumentalism, positivism) stagnates scientific criticism and progress.

Pluralistic realism is a *methodological prescription for scientists*: scientists should adopt a realistic interpretation of theories (and a realist interpretation of *conflicting* theories) because this *attitude* towards scientific theories promotes scientific criticism and growth. In terms of the scientific realism debates in philosophy of science,[3] pluralistic realism amounts to the methodological suggestion that the theoretical terms (or 'unobservable entities') in scientific theories (e.g., 'electrons,' 'genes,' 'force') should be interpreted *as literal descriptions of the world*, rather than instruments for generating predictions or summarizing empirical data. This anti-positivistic position agrees with the first tenet of Richard Boyd's definition of scientific realism "Theoretical terms' in scientific theories (non-observational terms) should be thought of as putatively referring expressions; scientific theories should be interpreted realistically" (Boyd 1983, p. 45). Hasok Chang (2021) argues that Feyerabend's methodological realism meets the realist standard prescribed by Bas van Fraassen's definition of scientific realism (cf. Tsou 2003): "Science aims to give us, in its theories, a literally true story of what the world is like; and acceptance of a scientific theory involves the belief that it is true" (van Fraassen 1980, p. 8). Feyerabend's argument for a realistic interpretation of scientific theories provides a methodological justification (viz., a *realist attitude* towards theories promotes criticism and engagement with alternative theories) for preferring realism over instrumentalism.

Textual evidence that Feyerabend endorsed pluralistic realism in works as late as 1981 stems largely from the introduction that Feyerabend prepared as a general introduction for the first two volumes of his *Philosophical Papers* (Feyerabend 1981e, pp. ix–xiv, 1981d, pp. vii–xii), which summarizes the general themes and arguments that unify his earlier published works. The first volume of Feyerabend's Philosophical Papers (Feyerabend 1981e), *Realism, Rationalism and Scientific Method*, features Feyerabend's classic early papers arguing for a realistic interpretation of scientific theories (Feyerabend 1958, 1960, 1962, 1964, 1965a, 1965c). Significantly, in his introduction to volumes 1 and 2, Feyerabend (1981b) provides no indication that he no longer accepts his earlier arguments for realism over instrumentalism (this omission is significant given that Feyerabend often announced his changes of view). Therein, Feyerabend (1981b, p. ix) presents pluralistic realism (somewhat unhelpfully) with the argumentative schema:

criticism ➔ proliferation ➔ realism

Feyerabend (1981b) emphasizes that arrows in this chain do not represent logical implications, but rather "starting with the left hand side and adding [a proliferation

[3] Feyerabend's various discussions of realism (e.g., Feyerabend 1981a, 1981c, 1989) largely neglect and do not engage with the expansive philosophy of science literature on scientific realism that developed in the 1980s and 1990s (cf. Shaw and Bschir 2021, p. 5). For exceptions, Feyerabend (1989, 1993) cites Arthur Fine's discussion of NOA (Fine 1986) and Ian Hacking's discussion of experimental realism (Hacking 1983).

of alternative theories]… a dialectical debate will eventually arrive at the right hand side" (p. x). There are at least two ways to interpret this argumentative chain. First, it can be interpreted as an argument for the formulation of better theories of reality. If the aim of science is the realist goal of formulating true scientific theories, then one means to this goal is to criticize accepted scientific theories by testing them against a plurality of alternative theories. Put in terms of Feyerabend's schema: *criticism* of accepted theories leads to a *proliferation* of alternative theories, which leads (through dialectical debate) to the formulation of better theories of *reality*. Second, Feyerabend's argumentative chain can be interpreted in light of his argument for the realistic interpretation of scientific theories. On this reading, the argument starts with Feyerabend's Popperian assumption that science is an essentially a critical enterprise. If you accept that criticism is the most important feature of science, then you should be a pluralist and support the proliferation of alternative theories that conflict with accepted scientific theories. Criticism and proliferation is facilitated by a realistic interpretation of scientific theories insofar as a realist attitude promotes genuine criticism of accepted theories and robust (metaphysical) engagement with alternative theories. This second more qualified interpretation coincides better with the details of Feyerabend's arguments for realistic interpretation of scientific theories (see Feyerabend 1981b, p. ix); Brown and Kidd (2016, p. 3) and Chang (2021, p. 43) present Feyerabend's argument along these lines. While I assume that the second interpretation is a closer approximation to Feyerabend's intended meaning, I believe that Feyerabend could accept both interpretations presented above.

3.5 Pluralistic Realism in the 1970s

While Feyerabend's endorsement of pluralistic realism is evident in the 1960s, his position on realism in the 1970s is ambiguous at best. In 1970s publications, it is notable that Feyerabend only rarely discusses his 1960s arguments for realistic interpretation of scientific theories, and he no longer explicitly defends realism as a methodological principle that promotes critical science. More generally, Feyerabend's paradigmatic publications in the 1970s (Feyerabend 1970a, 1975a, 1978) differ in style (and content) from his publications in 1960s (Feyerabend 1962, 1963, 1965b). Whereas Feyerabend's writings in the 1960s are fairly narrow and technical papers arguing against logical empiricist views of science, his writings in the 1970s are more broadly-focused and polemical. If Preston is correct that there is a shift in Feyerabend's philosophical thinking *away from realism*, then there is prima facie textual evidence supporting the thesis that this shift appears in Feyerabend's published work starting sometime in the late 1960s (c. 1968–1970).

Whereas Feyerabend presents pluralistic realism as a methodological ideal that promotes scientific criticism and metaphysical engagement with alternative theories in the 1960s, he presents the realistic interpretation of theories as a necessary presupposition of incommensurability in the 1970s (Preston 1997a, Chap. 6; Oberheim and Hoyningen-Huene 1997, 2024, §4; Oberheim 2006, Chap. 6).[4] In the 1970s, this latter argument is presented in Feyerabend's "Against Method" book chapter (Feyerabend 1970a, pp. 86–88) and repeated in the first and second editions of *Against Method* (see Feyerabend 1975a, pp. 278–282; Feyerabend 1988, pp. 221–224).[5] In these works, Feyerabend emphasizes that *under their "customary" (i.e., realist) interpretation*, certain pairs of theories (e.g., classical mechanics and quantum mechanics, general relativity and classical mechanics) are incommensurable. Feyerabend (1975a) responds to the instrumentalist objection that incommensurability does not exist under an instrumentalist interpretation of theories:

> Theories can be interpreted in different ways. They will be commensurable in some interpretations, incommensurable in others. *Instrumentalism*, for example, *makes commensurable all those theories which are related to the same observation language and are interpreted on its basis*. A realist, on the other hand, wants to give a unified account, both of observable and of unobservable matters, and he will use the most abstract terms... in order either to *give* meaning to observation sentences or else to *replace* their customary interpretation... Against this, it is pointed out by almost all [logical] empiricists that theoretical terms receive their interpretation [and meaning] from being connected with a pre-existing observation language, or with another theory that has already been connected with such a language. (p. 279, emphasis added)

Feyerabend (1975a) presents the 'guiding idea' behind this instrumentalist objection to incommensurability as the prescription that "new and abstract languages cannot be introduced in a direct way, but must first be connected with an already existing, and presumably stable, observational idiom" (p. 280). Feyerabend (1975a, pp. 280–282) chastises and dismisses this logical empiricist prescription for its artificial and implausible picture of language acquisition. Curiously, Feyerabend does not raise his earlier objections to this Carnapian view on scientific theories. In his earlier papers, Feyerabend (1958, 1960, 1962) rejects Carnap's strategy of identifying the *meaningfulness* of scientific theories with their observational content (Carnap 1956), which privileges the empirical part of scientific theories with respect to *meaning*. Feyerabend assumes a radical version of the theory-ladenness of observation thesis that inverts this logical empiricist picture of meaning. For Feyerabend, the meaning of scientific theories does not derive indirectly through their connections with observation sentences, but the meaning of observation sentences is "fully theoretical" in the sense of being determined by what a scientific theory says about the world (Feyerabend 1965a, 1965b, 1965c). Retrospectively, Feyerabend (1993) summarizes his early views on incommensurability as follows: "[T]he interpretation of an observation language is determined by the theories which we use to explain what we observe, and it changes as soon as these theories change...

[4] Feyerabend's argument that incommensurability requires a realist interpretation of theories was present in his earlier papers (e.g., see Feyerabend 1962, 1965a, 1965b, 1965c). In his reply to Peter Achinstein (1964), Feyerabend (1965a) explicitly states that his account of incommensurability presupposes an "epistemological realism" (p. 268), i.e., a realistic interpretation of theories.

[5] Feyerabend's argument that incommensurability requires a realistic interpretation of theories is removed in the third edition of *Against Method* (Feyerabend 1993, pp. 207–208).

observation statements are not just theory-*laden*… but *fully theoretical*" (pp. 211–212, emphasis in original).

For the purposes of this chapter, it is significant that Feyerabend endorses pluralistic realism in the 1970s insofar as his account of incommensurability requires a realistic interpretation (see Preston 1997a, Chap. 6; Oberheim and Hoyningen-Huene 2024; Oberheim 2006, Chap. 6). This exegetical position is amenable to interpretations of *Against Method* as a realist work (e.g., Lloyd 1996; Preston 1997b; Tsou 2003). Eric Oberheim (2006) usefully connects Feyerabend's view in the 1970s that incommensurability requires a realistic interpretation to Feyerabend's earlier 1960s publications on realism:

> Feyerabend's… point is simply that if one of two theories (or both) are interpreted instrumentally, i.e. merely as tools and not as attempted descriptions, then *they cannot conflict as attempts to describe nature*. The result is that there is no competition between them, and consequently no need to improve them or to find a better theory. Conceptually incompatible tools can be used for different purposes, and there is nothing stopping us from using such tools at different times. However, conceptually incompatible attempted descriptions can and do conflict. *On a realistic interpretation, theories expressed with incommensurable concepts are mutually exclusive*. This promotes competition; and consequently, progress. (p. 190, emphasis added)

Oberheim provides an astute reconstruction of how Feyerabend's arguments for a realistic interpretation of theories forms the basis for Feyerabend's arguments for incommensurability. Compared to instrumentalism, a realistic interpretation of theories encourages genuine metaphysical conflict and engagement with alternative theories of reality, whereas instrumentalism discourages it. This is due to the way that *conflicting theories* are regarded by realists and instrumentalists respectively. Whereas realists take conflicting theories as an impetus to resolve this conflict, instrumentalists take conflicting theories to be an invitation to compare these theories in terms of predictive accuracy or some other empiricist standard. Like his earlier arguments for realism over instrumentalism, Feyerabend points to an epistemic benefit of a ('realist') *psychological attitude* towards scientific theories. While there is textual evidence that Feyerabend is a pluralistic realist as late as 1988, I limit my argument in this chapter to the claim that Feyerabend is a pluralistic realist as late as 1981.

3.6 Relativism and Feyerabend's Retreat from Empiricism

My argument that Feyerabend is a realist in published works as late as 1981 faces the exegetical objection that in *Science and a Free Society*, Feyerabend (1978) explicitly endorses 'relativism.' With respect to Feyerabend's defense of 'relativism' in the 1970s and 1980s (e.g., Feyerabend 1978, 1987), I argue—as suggested by Brown and Kidd (2016)—that the relativistic tendencies in Feyerabend's work after the first edition of *Against Method* (Feyerabend 1975a) should be regarded as a *shift of emphasis and strategy*, rather than a major change in philosophical positions. On this understanding, Feyerabend's arguments for relativism represent different interests (rather than a major shift of philosophical position) that Feyerabend pursued in the late 1970s and 1980s. Feyerabend's interest in relativism stems from a more fundamental philosophical shift that Feyerabend made around the mid-1960s (Feyerabend's 'retreat from empiricism'), wherein Feyerabend gave up the view that science is superior to non-scientific ways of knowing because of its empirically testable methods.

It is worth mentioning that Feyerabend's interest in relativism (Feyerabend 1978, 1987) coincides with a 'dark period' (c. 1975–1980) in his personal life, when Feyerabend was dealing with physical and mental health issues.[6] During this period, Feyerabend was distressed by the largely negative reception of *Against Method* (Feyerabend 1975a). In his autobiography, he reports:

> Somewhere in the middle of the commotion I grew rather depressed. The depression stayed with me for over a year… I had often warned my students not to identify with their work… I myself had followed this advice in the past, but now I was alone, sick with some unknown affliction; my private life was a mess, and I was without a defense. I often wished I had never written that fucking book. (Feyerabend 1995, p. 147)

During his dark period, Feyerabend was uncivil and aggressive in some works, most notably in his acerbic responses to critics of *Against Method* in *Science in a Free Society* (Feyerabend 1978, part 3). Feyerabend's widow, Grazia Borrini-Feyerabend, reports that in his later years, Feyerabend regarded *Science in a Free Society* as the work that he regretted the most (Preston 2020, §5.2). It is unclear how much explanatory work that Feyerabend's dark period can do in accounting for his endorsement of relativism, but it is a salient aspect of Feyerabend's personal life in the mid-to-late 1970s that undoubtedly played some role.[7]

---

[6] I date the beginning of Feyerabend's dark period in 1975 since it occurs after the publication of *Against Method* (Feyerabend 1975a). I date the end of Feyerabend's dark period in 1980 since Feyerabend (1995) describes the 1980s as "ten wonderful years" (p. 158), when he was able to split time between UC Berkeley and the Federal Institute of Technology at Zurich (Preston 2020, §5.3). As is well documented, Feyerabend was unhappy at UC Berkeley (e.g., see Feyerabend 1995, pp. 112–113). In the late 1960s and early 1970s, Feyerabend's colleagues in the philosophy department at Berkeley (viz., John Searle, Hans Sluga, Hubert Dreyfus) tried to get Feyerabend fired over his teaching practices (see Feyerabend 1995, p. 126; Lakatos and Feyerabend 1999, pp. 154, 177, 193, 311).

[7] As a reckless speculation, Feyerabend's move towards relativism and more provocative ('post-modern') modes of expression in the 1970s (Preston 2000) may be explained, in part, by his aspirations to become a public intellectual. In his autobiography, Feyerabend discusses 'intellectuals' with bitterness:

> Gradually I became acquainted with "intellectuals." They constitute a very special community. They write in a special way, have special sentiments, and seem to think of themselves as the only legitimate representatives of the human race, which in practice means other intellectuals… Thomas Nagel is one, Rorty another; even Searle turns up here and there, though he lacks the smooth ways of the true intellectual. This community now started taking a slight interest in me: it lifted me up to its own eye-level, took a brief look at me, and dropped me again. After making me appear more important than I ever thought I was, it enumerated my shortcomings and put me back in my place. That really confused me. (Feyerabend 1995, pp. 146–147).

During his dark period, Feyerabend's publications appear to shift towards more polemical and skeptical (i.e., 'anti-science') positions (Feyerabend 1975a, 1975b, 1977, 1978), including his infamous defense of relativism. Retrospectively, Feyerabend (1995) describes his views on relativism in his autobiography with typical self-deprecation[8]:

> Other armchair views [of mine] did not fare so well. I am referring to my "relativism," to the idea that cultures are more or less *closed entities with their own criteria and procedures*, that they are intrinsically valuable and should not be interfered with… But cultures interact, they change, they have resources that go beyond their stable and objective ingredients… Considering how much cultures have learned from each other… I have come to the conclusion that *every culture is potentially all cultures* and that special cultural features are changeable manifestations of *a single human* nature (pp. 151–152, emphasis added)

As alluded to in this passage, the relativism that Feyerabend defended in the 1970s and 1980s is closely related to his incommensurability thesis: particular cultures ('forms of thought, 'forms of life,' etc.) are "closed entities," which cannot meaningfully engage with other cultures. One thing to notice about Feyerabend's particular view of relativism is *that it is entirely consistent with his methodological ideal of pluralistic realism*. The idea that cultures are closed systems ('paradigms,' 'conceptual schemes,' 'linguistic frameworks,' 'forms of life') with their own internal criteria and knowledge-producing procedures is not inconsistent with Feyerabend's methodological recommendation for a realistic interpretation of theories. By the logic of his earlier arguments for pluralistic realism, a plurality of alternative forms of life ('interpreted realistically') will lead, through dialectical debate, to new and better forms of life. Hence, Feyerabend's arguments for pluralism in society are not only consistent with, but they may *require* pluralistic realism (i.e., a realistic interpretation of forms of life).

Feyerabend's path to relativism can be traced to a significant change in his philosophical thinking, around the mid-1960s, regarding the demarcation problem and the *meaningfulness* ('cognitive content') of alternative theories used to test and criticize accepted scientific theories. In publications prior to 1966, Feyerabend (1962, 1965b, 1965c) assumed that alternative theories (e.g., myths, rejected worldviews, metaphysical speculations, religious cosmologies) are important resources for criticizing accepted theories, but alternative theories themselves did not possess cognitive (empirical) content. In a 1976 letter to the philosopher of science Zev Bechler, Feyerabend describes a change in his philosophical thinking (that occurred after the mid-1960s) regarding the *cognitive content* of alternative theories used to criticize science:

> I had tried to show [see Feyerabend 1962, 1965b, 1965c] that a separation of the *content* of science and non science would be detrimental to the former. Metaphysical views, ancient myths, the cosmologies inherent in the various religions are not only valuable reservoirs of ideas, they are possible *test cases of the status quo* and must be preserved and developed in close connexion with science rather than abandoned… All this means that the separation of

[8] For a more comprehensive discussion of Feyerabend's changing and somewhat unorthodox views on relativism, see Heller (2016), Kusch (2016), and Sankey (2020).

> science and non-science which plays such an important role in the development of modern thought must cease. However, I still assumed that non-scientific views are (A) not on par with science – they must be *changed until they become more definite and receive empirical content*... My studies of myth and of the theatre [see Feyerabend 1967a, 1967b, 1969c, 1975c], then made me very doubtful of assumption (A). We know now that the inventors of myth in astronomy, botany, zoology, biology, medicine, sociology, theology... *tested* this knowledge in laboratories and 'observatories'... In the course of my research I started suspecting that *myths have cognitive content as well.* Moreover... I suspected that there might be cases where science and myth are in conflict, but the myth is right, and science is not. *I suspected that myths are fully-fledged alternatives to science with a content and a method of presentation of their own.* (Feyerabend 1976, pp. 36–37, emphasis added)

When Feyerabend adopts the view that myths are 'fully-fledged alternatives to science' in the late-1960s, he has given up the view that *empiricism* is the best methodology for acquiring scientific knowledge and he rejects any empiricist demarcation criterion for science. Whereas Feyerabend's criticisms of empiricism in the early 1960s (Feyerabend 1962, 1963, 1965b) were arguments for *non-dogmatic (i.e., pluralistic) forms of empiricism*, his publications after 1966 reject empiricism entirely (Preston 2020, §4.4). Feyerabend's anti-empiricist stance is most explicitly announced in his 1969 paper, "Science without Experience," wherein Feyerabend (1969c) rejects the idea that scientific facts are *necessarily empirical.* In post-1966 works, Feyerabend begins to regard myths and metaphysical views as "fully fledged alternatives to science with a form and content of their own ... [that] provide knowledge not contained in, and perhaps even denied by science" (Feyerabend 1976, p. 38).[9] Feyerabend's argument for the cognitive (or *epistemic*) content of alternative theories is crystalized in Chap. 4 of *Against Method* (see Feyerabend 1976, p. 37), wherein Feyerabend (1975a) argues that "[t]here is no idea, however ancient and absurd, that is not capable of improving our knowledge" (p. 47, emphasis removed).

With respect to Feyerabend's alleged shift from realism to relativism (Preston 1997b; Brown and Kidd 2016), I contend that realism/relativism (or realism/antirealism) is not a useful variable for classifying different phases of Feyerabend's work. As indicated above, I believe that Feyerabend was a pluralistic realist in his publications in the 1970s and early 1980s (even as late as 1988), given that his account of incommensurability explicitly requires a realistic interpretation of theories (Feyerabend 1970a, pp. 86–88; Feyerabend 1975a, pp. 278–282; Feyerabend 1988, pp. 221–224). Moreover, there are not compelling reasons for regarding the 'relativism' that Feyerabend endorsed in the 1970s and 1980s (i.e., cultures are closed epistemic systems) as being *inconsistent* with pluralistic realism. Indeed, there are good reasons for thinking that Feyerabend's arguments for pluralism in society *require* a realistic interpretation of theories/forms of life (viz., a realistic interpretation encourages the proliferation of alternative theories and the genuine *testing* of theories against one another). For these reasons, I reject characterizations of

[9] For a fascinating analysis of Feyerabend's later views on art and science, see Ambrosio (2021).

Feyerabend's philosophical shift in the 1970s as a shift from scientific realism to relativism (Preston 1997a, 1997b; Brown and Kidd 2016).

Regarding Feyerabend's alleged 'retreat from realism' in the mid-1970s (Preston 1997b), it would be more accurate to say that Feyerabend made a 'retreat from empiricism' in the mid-1960s (cf. Preston 2020, §4.4), which paved the way for Feyerabend's more radical ('anti-science') arguments in the 1970s. In works prior to 1966, Feyerabend assumes that scientific knowledge is superior to other ways of knowing because of its *empirical testability*, and he argues that theoretical pluralism is a means to maximize the empirical content used as the basis for any testing situation. In works after 1966, Feyerabend assumes that non-scientific theories (e.g., myths, metaphysical speculations, religious beliefs) possess cognitive content and are *genuine epistemic alternatives to science*. This fundamental change in Feyerabend's thinking in the mid-1960s is more important for tracking changes in his overall views and more useful for classifying different phases of his work than any position that Feyerabend defended about realism. Feyerabend's retreat from empiricism and rejection of an empiricist demarcation criterion for science forms the basis for the key positions (e.g., the defense of society *from* science, relativism, the value of non-scientific ways of knowing) in his pluralism in society phase.[10]

Although I have argued that Feyerabend was a pluralistic realist in the 1970s and 1980s, it is important to acknowledge how unorthodox and narrowly focused his methodological ideal of realism is formulated, i.e., the recommendation to interpret theories realistically rather than instrumentally. Feyerabend's discussions of realism are always somewhat disengaged with the philosophy of science literature on scientific realism that emerged in the 1980s and 1990s, and his resulting realist positions were thin (and idiosyncratic) at best. I take this to be the reasonable core of Oberheim's (otherwise objectional) argument that Feyerabend was never a scientific realist (Oberheim 2006, Chap. 6): it is difficult to locate Feyerabend's methodological and highly contextualized realist position (viz., realism is preferable to instrumentalism for promoting *critical scientific practices*) in the context of the philosophy of science literature on scientific realism. Moreover, it is significant that while Feyerabend explicitly defends realism (versus instrumentalism) in publications prior to 1970 (Feyerabend 1958, 1964), in post-1970s publications, he no longer presents explicit arguments for realism. In terms of *emphasis and strategy*,

[10] In correspondence, Oberheim has chided my presentation of Feyerabend's "retreat from empiricism." Oberheim (2025) argues that the most important change that separates Feyerabend's early from later works is a "reversal" of Feyerabend's stance towards scientism. On Oberheim's picture (cf. Feyerabend 1981b, pp. xiii–xiv; Oberheim 2006, pp. 268–269; Kidd 2021; Shaw 2021, pp. 125–131), Feyerabend's early works are characterized by an endorsement of scientism (i.e., science should be employed to correct common sense), whereas his later works are characterized by a (Wittgensteinian) rejection of scientism (i,e., common sense should be employed to correct science). In response to Oberheim, I would argue that Feyerabend's retreat from empiricism (i.e., his rejection of the view that empirical science is the best source of knowledge) is closely related to his changed stance towards scientism (i.e., his rejection of the view that science should be used to correct common sense) insofar as the latter presupposes the former.

Feyerabend does not appear to be interested in presenting himself as a realist.[11] In the 1970s and 1980s, Feyerabend's pluralistic realism is reduced to the extremely narrow argument that some theories (interpreted realistically) are incommensurable (cf. Feyerabend 1981a, 1981b, 1981c). Given the paucity of textual evidence supporting a substantial realist reading of Feyerabend in the 1970s, I contend that—as a potential variable for classifying different phases of Feyerabend's work—realism-relativism is a red-herring at best.[12]

3.7 Feyerabend's Expansion of Pluralism: Pluralism in Society (c. 1970–1989)

Although I reject realism-relativism as a useful variable for clarifying different phases of Feyerabend's work, Brown and Kidd's distinction between pluralism within science and pluralism in society is an excellent variable for distinguishing the early and middle phases of Feyerabend's published works. Feyerabend's retreat from empiricism (as discussed in Sect. 3.6) in the mid-1960s provides the philosophical basis for Feyerabend's shift to arguments for pluralism in society. This shift can be understood as Feyerabend's expansion of pluralism: Feyerabend expands his earlier arguments for pluralism within science to apply more broadly to *knowledge produced in society*, wherein science is regarded as one ideology among many that can contribute to knowledge. Rather than seeing this shift as a movement from arguments *for science* to arguments *against science*, Feyerabend's shift should

---

[11] It is worth noting that in works published in the late 1960s and 1970s, Feyerabend appears to generalize his earlier arguments for a realist interpretation of theories to an argument for the "principle of tenacity," which suggests that scientists should remain committed to promising theories despite their difficulties (see Feyerabend 1969b, 1970b). Feyerabend (1970b) presents the principle of tenacity as fulfilling a similar scientific function as Kuhn's concept of normal science (Kuhn 1962): it captures the *necessary dogmatism* required for the formulation and elaboration of scientific theories. In contrast to Kuhn's idea that a tradition of normal science should develop *until* a new paradigm emerges that can supplant it in terms of puzzle-solving power, Feyerabend (1970b) argues that scientific development should be characterized by the *constant interaction among tenaciously held scientific theories*:

[T]he interplay between tenacity and proliferation… is… an essential feature of the actual development of science. It seems that it is not the puzzle-solving activity that is responsible for the growth of our knowledge but the active interplay of various tenaciously held views. (p. 209)

Feyerabend's argument in the 1960s that a realistic interpretation of theories promotes proliferation and metaphysical engagement with alternative views appears to be generalized to the principle of tenacity in the 1970s. For a more comprehensive discussion of the principle of tenacity, see Preston (1997a, Chap. 7), Oberheim (2006, Chap. 7), Tambalo (2015), and Shaw (2017).

[12] While I am sympathetic to Chang's argument that Feyerabend was a realist throughout his career (Chang 2021), I cannot completely follow his argument that Feyerabend's pluralistic realism in the 1960s and 1970s is closely related to (or 'coheres with') Feyerabend's defense of 'abundance realism' (see Brown 2016) in the 1990s.

be understood as *broadening the scope of application* of his arguments for pluralism within science (wherein empirical science is assumed to be the best source of knowledge) to arguments for pluralism for knowledge considered more broadly (wherein empirical science is assumed to be one ideology among many). Feyerabend's arguments for pluralism in society are exemplified by his arguments for defending society from science, his rejection of empirical science as the best methodology for attaining knowledge, his endorsement of relativism, and his defense of non-scientific ways of knowing (e.g., witchcraft, art). I contend that a defensible beginning date for Feyerabend's pluralism in society phase is 1970.

Brown and Kidd (2016) date the beginning of Feyerabend's shift to pluralism in society in 1978 on the historical ground that, by the publication of *Science in a Free Society* (Feyerabend 1978), Feyerabend has *definitively shifted away from arguments for pluralism within science to arguments for pluralism in society*. They write:

> The shift is definitely complete by 1978, with the publication of *Science in a Free Society*, which is firmly situated in the themes of Feyerabend's middle work. We suggest that one can read the 1975 edition of *Against Method* as a transitional work, mainly occupied with the concerns of the early work, but moving in the direction of the middle period (Brown and Kidd 2016, p. 3)

The historical methodological reasons cited by Brown and Kidd for dating the start of Feyerabend's pluralism in society phase in 1978 are sensible. However, by their own standard, the beginning of Feyerabend's pluralism in society phase could be dated much earlier. I contend that a more appropriate starting date for Feyerabend's arguments for pluralism in society and expansion of pluralism is 1970, with the publication of the book chapter version of "Against Method" (Feyerabend 1970a).

Leaving the variable of realism-relativism to the side, when did Feyerabend's arguments for pluralism in society appear in his publications? Below is a list of important publications that firmly belong in the tradition of Feyerabend's arguments for pluralism in society:

- "On the Improvement of the Sciences and the Arts, and the Possible Identity of the Two" (book chapter in *Boston Studies*): Feyerabend (1967a) argues that science and art are similar insofar as they both assume a certain form (or 'style') that influences people's attitudes (e.g., critical or dogmatic) towards the knowledge presented (cf. Feyerabend 1976, pp. 37–38).
- "Science, Freedom, and the Good Life": Feyerabend (1968b) argues that pluralism is a remedy for conservatism in science and conservatism in society.
- "Against Method" (book chapter of *Minnesota Studies*): Feyerabend (1970a) cites John Stuart Mill's pluralism in *On Liberty* as a means to combat dogmatism in science, facilitate the psychological development of individuals in society, and improve civilization (pp. 27–30). Besides promoting progress in science, pluralism and proliferation are essential ingredients for a "humanitarian outlook" that seeks to improve society and culture (pp. 26–27).

- "Experts in a Free Society": Feyerabend (1970c) defends society from science by arguing against the legitimacy of scientific experts.
- *Against Method* (first edition of the book): Feyerabend (1975a) argues that non-scientific ideas are meaningful and can improve human knowledge (Chap. 4) and that science and the state should be separated because of the authoritative status of science in society (Chap. 18).
- "How to Defend Society from Science": Feyerabend (1975b) aims "to defend society and its inhabitants from all ideologies, science included" (p. 4).
- *Science in a Free Society*. Feyerabend (1978, part 2) argues that modern science is not superior to non-scientific ways of knowing. Feyerabend defends relativism and argues that science is one ideology among many. Because of its authoritative status, science should be supervised by the public and democratic judgment should overrule expert opinions on science.
- "Democracy, Elitism, and Scientific Method": Feyerabend (1980) argues for a "democratic relativism" to counter "intellectual elitism" (e.g., the authority of science).
- *Farewell to Reason*: Feyerabend (1987) defends a specific form of cultural relativism ('practical relativism' or 'opportunism') against rationalistic forms of thought. Cultural diversity promotes human happiness and freedom, while cultural uniformity is detrimental for attaining these goods.
- *Against Method* (revised edition of the book): Feyerabend (1988) adds new chapters that incorporate materials from *Science in Free Society* (Feyerabend 1978).

Brown and Kidd date the beginning of Feyerabend's pluralism in society phase in 1978 because Feyerabend's examination of the political implications of *Against Method* (Feyerabend 1975a) in *Science in a Free Society* (Feyerabend 1978) marks a clear shift away from his arguments for pluralism in science. However, Brown and Kidd's proposed start date for Feyerabend's shift to arguments for pluralism in society (and hence his 'middle works') is misleading insofar as they date the *beginning* of Feyerabend's interest in pluralism in society *a full decade after* his first published works on defending society from science (Feyerabend 1967a, 1968b).

A defensible start date for Feyerabend's shift to arguments for pluralism in society and expansion of pluralism is 1970, which coincides with Feyerabend's publication of "Against Method" (Feyerabend 1970a) and "Experts in a Free Society" (Feyerabend 1970b). There are multiple historical and practical reasons for identifying this date as marking the beginning of Feyerabend's pluralism in society phase, and hence the start of his 'middle works.' First, this date accommodates significant changes in Feyerabend's philosophical thinking that started around the mid-1960s. In terms of his publications, the mid-to-late 1960s (c. 1966–1969) were a volatile time in Feyerabend's thought, when he changed his mind on various positions and his views evolved in multiple ways. This was a period when Feyerabend reassessed the significance of Bohr's 'instrumentalist' interpretation of quantum mechanics,[13] and he was increasingly distancing himself from Popper's school of philosophy of science (Farrell 2000; Tambalo 2015; Collodel 2016; Oberheim 2025). For the purposes of this chapter, two important changes that occurred around the mid-1960s are: (1) Feyerabend's retreat from empiricism, and (2) Feyerabend's incorporation of John Stuart Mill's arguments for pluralism into his own.

Feyerabend's retreat from empiricism, which appears in his post-1966 publications, marks a fundamental philosophical shift from a position wherein empirical science is viewed as the best source of knowledge to a position wherein science and non-science are placed on equal footing with respect to the legitimacy of their knowledge claims. From a historical perspective, Feyerabend's retreat from empiricism (and rejection of a demarcation criterion) provides the philosophical basis for his shift towards arguments for pluralism in society. It also provides the basis for Feyerabend's shift from arguments for pluralism within science in the 1960s to his more radicalized arguments for pluralism during his 'anarchism period' starting in the 1970.[14] Whereas Feyerabend arguments for pluralism in the 1960s can be summarized as an argument that a plurality of scientific theories promotes progress within science, in his arguments for anarchism, Feyerabend (1970a, 1975a, 1988, 1993) generalizes his previous arguments for a plurality of scientific theories to scientific methodology itself (including scientific standards and rules of practice). These arguments ('against method') are formulated in support of the claim that—with respect to *knowledge* (including scientific knowledge)—"theoretical anarchism is *more humanitarian* and *more likely to encourage progress* than its law-and-order alternatives" (Feyerabend 1975a, p. 17, emphasis added). Feyerabend's anarchistic dictum (Feyerabend 1975a, pp. 26–27; Feyerabend 1951a, p. 17) that pluralism encourages more 'humanitarian knowledge' (and more humanitarian scientific knowledge) marks a significant change from his 1960s pluralistic arguments for progress in science, in which progress is marked by scientific theories with 'greater empirical content' (or theories, when 'interpreted realistically,' produce 'better descriptions of reality').

---

[13] Feyerabend's reassessment of Bohr plays into his adoption of anarchism in 1970s. Early in his career, Feyerabend (1958) criticized the Copenhagen interpretation of quantum mechanics (on Popperian grounds) for its instrumentalist interpretation (Feyerabend 1958, 1963). In his papers in the late 1960s, Feyerabend (1968a, 1969a) defends Bohr's specific arguments for complementarity on *physical grounds* (Kuby 2021b). Feyerabend does not take this to imply a greater epistemological benefit of instrumentalism *versus* realism, but that Bohr's *physical arguments* for complementarity should not be regarded as an unsophisticated form of instrumentalism or positivism because it is driven by *scientific reasons* rather than *a priori* philosophical reasons. For a more comprehensive discussion of this shift, see Farrell (2001), Shaw (2020), van Strien (2020) and Kuby (2021b). Van Strien (2020) presents Feyerabend's reevaluation of Bohr as an important precursor to Feyerabend's anarchism (c. 1970–1993) insofar as it solidified Feyerabend's position that universal *philosophical standards* of scientific methodology (e.g., universal, rational principles) should not be imposed on science from outside sources (e.g., philosophy of science).

[14] Feyerabend's anarchist slogan ("anything goes") first appears in his book chapters "Against Method" (Feyerabend 1970a, p. 26) and "Problems of Empiricism, Part II" (Feyerabend 1970d, p. 278). For discussion of Feyerabend's anarchism, see Lloyd (1996), Preston (1997a, Chap. 9), Tsou (2003), Oberheim (2006), and Shaw (2017).

Another significant historical reason for dating the start of Feyerabend's expansion of pluralism in 1970 is that it coincides with the publication of the book chapter "Against Method," which is the first work that Feyerabend (1970a) integrates Mill's arguments for pluralism into his own. Based on various accounts, Feyerabend read *On Liberty* (Mill [1859] 1977) sometime in the mid-to-late 1960s (c. 1968–1969) on the advice of the Popperian philosopher of science John Watkins (see Watkins 2000; Collodel 2016, pp. 51–52). In a book review of *Beyond the Edge of Certainty*, which included Feyerabend's "Problems of Empiricism" (Feyerabend 1965b), Watkins (1966) remarks that "a reference to J. S. Mill would have been in place ... where [Feyerabend] calls for an extension of his epistemological pluralism to extra-scientific walks of life" (p. 361). Based on various accounts, Feyerabend read *On Liberty* sometime around 1968 or early 1969, when he made it required reading for one of his courses at UC Berkeley (Watkins 2000, pp. 49–50). In a letter from Feyerabend to Watkins (dated January 17, 1969), Feyerabend states that he is more enthusiastic about Mill's *On Liberty* he has been about anything for a long time (quoted in Collodel 2016, 51). In a letter from Feyerabend with Lakatos (dated April 25, 1969), Feyerabend proclaims that *On Liberty* is his "new bible" (quoted in Collodel 2016, pp. 51–52), and Feyerabend discusses Mill's pluralism effusively in multiple letters to Lakatos in the late 1960s and early 1970s (Lakatos and Feyerabend 1999, pp. 169, 216, 238–242). In his publication of the book chapter of "Against Method," Feyerabend (1970a) presents Mill (and Hegel) as providing the philosophical arguments for anarchism. In publications after 1970, Mill is consistently presented by Feyerabend as the most important influence on his pluralism and anarchism. In 1981, Feyerabend presents Mill as the most important historical proponent of pluralistic realism (Feyerabend 1981b, 1981c).

Feyerabend's incorporation of Mill in publications starting in 1970 marks a change from his previous arguments for pluralism insofar as Mill provides the philosophical basis for applying pluralism to non-scientific forms of life (e.g., society, culture, civilization) and for Feyerabend's somewhat cryptic claim in *Against Method* that "theoretical anarchism is *more humanitarian* and more likely to encourage progress than its law-and-order alternatives" (Feyerabend 1975a, p. 17, emphasis added). In "Against Method," Feyerabend (1970a) presents Mill's methodological pluralism as a means for advancing knowledge, facilitating the free development of individuals, and improving civilization (p. 27). Feyerabend (1970a) is particularly impressed that Mill's pluralism is not applied narrowly to questions about knowledge and truth (e.g., how to improve scientific methods), but it is applied broadly to problems of life, wherein ideals of 'scientific method' are closely related to humanitarian ideals (i.e., ethical ideals regarding how to live the good life, promote individuality, or improve society):

> Proliferation is introduced [by Mill] as *the solution to a problem of life*: how can we achieve full consciousness; how can we learn what we are capable of doing; how can we increase our freedom so that we are able to decide... the manner in which we want to use our talents?... Today, the only question is how science can improve its own resources, no matter what the human effect of its methods and of its results. For Mill the connection still exists. *Scientific method is part of a general theory of man*. It receives its rules from this theory and is *built up in accordance with our ideas of a worthwhile human existence*.... Thus method-

> ological and humanitarian arguments are intermixed in every part of Mill's essay, and it is on *both* grounds that a pluralistic epistemology is defended. (p. 29)

Feyerabend appreciates the broad vision of Mill's pluralism that understands specific human ideas (e.g., 'scientific method') as connected to broader (and unstable) humanitarian or ethical ideals (e.g., ideals regarding how to live a valuable human life). Whereas Feyerabend's arguments in his 1960s papers are confined to the claim that pluralism will promote scientific progress and better empirical scientific theories, in publications starting in 1970, Feyerabend expands his arguments for pluralism (or 'anarchism') to apply to broader (ethical and humanitarian) ideals of knowledge (cf. Lloyd 1997; Shaw 2017).[15] Matteo Collodel (2016) reconstructs Feyerabend's shift in thinking in the late 1960s as follows: "On the basis of its ethical underpinning, then, pluralism was not to be recommended only within the limited precincts of scientific methodology, but had to be extended across the board to every aspect of human life" (p. 51). In appreciating the full philosophical power and extensive scope of Mill's arguments for pluralism, Feyerabend expanded (or broadened) the scope of his own arguments for pluralism in science to arguments for pluralism in society. Feyerabend's incorporation of Millian arguments for pluralism (Feyerabend 1970a) is the most compelling historical reason for dating the start of his pluralism in society phase in 1970.

The argument articulated in this chapter suggests revising Brown and Kidd's specific periodization of Feyerabend's early and middle works. I suggest the following the alternative manner of distinguishing Feyerabend's early works and middle works:

- *Early Work* (c.1951—1969): Feyerabend argues for pluralism within science, and his main goal is to defend science against philosophies of science that would damage it. Exemplary works of this period include Feyerabend (1958, 1962, 1963, 1965b)
- *Middle Work* (c.1970—1987): Feyerabend argues for pluralism in society, with science understood as one social tradition amongst others, and his main goal is to defend society from the undue authority of science. Exemplary works of this period include Feyerabend (1970a, 1975a, 1975c, 1978, 1987)

As argued above, this periodization is advantageous for distinguishing various important differences in Feyerabend's ('mature') middle works and his earlier works. First, this periodization captures the qualitative differences in the style and content of Feyerabend's (narrowly focused and technical) works in the 1960s compared to his (broader and more polemical) works in 1970s. On this view, Feyerabend's early works can be informally understood as his arguments *for science*, while his middle works can be seen as Feyerabend's arguments *against science* (and the source of the accusation that Feyerabend was the 'worst enemy of science'). Second, this periodization is usefully tracks Feyerabend's 'retreat from empiricism' and

[15] My generalization is a simplification insofar as Feyerabend provides quasi-ethical and humanitarian arguments for pluralism in the early 1960s (see Feyerabend 1961, 1963; for discussion, see Shaw 2017, p. 15).

other significant changes in his philosophical views (e.g., his reevaluation of Bohr, his break from Popper's philosophy) that culminated in the mid-to-late 1960s. Third, dating the beginning of Feyerabend's 'middle works' with the publication of "Against Method" (Feyerabend 1970a) coincides with Feyerabend's 'anarchism period' (c. 1970–1993) when he expanded and broadened the scope of his earlier arguments for a plurality of scientific theories to apply to scientific methodology itself. This anarchistic stance is related to Feyerabend's retreat from empiricism and rejection of attempts to demarcate science from non-science (c. 1966–1969). This periodization identifies the beginning of Feyerabend's middle works with his first citation and discussion of Mill's *On Liberty* (Feyerabend 1970a). My suggested revision can accommodate Brown and Kidd's argument that *Against Method* (Feyerabend 1975a) is a "transitional work." As Feyerabend often said, *Against Method* is a "collage" (Feyerabend 1995, pp. 139–140): it is largely constructed by patching together arguments from works that Feyerabend published in 1960s and early 1970s (Feyerabend 1963, 1965b, 1970a, 1970d). According to the analysis of this chapter, the parts of *Against Method* (Feyerabend 1975a) that summarize his arguments for pluralism within science from the 1960s render it an early work; the parts that argue that pluralism is a means for achieving more humanitarian knowledge, that non-scientific ways of knowing are valuable (Chap. 4), and that society needs to be defended from science (Chaps. 17, 18) render it a middle work.

3.8 Conclusion

In this chapter, I articulated two fairly narrow arguments against Brown and Kidd's framework for classifying Feyerabend's publications: (1) realism-relativism is not a useful variable for classifying different phases of Feyerabend's published work, and (2) Feyerabend's middle-works, distinguished by Feyerabend's interest in arguments for pluralism in society, should be dated earlier than they suggest (viz., 1970 rather than 1978). My analysis was motivated to clarify the nature of a philosophical shift in Feyerabend's thinking that occurred in the 1970s. I presented Feyerabend's shift—from arguments for pluralism within science to arguments for pluralism in society—as Feyerabend's *expansion of pluralism*. In publications starting in 1970, Feyerabend expands the scope of arguments for pluralism to apply to scientific methods, ethical and humanitarian ideals of knowledge, and society more generally.

While I argued that the variable of realism-relativism is a red-herring for distinguishing different phases of his published work, my analysis of Feyerabend's pluralistic realism is not merely of historical interest.[16] Like many contemporary philosophers of science, I endorse both realism and pluralism as fruitful regulative

[16] In this regard, I confess that I share Preston's sympathies towards Feyerabend's early philosophical work in the 1960s, which provided careful and ingenious philosophical arguments that were articulated through a close engagement with the history of science. I am less sympathetic towards Feyerabend's more a priori (armchair) analyses in the mid-to-late 1970s, which deployed fairly trite observations to argue for views such as relativism.

ideals. In the 1960s and 1970s, Feyerabend articulated a novel and a promising way to combine pluralism and realism, although it must be said that he never explicated a comprehensive nor detailed account.[17] The desideratum to combine pluralism with realism is a general topic that has only received limited attention among historians and philosophers of science (see Ludwig and Ruphy 2021, §4). This problem was addressed in the pluralist and deflationary accounts of realism articulated by members of the Stanford School of pluralism (e.g., Duprè 1981, 1993; Cartwright 1983; Hacking 1983).[18] While pluralism and realism are now platitudes in contemporary history and philosophy of science, it is surprising how little work has been explicitly devoted to clarifying their relationship (cf. Chang 2018). Feyerabend's pluralistic position on the methodological benefit of realism for scientific practice ('pluralistic realism') provides a fruitful basis for the defense and elaboration of pluralistic accounts of realism.

Acknowledgments I am grateful to Jamie Shaw for his valuable and extensive feedback on earlier versions of this paper. I also owe thanks to Eric Oberheim, Jacob Stegenga, Matthew Brown, Ian Kidd, and an anonymous reviewer for helpful comments and discussion. A draft of this chapter was presented at the 11th Values in Medicine, Science, and Technology Conference (VMST-11) at the University of Texas at Dallas in May 2023, which celebrated Matt Brown's contributions to the Center for Values in Medicine, Science, and Technology (CVMST) as its former director (2011–2022). I thank the participants of this conference for feedback and suggestions.

References

Achinstein, Peter. 1964. On the Meaning of Scientific Terms. *Journal of Philosophy* 61 (17): 497–509.

Ambrosio, Chiara. 2021. Feyerabend on Art and Science. In *Interpreting Feyerabend: Critical Essays*, ed. Karim Bschir and Jamie Shaw, 11–39. New York: Cambridge University Press.

Boyd, Richard N. 1983. The Current Status of Scientific Realism. *Erkenntnis* 19 (1–3): 45–90.

Brown, Matthew J. 2009. Models and Perspectives on the Stage: Remarks on Giere's *Scientific Perspectivalism*. *Studies in History and Philosophy of Science* 40:213–220.

Brown, Matthew J. 2016. The Abundant World: Paul Feyerabend's Metaphysics of Science. *Studies in History and Philosophy of Science* 57:142–154.

Brown, Matthew J., and Ian James Kidd. 2016. Introduction: Reappraising Paul Feyerabend. *Studies in History and Philosophy of Science* 57:1–8.

[17] Matthew Brown (2009) suggests a promising line of exegesis when he notes similarities between Feyerabend's pluralism and Ronald Giere's perspectivalism (Giere 2006). Giere (2016) agrees that Feyerabend's views on science can be characterized as perspectivalist, but he expresses disappointment that Feyerabend's later perspectivalism lacked the critical (Popperian) character of Feyerabend's earlier views.

[18] Elsewhere, I suggest that the methodological assumptions (i.e., pluralism, disunity of science, anti-reductionism) of the HPS community in the 1980s and 1990s (associated with the Stanford and Minnesota schools of pluralism) render post-Kuhnian normative philosophy of science more Feyerabendian in character than Kuhnian (Tsou 2024). For a more comprehensive discussion of the Stanford and Minnesota schools of pluralism, see Kellert et al. (2006), Solomon and Richardson (2005), Richardson (2006), Ludwig and Ruphy (2021), and Cat (2024). For discussion of the Stanford School in connection with Feyerabend, see Cat and Shaw (2024) and Shaw and Tsou (2025).

Carnap, Rudolf. 1956. The Methodological Character of Theoretical Concepts. In *The Foundations of Science and the Concepts of Psychology and Psychoanalysis*, Minnesota Studies in the Philosophy of Science, ed. Herbert Feigl and Michael Scriven, vol. 1, 38–76. Minneapolis: University of Minnesota Press.

Cartwright, Nancy N. 1983. *How the Laws of Physics Lie*. Oxford: Oxford University Press.

Cat, Jordi, and Jamie Shaw. 2024. From Quantum Holism to the Disunity of Science and Social Activism: The Cat-Feyerabend Correspondence. *International Studies in the Philosophy of Science* 37 (4): 243–290.

Cat, Jordi. 2024. The Unity of Science. In *The Stanford Encyclopedia of Philosophy* (Summer 2024 Edition), ed. Edward N. Zalta and Uri Nodelman. https://plato.stanford.edu/archives/sum2024/entries/scientific-unity/

Collodel, Matteo. 2016. Was Feyerabend a Popperian? Methodological Issues in the History of the Philosophy of Science. *Studies in History and Philosophy of Science* 57:27–56.

Collodel, Matteo, and Eric Oberheim, eds. 2020. *Feyerabend's Formative Years. Volume 1. Feyerabend and Popper: Correspondence and Unpublished Papers*, Vienna Circle Institute Library. Vol. 5. Cham: Springer.

Chang, Hasok. 2018. Is Pluralism Compatible with Scientific Realism? In *The Routledge Handbook of Scientific Realism*, ed. Juha Saatsi, 176–186. London: Routledge.

Chang, Hasok. 2021. The Coherence of Feyerabend's Pluralist Realism. In *Interpreting Feyerabend: Critical Essays*, ed. Karim Bschir and Jamie Shaw, 40–56. Cambridge: Cambridge University Press.

Duprè, John. 1981. Natural Kinds and Biological Taxa. *Philosophical Review* 90 (1): 66–90.

Duprè, John. 1993. *The Disorder of Things: Metaphysical Foundations of the Disunity of Science*. Cambridge: Harvard University Press.

Farrell, Robert P. 2000. Will the Popperian Feyerabend Please Step Forward: Pluralistic, Popperian Themes in the Philosophy of Paul Feyerabend. *International Studies in the Philosophy of Science* 14 (3): 257–266.

Farrell, Robert P. 2001. Feyerabend's Metaphysics: Process-realism or Voluntarist-idealism? *Journal for General Philosophy of Science* 32 (2): 351–369.

Feyerabend, Paul K. 1958. An Attempt at a Realistic Interpretation of Experience. *Proceedings of the Aristotelian Society* 58:143–170. (Reprinted in Feyerabend 1981e).

Feyerabend, Paul K. 1960. On the Interpretation of Scientific Theories. In *Atti del XII Congresso Internazionale di filosofia*, Vol. 5, *Logica, gonseologica, filosofia della scienza, filosofia del linguaggio*. 151–159. Florence: Sansoni. (Reprinted in Feyerabend 1981e).

Feyerabend, Paul K. 1961. *Knowledge Without Foundations: Two Lectures Delivered on the Nellie Heldt Lecture Fund*. Oberlin: Oberlin College. (Reprinted in Preston 1999).

Feyerabend, Paul K. 1962. Explanation, Reduction, and Empiricism. In *Scientific Explanation, Space, and Time*, Minnesota Studies in the Philosophy of Science, ed. Hebert Feigl and Grover Maxwell, vol. 3, 28–97. Minneapolis: University of Minnesota Press. (Reprinted in Feyerabend 1981e).

Feyerabend, Paul K. 1963. How to be a Good Empiricist—A Plea for Tolerance in Matters Epistemological. In *Philosophy of Science: The Delaware Seminar*, ed. Bernard Baumrin, vol. 2, 3–39. New York: Interscience Publishers. (Reprinted in Preston 1999).

Feyerabend, Paul K. 1964. Realism and Instrumentalism: Comments on the Logic of Factual Support. In *Critical Approaches to Science and Philosophy*, ed. Mario Bunge, 280–309. Glencoe: The Free Press. (Reprinted in Feyerabend 1981e).

Feyerabend, Paul K. 1965a. On the 'Meaning' of Scientific Terms. *Journal of Philosophy* 62 (10): 266–274. (Reprinted in Feyerabend 1981e).

Feyerabend, Paul K. 1965b. Problems of Empiricism. In *Beyond the Edge of Certainty: Essays in Contemporary Science and Philosophy*, University of Pittsburgh Series in Philosophy of Science, ed. Robert G. Colodny, vol. 2, 145–206. Englewood Cliffs: Prentice-Hall.

Feyerabend, Paul K. 1965c. Reply to Criticism: Comments on Smart, Sellars, and Putnam. In *Proceedings of the Boston Colloquium for the Philosophy of Science 1962–64: In Honor of Philipp Frank*, Boston Studies in the Philosophy of Science, ed. Robert S. Cohen and Marx W. Wartofsky, vol. 2, 223–261. New York: Humanities Press. (Reprinted in Feyerabend 1981e).

Feyerabend, Paul K. 1967a. On the Improvement of the Sciences and the Arts, and the Possible Identity of the Two. In *Proceedings of the Boston Colloquium for the Philosophy of Science, 1964–66: In Memory of Norwood Russell Hanson*, Boston Studies in the Philosophy of Science, ed. Robert S. Cohen and Marx W. Wartofsky, vol. 3, 387–415. Dordrecht: Springer.

Feyerabend, Paul K. 1967b. The Theatre as an Instrument of the Criticism of Ideologies: Notes on Ionesco. *Inquiry* 10 (3): 298–312.

Feyerabend, Paul K. 1968a. On a Recent Critique of Complementarity: Part I. *Philosophy of Science* 35 (4): 309–331. (Reprinted in Feyerabend 1981e).

Feyerabend, Paul K. 1968b. Science, Freedom, and the Good Life. *Philosophical Forum* 1 (2): 127–135.

Feyerabend, Paul K. 1969a. On a Recent Critique of Complementarity: Part II. *Philosophy of Science* 36 (1): 82–105. (Reprinted in Feyerabend 1981e).

Feyerabend, Paul K. 1969b. Outline of a Pluralistic Theory of Knowledge and Action. In *Planning for Diversity and Choice*, ed. Stanford Anderson, 275–284. Cambridge: MIT Press. (Reprinted in Preston 1999).

Feyerabend, Paul K. 1969c. Science Without Experience. *Journal of Philosophy* 66 (22): 791–794. (Reprinted in Feyerabend 1981e).

Feyerabend, Paul K. 1970a. Against Method: Outline of an Anarchistic Theory of Knowledge. In *Analyses of Theories and Methods of Physics and Psychology*, Minnesota Studies in the Philosophy of Science, ed. Michael Radner and Stephen Winokur, vol. 4, 17–130. Minneapolis: Minnesota University Press.

Feyerabend, Paul K. 1970b. Consolations for the Specialist. In *Criticism and the Growth of Knowledge: Proceedings of the International Colloquium in the Philosophy of Science*, ed. Imre Lakatos and Alan Musgrave, vol. 4, 197–230. Cambridge: Cambridge University Press. (Reprinted in Feyerabend 1981d).

Feyerabend, Paul K. 1970c. Experts in a Free Society. *The Critic* 29 (2): 58–69. (Reprinted in Preston 1999).

Feyerabend, Paul K. 1970d. Problems of Empiricism, Part II. In *The Nature and Function of Scientific Theories: Essays in Contemporary Science and Philosophy*, University of Pittsburgh Series in Philosophy of Science, ed. Robert G. Colodny, vol. 4, 275–353. Pittsburgh: University of Pittsburgh Press.

Feyerabend, Paul K. 1974. Popper's *Objective Knowledge*. Inquiry 17(1–4): 475–507. (Reprinted in Feyerabend 1981d).

Feyerabend, Paul K. 1975a. *Against Method: Outline of an Anarchistic Theory of Knowledge*. London: New Left Books.

Feyerabend, Paul K. 1975b. How to Defend Society from Science. *Radical Philosophy* 11 (1): 3–9. (Reprinted in Preston 1999).

Feyerabend, Paul K. 1975c. Let's Make More Movies. In *The Owl of Minerva: Philosophers on Philosophy*, ed. Charles J. Bontempo and S. Jack Odell, 201–210. New York: McGraw-Hill. (Reprinted in Preston 1999).

Feyerabend, Paul K. 1976. Autobiographical Remarks (1976). In Collodel and Oberheim 2020, 19–44.

Feyerabend, Paul K. 1977. Changing Patterns of Reconstruction. *British Journal for Philosophy of Science* 28 (4): 351–369.

Feyerabend, Paul K. 1978. *Science in a Free Society*. London: Verso.

Feyerabend, Paul. 1980. Democracy, Elitism, and Scientific Method. *Inquiry* 23 (1): 3–18.

Feyerabend, Paul K. 1981a. Introduction: Scientific Realism and Philosophical Realism. In Feyerabend 1981e, 3–16.

Feyerabend, Paul K. 1981b. Introduction to Volumes 1 and 2. In Feyerabend 1981e, ix–xiv.

Feyerabend, Paul K. 1981c. Introduction: Proliferation and Realism as Methodological Principles. In Feyerabend 1981e, 139–145.

Feyerabend, Paul K. 1981d. *Problems of Empiricism: Philosophical Papers*. Vol. 2. Cambridge: Cambridge University Press.

Feyerabend, Paul K. 1981e. *Realism, Rationalism and Scientific Method. Philosophical Papers.* Vol. 1. Cambridge: Cambridge University Press.
Feyerabend, Paul K. 1987. *Farewell to Reason.* London: Verso.
Feyerabend, Paul K. 1988. Against Method, Revised edition. London: Verso.
Feyerabend, Paul K. 1989. Realism and the Historicity of Knowledge. *Journal of Philosophy* 86 (8): 393–406.
Feyerabend, Paul K. 1993. *Against Method.* 3rd ed. London: Verso.
Feyerabend, Paul K. 1995. *Killing Time.* Chicago: University of Chicago Press.
Fine, Arthur. 1986. *The Shaky Game: Einstein, Realism and the Quantum Theory.* Chicago: University of Chicago Press.
Giere, Ronald N. 2006. *Scientific Perspectivalism.* Chicago: University of Chicago Press.
Giere, Ronald N. 2016. Feyerabend's Perspectivalism. *Studies in History and Philosophy of Science* 57:137–141.
Hacking, Ian. 1983. *Representing and Intervening: Introductory Topics in the Philosophy of Science.* Cambridge: Cambridge University Press.
Heller, Lisa. 2016. Between Relativism and Pluralism: Philosophical and Political Relativism in Feyerabend's Late Work. *Studies in History and Philosophy of Science* 57:96–105.
Horgan, John. 1996. *The End of Science: Facing the Limits of Knowledge in the Twilight of the Scientific Age.* New York: Broadway.
Kellert, Stephen H., Helen E. Longino, C. Kenneth, and C. Waters. 2006. Introduction: The Pluralist Stance. In *Scientific Pluralism*, Minnesota Studies in the Philosophy of Science, ed. Stephen H. Kellert, Helen E. Longino, and C. Kenneth Waters, vol. 19, vii–xxix. Minneapolis: University of Minnesota Press.
Kidd, Ian James. 2021. Feyerabend, Science and Scientism. In *Interpreting Feyerabend: Critical Essays*, ed. K. Bschir and J. Shaw, 172–190. New York: Cambridge University Press.
Kuby, Daniel. 2021a. Decision-Based Epistemology: Sketching a Systematic Framework of Feyerabend's Metaphilosophy. *Synthese* 199 (1–2): 3271–3299.
Kuby, Daniel. 2021b. Feyerabend's Re-evaluation of Scientific Practice: Quantum Mechanics, Realism and Niels Bohr. In *Interpreting Feyerabend: Critical Essays*, ed. Karim Bschir and Jamie Shaw, 132–156. New York: Cambridge University Press.
Kuhn, Thomas S. 1962. *The Structure of Scientific Revolutions.* Chicago: University of Chicago Press.
Kusch, Martin. 2016. Relativism in Feyerabend's Later Writings. *Studies in History and Philosophy of Science* 57:106–113.
Lakatos, Imre, and Paul K. Feyerabend. 1999. In *For and Against Method: Including Lakatos' Lectures on Scientific Method and the Lakatos-Feyerabend Correspondence*, ed. Matteo Motterlini. Chicago: University of Chicago Press.
Lloyd, Elisabeth A. 1996. The Anachronistic Anarchist. *Philosophical Studies* 81 (2–3): 247–261.
Lloyd, Elisabeth A. 1997. Feyerabend, Mill, and Pluralism. *Philosophy of Science* 64 (S4): S396–S407.
Ludwig, David, and Stèphanie Ruphy. 2021. Scientific Pluralism. In *The Stanford Encyclopedia of Philosophy* (Winter 2021 Edition), ed. Edward N. Zalta. https://plato.stanford.edu/archives/win2021/entries/scientific-pluralism/.
Mill, John S. 1859/1977. On Liberty. In *Essays on Politics and Society.* Collected Works of John Stuart Mill, ed. J. M. Robson, vol. 18, 213–310. Toronto: University of Toronto Press.
Oberheim, Eric. 2006. *Feyerabend's Philosophy.* Berlin: Walter de Gruyter.
Oberheim, Eric. 2025. Paul Feyerabend on Meaning and Method: From the Limited Validity of Falsificationism to 'Anything Goes.' *HOPOS: The Journal of the International Society for the History and Philosophy of Science* 15 (2).
Oberheim, Eric, and Paul Hoyningen-Huene. 1997. Incommensurability, Realism and Meta-incommensurability. *Theoria* 12 (3): 447–465.
Oberheim, Eric, and Paul Hoyningen-Huene. 2024. The Incommensurability of Scientific Theories. In *The Stanford Encyclopedia of Philosophy* (Summer 2024 edition), ed. Edward N. Zalta and Uri Nodelman. https://plato.stanford.edu/archives/sum2024/entries/incommensurability/.

Popper, Karl R. 1963. *Conjectures and Refutations: The Growth of Scientific Knowledge*. London: Routledge.
Popper, Karl R. 1972. *Objective Knowledge: An Evolutionary Approach*. Oxford: Oxford University Press.
Preston, John. 1997a. *Feyerabend: Philosophy, Science and Society*. Cambridge: Polity Press.
Preston, John. 1997b. Feyerabend's Retreat from Realism. *Philosophy of Science* 64 (Proceedings): S421–S431.
Preston, John, ed. 1999. *Paul K. Feyerabend: Knowledge, science, and relativism: Philosophical papers*, vol. 3. Cambridge: Cambridge University Press.
Preston, John. 2000. Science as Supermarket: 'Postmodern' Themes in Paul Feyerabend's Later Philosophy of Science. In *The Worst Enemy of Science? Essays in Memory of Paul Feyerabend*, ed. John Preston, Gonzalo Munèvar, and David Lamb, 80–101. New York: Oxford University Press.
Preston, John. 2020. Paul Feyerabend. In *The Stanford Encyclopedia of Philosophy* (Fall 2020 edition), ed. Edward N. Zalta. https://plato.stanford.edu/archives/fall2020/entries/feyerabend/.
Richardson, Alan W. 2006. The Many Unities of Science: Politics, Semantics, and Ontology. In *Scientific Pluralism*, Minnesota Studies in the Philosophy of Science, ed. Stephen H. Kellert, Helen E. Longino, and C. Kenneth Waters, vol. 19, 1–25. Minneapolis: University of Minnesota Press.
Sankey, Howard. 2020. The Relativistic Legacy of Kuhn and Feyerabend. In *The Routledge Handbook of the Philosophy of Relativism*, ed. Martin Kusch, 379–387. Abingdon: Routledge.
Shaw, Jamie. 2017. Was Feyerabend an Anarchist? The Structure(s) of 'Anything Goes'. *Studies in History and Philosophy of Science* 17:11–21.
Shaw, Jamie. 2020. The Revolt Against Rationalism: Feyerabend's Critical Philosophy. *Studies in History and Philosophy of Science* 80:110–122.
Shaw, Jamie. 2021. Feyerabend was Never an Eliminative Materialist: Feyerabend's Meta-philosophy and the Mind-body Problem. In *Interpreting Feyerabend: Critical Essays*, ed. K. Bschir and J. Shaw, 114–131. New York: Cambridge University Press.
Shaw, Jamie, and Karim Bschir. 2021. Introduction: Paul Feyerabend's Philosophy in the Twenty-first Century. In *Interpreting Feyerabend: Critical Essays*, ed. K. Bschir and J. Shaw, 1–10. New York: Cambridge University Press.
Shaw, Jamie, and Jonathan Y. Tsou. 2025. Feyerabend's Metaphysical Turn and the Stanford School of Pluralism. *HOPOS: The Journal of the International Society for the History of Philosophy of Science* 15(2).
Solomon, Miriam, and Alan Richardson. 2005. A Critical Context for Longino's Critical Contextual Empiricism. *Studies in History and Philosophy of Science* 36:211–222.
Tambalo, Luca. 2015. A Tale of Three Theories: Feyerabend and Popper on Progress and the Aim of Science. *Studies in History and Philosophy of Science* 51:33–41.
Tsou, Jonathan Y. 2003. Reconsidering Feyerabend's 'Anarchism'. *Perspectives on Science* 11 (2): 208–235.
Tsou, Jonathan Y. 2024. The Ambiguous Legacy of Kuhn's *Structure* for Normative Philosophy of Science. In *Kuhn's The Structure of Scientific Revolutions at 60*, ed. K. Brad Wray, 217–234. Cambridge: Cambridge University Press.
van Fraasen, Bas C. 1980. *The Scientific Image*. Oxford: Clarendon Press.
van Strien, Marij. 2020. Pluralism and Anarchism in Quantum Physics: Paul Feyerabend's Writings on Quantum Physics in Relation to his General Philosophy of Science. *Studies in History and Philosophy of Science* 80:72–81.
Watkins, John. 1966. Book Review of *Beyond the Edge of Certainty: Essays in Contemporary Science and Philosophy*, edited by Robert G. Colodny. Philosophy 41(158): 359–362.
Watkins, John. 2000. Feyerabend Among the Popperians, 1948–1978. In *The Worst Enemy of Science? Essays in Memory of Paul Feyerabend*, ed. John Preston, Gonzalo Munèvar, and David Lamb, 47–57. New York: Oxford University Press.

Part II
Dewey, Pragmatism, and Communities

Chapter 4
Quine, Dewey, and the Pragmatist Tradition in American Philosophy of Science

Don Howard

Abstract This chapter provides an answer to the question: Who picked up Dewey's pragmatist mantle in the second half of the twentieth century? The answer provided is W. V. O. Quine. Despite insistences to the contrary, Quine's commitment to pragmatism can be understood by properly appreciating his views on naturalism, holism and underdetermination, behavioral semantics, radical untranslatability, ontological commitment, relativism, truth, and realism. Here, it is shown that Quine consciously developed a pragmatic alternative to a foundationalist empiricism that he repeatedly renounced.

Keywords W. V. O. Quine · John Dewey · B. F. Skinner · Truth · Underdetermination · Holism · Naturalism

4.1 Introduction

How and where pragmatism lived and lives on in philosophy after John Dewey's death in 1952 are contested issues. Richard Rorty proudly proclaimed himself a pragmatist (Rorty 1979, 1982). Hilary Putnam, Robert Brandom, and Jürgen Habermas are sometimes cited as having carried on aspects of the pragmatist tradition (Legg and Hookway 2021). But the obvious diversity of these thinkers' views makes clear that speaking of a well-defined pragmatist legacy is problematic. The challenge grows only greater as we survey the diverse theorists and issues reflecting a so-called "revival of pragmatism" (Dickstein 1998) or the work of philosophers termed "new pragmatists" (Misak 2007).

D. Howard (✉)
Department of Philosophy, Reilly Center for Science, Technology and Values, University of Notre Dame, Notre Dame, IN, USA
e-mail: dhoward1@nd.edu

J. Y. Tsou et al. (eds.), *Values, Pluralism, and Pragmatism: Themes from the Work of Matthew J. Brown*, Boston Studies in the Philosophy and History of Science 347, https://doi.org/10.1007/978-3-031-92958-8_4

A more sharply focused question might prove more tractable: Who has the best claim to have carried forward the Deweyan tradition in the philosophy of science? In the first half of the twentieth century, Dewey's theory of science was dominant in the United States:

> Dewey's... philosophy of science was one of the most important players in the formation of that area of specialization, and it wouldn't be an exaggeration to suggest that, as in most areas of intellectual culture, no public discussion of science in the inter war period passed without seeking the input of Dewey and his students. (Brown 2009, p. 2)

Who picked up the mantle? One still too commonly heard answer to that question is that, after the mid-1930s and the arrival on American shores of emigre logical empiricists, pragmatism and positivism were melded into a distinctively American alloy. But, in fact, the European newcomers and a few American allies, most prominently Ernest Nagel, launched a campaign to displace Dewey's theory of science, a campaign that succeeded so well that Ronald Giere could ask in 1996:

1. How, between 1930 and 1960, did a dissident European movement advocating the replacement of much established German philosophy by *Wissenschaftliche Philosophie* transform itself into the dominant tradition for philosophy of science in North America? (Giere 1996, p. 336)
2. How did a naturalistic pragmatism incorporating an empirical theory of inquiry get replaced by a philosophy that regarded induction as a formal relationship between evidence and hypothesis? (Giere 1996, p. 347) [As quoted in Brown 2009, p. 12]

The full story of the displacement includes direct critiques of Dewey's theory of science (Reichenbach 1939; Nagel 1950), the seizure of control over journals like *Philosophy of Science* and organizations like the Philosophy of Science Association, and the placement of new PhDs in strategically important university positions (Howard 2003). But if Dewey's theory of science was so totally eclipsed, how can there be talk of the tradition's having been carried on? The answer is that it was carried on by a philosopher of science who contributed as much or more than Thomas Kuhn, Norwood Russell Hanson, Stephen Toulmin, or Paul Feyerabend to the ultimately successful effort to undermine the post-WWII neo-positivist hegemony, Willard Van Orman Quine, whose "Two Dogmas of Empiricism" (Quine 1951) was the first shot fired in the pragmatist counterattack. For at the heart of Quine's project is a philosophy of science with roots firmly planted in the Deweyan pragmatist tradition.

There are obvious obstacles to a pragmatist reading of Quine. Start with the fact that he is commonly remembered as part of the larger twentieth-century story of analytic philosophy, even though, ironically, "Two Dogmas of Empiricism" was and remains one of the most devastating attacks on the central legitimating premise of the analytic project, namely, the analytic-synthetic distinction. The irony grows as we remember that Quine was also one of the twentieth century's most vocal defenders of epistemological naturalism, a stance more or less diametrically opposed to analytic philosophy. The attack on analyticity and the naturalism are, of course, central to Quine's version of pragmatism.

Another obstacle to the pragmatist reading of Quine, especially one foregrounding his philosophy of science, is that he is remembered more commonly as a logician and philosopher of language than as a philosopher of science. When we think about modern, analytic philosophy of language or modern, formal logic, we do not think of William James and Dewey (though Peirce gets some of our attention). More's the pity, because logic and philosophy of language were major interests of Dewey's. But so it is. Quine *was* a logician and philosopher of language from the beginning, this being the font from which all else flowed. Nonetheless, the way of remembering Quine that foregrounds logic and philosophy of language to the exclusion of philosophy of science has much to do with the historical accidents of who chose to be his principal interlocutors. The fact that Donald Davidson, Jerrold Katz, and Saul Kripke were not ever or in the first instance philosophers of science implies nothing, however, about Quine's understanding of his own project. In fact, Quine wrote extensively about science and about core issues in the philosophy of science, such as conventionalism, induction, ontology, the theory-ladenness of observation, inter-theory relations, and incommensurability. Indeed, the engagement with philosophical issues about science is pervasive. Recall, also, Quine's own maxim: "Philosophy of science is philosophy enough."[1]

More importantly, the proper way of setting Quine in the context of twentieth-century philosophy, especially his early debts to C. I. Lewis, the younger Rudolf Carnap, and Otto Neurath, highlights Quine's central place in the century's major wrangles in the philosophy of science, such as the Vienna Circle's protocol sentence debate, to which one cannot help but read "Two Dogmas" as almost a direct response, even if Quine omitted the relevant footnotes, just as he managed to forget having been part of an argument about Pierre Duhem's brand of conventionalism when he first published "Two Dogmas" in the *Philosophical Review* in 1951.[2]

Shifting our glance forward in time, Quine's work from "Two Dogmas" in 1951 to *Word & Object* in 1960 must been seen in the context of pre-Kuhn (Kuhn 1962) anti-foundationalist challenges to neo-positivist orthodoxy in the philosophy of science. Other such include Nelson Goodman's "The New Riddle of Induction"

---

[1] It seems that this oft-quoted remark appears nowhere in Quine's published writings, having, instead been first attributed to Quine by P. F. Strawson in a review of Quine's *Theories and Things* (Quine 1981b). The review begins thusly: "In earlier essays, not reproduced in this volume, Quine wrote, 'Philosophy, or what appeals to me under that head, is continuous with science'; and, more bluntly: 'Philosophy of science is philosophy enough.'" (Strawson 1982). See Isaacson (2004, p. 263).

[2] The story, in brief, is that the 1951 *Philosophical Review* version of "Two Dogmas" lacks the well-known footnote about the similarity to Duhem that was added for the 1953 reprinting in *From a Logical Point of View*. Quine reported that he had been unaware of the similarity to Duhem until that was pointed out to him by Philip Wiener and Paul Oppenheim (Quine, personal communication), at which point he cast about for English-language sources on Duhem to mention in a footnote. It would appear that Wiener's 1954 translation of Duhem's *Aim and Structure of Physical Theory* (Duhem 1906) was a direct consequence of this interaction. I always found incredible Quine's claim not to have known of or to have remembered Duhem's work in 1951. Debates about Duhem's holism and under-determinationist conventionalism were so prominent a part of the landscape, especially during Quine's year in Vienna and Prague in the early 1930s, and especially among the people with whom Quine was working then and there – Rudolf Carnap, Otto Neurath, Moritz Schlick – that it is impossible for Duhem's work to have escaped Quine's notice.

(Goodman 1955), Hanson's *Patterns of Discovery* (Hanson 1958), Toulmin's *The Philosophy of Science: An Introduction* (Toulmin 1953) and *Foresight and Understanding* (1961), and Feyerabend's "Explanation, Reduction, and Empiricism" (Feyerabend 1962). Looking backward from the viewpoint of later twentieth-century discussions of Quine, he might appear mainly as a philosopher of language. Looking forward, however, from the viewpoint of the 1930s through the early 1960s, Quine looks much more like a philosopher of science. That Quine was once seen as a philosopher of science is evidenced by the first collection of critical essays on Quine's work, *Words and Objections* (Davidson and Hintikka 1969), the first contribution to which, by J. J. C. Smart, is entitled, simply, "Quine's Philosophy of Science" (Smart 1969). Anachronism is a vice.

A third obstacle to the reading of Quine as a pragmatist, at least in the Dewey tradition, is that he tendered no social philosophy and, unlike Neurath (Neurath 1913; Howard 2009) or Quine's Harvard colleague, Philipp Frank (Frank 1957), he saw little place for talk of values in science, or for ethics more generally in his naturalistic empiricism (Misak 2013, pp. 207–208). There might, however, have been an opening for a kind of weak normativity, for while Quine explicitly disavowed a justificatory task for epistemology (Quine 1969a), he did not altogether disavow epistemic norms, arguing, instead that such norms should be understood naturalistically, which maneuver is similar in part to Dewey's way regarding norms:

> The normative is naturalized, not dropped. The crowning normative principle of naturalized epistemology is nothing less than empiricism itself; for empiricism is both a rule of scientific method and a scientific discovery. It is natural science that tells us that our information about the world comes only through impacts on our sensory surfaces. And it is conspicuously normative, counseling us to mistrust soothsayers and telepathists. (Quine 1990, p. 229; as quoted in Misak 2013, p. 204)

But if epistemic norms may be so regarded, why not other norms? Experience informs improved performance as much with an aspiring athlete or a moral learner as with an inquiring mind. But such was not Quine's way. Dewey was as much a political activist as a philosopher (Ryan 1995, pp. 154–199). Quine's Nixonian Republican politics were well known, but they left no trace on his philosophy, which is just as well. Quine's claim to a place in the pragmatist tradition rests all the more crucially, for this reason, on his philosophy of science and its filiations to Deweyan pragmatism.

4.2 Quine on His Debt to Dewey and Pragmatism

That Quine understood himself to stand in the pragmatist tradition is clear from the concluding words of "Two Dogmas":

> Carnap, Lewis, and others take a pragmatic stand on the question of choosing between language forms, scientific frameworks; but their pragmatism leaves off at the imagined boundary between the analytic and the synthetic. In repudiating such a boundary I espouse a more thorough pragmatism. Each man is given a scientific heritage plus a continuing bar-

> rage of sensory stimulation; and the considerations which guide him in warping his scientific heritage to fit his continuing sensory promptings are, where rational, pragmatic. (Quine 1951, p. 46)

That Quine owed a debt specifically to Dewey is something that Quine, himself, proclaimed when he delivered the inaugural John Dewey lecture at Columbia University in 1968, the lecture that we know as "Ontological Relativity." Permit me to quote at length:

> I listened to Dewey on *Art as Experience* when I was a graduate student in the spring of 1931. Dewey was then at Harvard as the first William James Lecturer. I am proud now to be at Columbia as the first John Dewey Lecturer.
>
> Philosophically I am bound to Dewey by the naturalism that dominated his last three decades. With Dewey I hold that knowledge, mind, and meaning are part of the same world that they have to do with, and that they are to be studied in the same empirical spirit that animates natural science. There is no place for a prior philosophy.
>
> When a naturalistic philosopher addresses himself to the philosophy of mind, he is apt to talk of language. Meanings are, first and foremost, meanings of language. Language is a social art which we all acquire on the evidence solely of other people's overt behavior under publicly recognizable circumstances. Meanings, therefore, those very models of mental entities, end up as grist for the behaviorist's mill. Dewey was explicit on the point: "Meaning… is not a psychic existence; it is primarily a property of behavior" [Dewey 1925, p. 179].
>
> Once we appreciate the institution of language in these terms, we see that there cannot be, in any useful sense, a private language. This point was stressed by Dewey in the twenties. "Soliloquy," he wrote, "is the product and reflex of converse with others" [Dewey 1925, p. 170]. Further along he expanded the point thus: "Language is specifically a mode of interaction of at least two beings, a speaker and a hearer; it presupposes an organized group to which these creatures belong, and from whom they have acquired their habits of speech. It is therefore a relationship" [Dewey 1925, p. 185]. Years later, Wittgenstein likewise rejected a private language. When Dewey was writing in this naturalistic vein, Wittgenstein still held his copy theory of language. (Quine 1968, pp. 185–186).

Quine is, himself, here quoting from Dewey's *Experience and Nature* (1925).

The pragmatism in Quine's philosophy of science, at least at the time of "Two Dogmas," was noted by Smart in his above-mentioned contribution to *Words and Objections* (Smart 1969, pp. 8–9), coupling it, as once was common and right, with a version of instrumentalism about theories and theoretical terms. Smart thought that he detected a shift toward realism by the time of *Word & Object*, but Quine, in his reply, chides Smart for confusion on this point, claiming no change in doctrine, only a change in formulation (Quine 1969b, p. 293).

Six years later, Ernest Gellner described Quine as "The Last Pragmatist" (Gellner 1975). The immediate context for Gellner is ontology, Quine's views on which were famously expressed in the formula, "to be is to be the value of a bound variable," in "On What There Is" (Quine 1948). Gellner brands Quine as a pragmatist for his eschewing ontological absolutes:

> Quine is indeed a pragmatist philosopher. A number of themes are generally involved in pragmatism, but two are specially relevant: a joyful acceptance of change, trial and error, impermanence (as opposed to the old pursuit of absolute repose), and a vision which considers human activities and cognition in a biological perspective. Each of these themes is conspicuously present in Quine… But over and above exemplifying pragmatism in some

> measure, Quine has also elaborated an overt epistemological doctrine which actively preaches and justifies this attitude, which repudiates "Cartesianism" in so many words. (Gellner 1975, pp. 848–849)

Gellner goes on to describe Quine's "free-and-easy pragmatism" as emphasizing two continuities: the continuity between man and nature and the continuity between philosophy and other forms of inquiry.

It is instructive that those reading Quine at a time when Dewey and his work were still living memories remark upon the link to pragmatism as something not at all surprising, no special argument being needed to secure Quine's place in the tradition. But opinion in the secondary literature shifted in later years. An essay of Quine's, "The Pragmatists' Place in Empiricism," is cited as evidence. Several authors quote this passage[3]:

> It is not clear to me what it takes to be a pragmatist. It is not clear in what ways the philosophers who have been called pragmatists are nearer in outlook to one another than to philosophers who are not so called. I suspect that the term "pragmatism" is one we could do without. It draws a pragmatic blank. However, we have the term, and we can make some sense of it by enumeration. Peirce, James, Schiller, Mead, and Dewey have been called pragmatists and have owned the soft impeachment… All of these professing… pragmatists (Peirce, James, Schiller, Mead, Dewey) belong, it seems, to the empiricist tradition. (Quine 1981a, p. 23)

They quote, as well, this later remark:

> In limiting myself to the card-carriers, I have found little in the way of shared and distinctive tenets. The two best guesses seemed to be behavioristic semantics, which I heartily approve, and the doctrine of man as truth-maker, which I share in large measure. (p. 37)

That Quine is not as good a historian of pragmatism as, say, Louis Menand (2001) is indisputable, but, still, he allies himself clearly with what he takes to be core pragmatist tenets: behavioral semantics and man the truth-maker.

Matthew Brown thinks that Quine's "ham-fisted" history of pragmatism is so bad—mainly for want of a recognition of the pragmatists' testy relationship to empiricism—as to constitute "good reason to dispense with the myth of "Quine, the Last Pragmatist"" (Brown 2006, p. 340). But Quine never pretended to be a serious historian. I think that the right way of appreciating this and other of Quine's persuasive histories is as telling us more about Quine's view of his own place in history than about the thinkers and movements that here received a creative reading.[4] That Quine emphasizes behavioral semantics, as he did also in his hymn to Dewey in

[3] See, for example, Brown (2006) and Gibson (2009).

[4] A similar point is to be made about Quine's shaky grasp on the history of phenomenalism. In "Two Dogmas" (Quine 1951), "Epistemology Naturalized" (Quine 1969a), and other essays, Quine seemed to forget that Carnap's official doctrine in the *Aufbau* (Carnap 1928) was one of neutrality on the choice between a physicalist or phenomenalist protocol language. I think that Quine rightly discerned phenomenalist leanings in Carnap in 1928, but here too it is Quine's place in history, not Carnap's, that is relevant. Quine emphasized Carnap's phenomenalist side because that's what he, Quine, wanted to dispute.

"Ontological Relativity," is the crucial point. I will argue below that this is, indeed, the starting point of Quine's pragmatic philosophy of science.

Remember that we are discussing Quine the philosopher of science as part of the pragmatist tradition. One might still stumble over the fact that Quine's own explanation of his debt to Dewey speaks mainly to the philosophy of language, so it would seem, not philosophy of science. We are helped here by recalling some pertinent history.

The historiography of twentieth-century philosophy of science and the historiography of analytic philosophy remain curiously disconnected from one another. I say "curiously", because so much of the history of analytic philosophy, stressing logic and philosophy of language, seems to forget how most of the central questions in this arena grew up originally as part of the philosophy of science. Surely this is true for Gottlob Frege, Bertrand Russell, and Rudolf Carnap, to name only three of the core figures in the tradition. Consider the case of Carnap. His earliest work, from his dissertation (Carnap 1921) through the *Aufbau* (Carnap 1928), was almost exclusively devoted to philosophy of physics and general philosophy of science (see Howard 1996). It is in this context that his interest in the philosophy of language matures, the questions all being about how the language of science acquires and expresses empirical content, mainly so as to be able to assess the empirical credentials of Einstein's theory of general relativity. Such remains the focus in the context of the protocol-sentence debate (Carnap 1932a, 1932b), namely, what should be the observation language of science and how that language relates to the world. And when, by the time of the *Logische Syntax der Sprache* (Carnap 1934), the focus turns more directly on the *languages* of science, with questions about analyticity and conventionalism, the distinction between internal and external questions, and so forth, the context is still that of the languages of *science*.

These were the questions with which Carnap and his Vienna Circle friends were wrangling when Quine visited them in Vienna and Prague during his European *Wanderjahr* of 1932–33. It was out of the conversations that took place there that early essays like "Truth by Convention" (Quine 1936) were born, again, the context being philosophy of language as it informs philosophy of science.

Another fact about the debates of the 1930s is relevant to appreciating Quine's pragmatism. A common North American response to the philosophy of language emerging from the Vienna Circle, especially the work of Carnap, was that syntax and semantics had to be supplemented by pragmatics. The person who pressed this point with special vigor and made it well known was Charles Morris, a student of George Herbert Mead who was a kind of pragmatist emissary to the Vienna Circle during his *Wanderjahr* of 1934–35. It was Morris who later brokered Carnap's move to Chicago, and he played a major role in the gestation and editing of the *International Encyclopedia of Unified Science*. In a series of papers explaining pragmatism to the Viennese and logical empiricism to an American audience (Morris 1935a, 1935b, 1935c, 1936, 1937a, 1937b), Morris the semiotician argued that Carnap's philosophy of language devolves into an empty conventionalism if not supplemented with an appreciation of the pragmatic aspects of language in use. When Carnap, himself, later spoke of external questions as "pragmatic" (Carnap 1950), he had learned that way of speaking from Morris.

Morris's championing the addition of pragmatics to syntax and semantics affords the context in which to understand yet another late remark by Quine sometimes cited in more recent secondary literature to question Quine's self-identification as a pragmatist (see, for example, Isaacson 2004). Recall the above-quoted, closing paragraph from "Two Dogmas", where Quine mentioned the pragmatism of Carnap and Lewis and described himself as espousing "a more thorough pragmatism." Years later, in "Two Dogmas in Retrospect," Quine wrote that this remark had "unforeseen consequences," adding:

> I suspect it is responsible for my being widely classified as a pragmatist. I don't object, except that I am not clear on what it takes to qualify as a pragmatist. I was merely taking the word from Carnap and handing it back; in whatever sense the framework for science is pragmatic, so is the rest of science. (Quine 1991, p. 272)

Carnap learns this way of speaking from Morris. Quine knew the pragmatist tradition perfectly well and identified one of its defining themes as involving the continuities between and among science, the philosophy of science, philosophy more generally, and the rest of human activity. Here Quine simply and rightly says to Carnap that a genuine pragmatism sees the pragmatic moment as reaching beyond the choice of a linguistic framework for science to the whole of science.

Daniel Isaacson prefaces his quotation of, first, the passage from "Two Dogmas" and then, in a footnote, the just-quoted remark from "Two Dogmas in Retrospect" with the following peculiar comment:

> In "Two Dogmas," Quine formulates explicitly his rejection of Carnap and the Vienna Circle. They are not sufficiently empiricist. He makes this point in the final paragraph of "Two Dogmas," though the point is a little obscured by his use of the word "pragmatism" instead of "empiricism." (Isaacson 2004, pp. 250–251).

What kind of strong hermeneutic is it that ignores both the context and the manifest content of Quine's own words? Quine said that he has been labeled a pragmatist and that he does not object. He does not say in "Two Dogmas in Retrospect" that he meant to write or should have written "empiricism" instead of "pragmatism." He said, "I don't object." His saying that he was a bit unclear on what qualifies one, in general, as a pragmatist does not imply and was not meant to imply that he thought himself something other.

No. Quine's place in the pragmatist tradition and, specifically, his debt and intellectual kinship to Dewey, is clear and was, in the day, recognized by everyone who remembered and appreciated Dewey's work. Was Quine "the last pragmatist," as Gellner described him? I hope not, but this only because I choose to think of pragmatism as a still living tradition, not a dead one, with Quine's variety of pragmatist philosophy of science pointing the way.

4.3 The Principal Features of Quine's Pragmatist Philosophy of Science

I turn now to a sketch of the principal features of Quine's pragmatist philosophy of science. Space being scant, what follows is more an outline of main points, not a densely analytic or critical exegesis. The sketch will likewise say little about the historical development Quine's views, interesting though it is. I aim for synopsis, a kind of rational reconstruction, if you will excuse the term, one organized around a set of core themes.

4.3.1 Naturalism

In "Ontological Relativity," Quine wrote that he shared with Dewey, first and foremost, "the naturalism that dominated his last three decades." "Naturalism" means different things in different contexts. To Quine, as to Dewey, it meant this:

> With Dewey I hold that knowledge, mind, and meaning are part of the same world that they have to do with, and that they are to be studied in the same empirical spirit that animates natural science. There is no place for a prior philosophy. (Quine 1968, p. 185)

There is a long, historical context in North American philosophy for this version of naturalism.[5] A more commodious name for the larger tradition is "evolutionary naturalism." That human cognition is to be understood as a capacity with biological roots reaching deep into the evolutionary history of homo sapiens was a view shared by all of the major figures in the pragmatist movement. But it was a view shared well beyond, even by some of the pragmatist movement's chief antagonists, among them critical realists, like Roy Wood Sellars, and critical rationalists, like Morris Raphael Cohen.

Quine's most famous elaboration of his specific variant of naturalism is "Epistemology Naturalized" (Quine 1969a). The argument one finds there is an indirect one. Quine holds up the reductionist phenomenalism that he attributes to the Carnap of the *Aufbau* period as the last, best attempt to construct a justificationist, empiricist epistemology. The failure of that variety of foundationalist empiricism leave us with little choice but to settle, instead, for an empirical, descriptive epistemology, one that turns to psychology as the framing empirical science, for the purpose of generating a story about how input in the form of sensory stimulations becomes output in the form of verbal behavior. To the objection that such a project is viciously circular because it uses empirical science to ground empirical science Quine replies that, if we have eschewed justification as an aim, settling for

[5] Kim (2003) and Tsou (forthcoming) are good sources on the history of North American philosophical naturalism in the twentieth century.

description, instead, then a vice becomes a virtue and objectionable circularity becomes commendable theoretical closure.

Quine's argument here is curious in at least three respects, the first being that naturalism is made to appear a kind of second best, a stance to which we retreat only after the failure of foundationalist empiricism. Dewey would have urged naturalism as the preferred alternative from the start, as did Quine, himself, in many other places. Of a piece with Dewey's famous rejection of Cartesian first philosophy in writings like *The Quest for Certainty* (Dewey 1929), Quine famously rejected the idea of an Archimedean point from which to leverage a justificatory epistemology. It's not that we tried the foundationalist project and reverted to naturalism *because* foundationalism failed. Quine and Dewey both believed that, *as a consequence of naturalism*, foundationalism was fated to fail.

Curious, as well, is Quine's taking psychology alone or in the main as the framing empirical science for a naturalized epistemology. Dewey, as well as Quine in many other places, stressed the ineluctably social nature of language, meaning, and, therefore, the science expressed in language and expressive of such meaning. Above I quoted Quine quoting Dewey:

> Language is specifically a mode of interaction of at least two beings, a speaker and a hearer; it presupposes an organized group to which these creatures belong, and from whom they have acquired their habits of speech. It is therefore a relationship. (Quine 1968, p. 186; quoting Dewey 1925, p. 185)

Why, then, not accord sociology a comparably fundamental role in a naturalized epistemology? For that matter, biology deserves mention, obviously and too. But Quine's privileging the psychological perspective in spite of his explicitly social view of language was a consistent feature of his philosophy from early to late. Some commentators see here a vestige of positivism in the form of the unexamined assumption that all of the relevant questions concern the manner in which input becomes output in the behavior of the individual knower.

Vestigial positivism might be the explanation for the third curious feature of Quine's naturalism, which is his making the question one about how input *in the form of sensory stimulation* becomes verbal output. Why sensory stimulation? The very same behaviorist psychologists whose lead Quine claimed to be following would have said to him that proximal stimuli such as states of affairs on receptor surfaces turn out not to be the relevant controlling variables for most behaviors and certainly not for verbal behavior. Quine's good friend and Harvard colleague, B. F. Skinner, made this a central point in his sadly neglected 1957 classic, *Verbal Behavior* (Skinner 1957), specifically in developing his behaviorist surrogate for the semantic notion of reference, the verbal operant that he dubs the "tact."[6] There is irony in Quine's failure to appreciate the point, for the fruits of a Skinnerian operant analysis of verbal behavior would have equipped Quine with additional arguments for hallmark theses such as referential inscrutability.

[6] See Howard (2012) for a more thorough discussion of the Skinnerian approach to reference.

There is also an historical point to make in passing. Legend has it that Skinner's place in the history of the philosophy of science in the twentieth century demonstrates what goes wrong when simple-minded positivism or operationalism invades the social sciences. That is an ungenerous story in any case, as a thoughtful scholarship has shown (see, for example, Smith 1986). But it is doubly unfortunate folklore, because it also obscures Skinner's debt to the functionalist tradition in American psychology, the tradition in which Dewey worked and which was prominently championed by Dewey's University of Michigan student and, later, University of Chicago colleague, James B. Angell. By obscuring Skinner's debt to functionalism, the folklore also obscures Skinner's debt to pragmatism and his place in that tradition.[7] Properly appreciating Skinner's place in the pragmatist tradition makes only clearer still Quine's place in that tradition.[8]

4.3.2 Holism and Underdetermination

Naturalism is one premise of Quine's philosophical project. The other major premises are the two related claims of theory holism and the underdetermination of theory choice by empirical evidence. Quine credited neither to Dewey. Later on, Quine often noted an affinity on this point with Neurath, making frequent use of Neurath's boat metaphor for theory holism (Neurath 1932; see, e.g., Quine 1960, pp. 3–4). Neurath, however, was clear about having learned this way of thinking from Duhem, to whom Quine also acknowledged a debt.

It is important, however, that, unlike Duhem, Neurath explicitly situated his argument for holism and underdetermination in a broadly naturalistic epistemological project. For Neurath, holism and underdetermination were not derived from a logic a theory testing. They were, instead, empirical facts about the behavior of

[7] For Angell's version of functionalism, see Angell (1904). See Hothersall (2004, pp. 139–174, pp. 361–392) for an overview of functionalism. Explaining how it is that Skinner is to be understood as a functionalist would take us beyond the bounds of this paper. Suffice it to note for now that Dewey, like Skinner, was a forceful critic of the older stimulus-response form of early behaviorism (see Dewey 1896), and that Skinner's fundamental unit of analysis, the operant, is defined as a three-term functional relationship among stimulus conditions, probability of response, and history of contingencies of reinforcement. Skinner also shared with Dewey an aversion to mentalism, but, as with Dewey, this was not a consequence of a stringent verificationism. For Skinner it was mainly a worry about the danger of too quickly adverting to contingently available analogues for modeling possible intervening variables in the analysis of behavior. One final bit of folklore that has long obscured the real history is that pragmatism and logical positivism were closely allied philosophical movements. How and why this is false is discussed at some length in Howard (2003).

[8] For more on Skinner's debt to pragmatism, specifically the pragmatists' way of regarding meaning, see Moxley (2004), who notes the specific influence on Skinner of Dewey and James during Skinner's studies at Harvard (p. 756) and the striking similarity between the Skinnerian notion of the operant and Dewey's way of modeling human action (p. 761). He also notes that Dewey and Arthur Bentley commend important elements of Skinner's analysis of verbal behavior in their 1947 *Journal of Philosophy* article on "Definition" (p. 744; Dewey and Bentley 1947).

individual human cognizers and communities of inquirers. Neurath made this explicit in the early and influential essay in which he premiered his version of holism and underdetermination, "Die Verirrten des Cartesius und das Auxiliarmotiv. Zur Psychologie des Entschlusses" ["The Lost Wanderers of Descartes and the Auxiliary Motive (On the Psychology of Decision)"] (Neurath 1913), where he deliberately chose the psychological language of "motives", not the logical language of "reasons" to name the factors guiding theory choice within the domain of underdetermination. In this and later writings, Neurath explicitly proposed our making the nature and function of such motives a subject for empirical, scientific inquiry: "It is an empirical question how the auxiliary motive meets the test in practice" (Neurath 1913, p. 9).[9]

Turn back to Neurath's epigone, Quine. It is too easy to misread Quine's holism as only a logical thesis about the structure and empirical interpretation of theories and to miss the fact that, for Quine, as for Neurath, holism and underdetermination are, themselves, *empirical* facts about, as mentioned, the verbal behavior of individuals and the practices of communities of inquirers. For Quine, it is an *empirical* fact about verbal behavior that meaning does not attach to terms and sentences in the manner imagined in the Fregean tradition. A behavioral semantics cannot sustain the abstractions of *Sinn and Bedeutung*. That there is no sharp analytic-synthetic distinction is, therefore, also an *empirical* fact about verbal behavior, and so, too, is the semantic holism and theory holism that follow from that empirical fact. Quine's employing in *Word & Object* an amateur anthropological linguistics in the argument for the indeterminacy of radical translation was not just for rhetorical effect. He knew that the details were fanciful, that his was just a thought experiment. But he meant us to conclude that the concept of meaning sustaining an analytic-synthetic distinction and the phenomenological reductionism that was its correlate were contradicted by a consequent epistemological naturalism.[10]

Writing in a different philosophical idiom, Dewey in effect endorsed semantic holism, theory holism, and underdetermination (Dewey's Hegelian background is relevant here), but none were focal in Dewey's theory of science nor were they claims prominently associated with American pragmatism in any of its canonical forms. That they are, nonetheless, integral to Dewey's theory of science, is obvious if we query the possibility of a Deweyan pragmatist's denying them. The denial would take the form of the assertion that an individual theoretical claim possesses determinate empirical content of such a kind that the truth or falsity of the claim is fixed by the possible experience corresponding to that empirical content. Asserting such implicates several sins from a pragmatist point of view, foremost among them

[9] See also, for example, Neurath (1936), where Neurath contrasts what can be achieved by a restricted logic of science with the fuller understanding of how theory choice proceeds that we attain through the history of science and the "behavioristics of scholars." "Social behaviorism" was Neurath's preferred name for the kind of sociology that constituted the framing science of his version of epistemological naturalism; see Neurath (1931b).

[10] De Rosa and Lepore (2004) make a similar point about the roots of Quine's holism in his naturalism.

the mortal sin of hypostatization, and this in two forms: hypostatizing "experience" and hypostatizing "meaning." About the latter, Dewey wrote:

> The ordered development of meanings in their relations to one another may become an engrossing interest. When this happens, implicit logical conditions are made explicit and then logical theory of some sort is born. It may be imperfect; it will be imperfect from the standpoint of the inquiries and symbol-meanings that later develop. But the first step, the one that costs and counts, was taken when someone began to reflect upon language, upon *logos*, in its syntactical structure and its wealth of meaning contents. Hypostatization of *Logos* was the first result, and it held back by centuries the development of inquiries of a kind that are competent to deal with the problems of the existent world. (Dewey 1938, pp. 57–58)

Of course, Dewey sought a theory of science "competent to deal with the problems of the existent world." Logical empiricist verificationism, or any theory of science that proffered an algorithmic approach to theory choice, like Karl Popper's falsificationism, would have failed Dewey's competence test and would have been derided by Neurath as "pseudorationalism" (Neurath 1913).[11]

Experience, likewise, shall not by hypostatized. Dewey famously defended a view according to which experience, or better, experiencing, is an activity, not a static content of some psychic state (Dewey 1925). As such, experience cannot be parsed in units like sense data, elements of sensation, or Carnapian elementary experiences. The kind of empiricism that assumes otherwise was regarded by Dewey as the philosophy of science of nineteenth-century, British shopkeepers and bookkeepers, not a theory of science competent to deal with the challenges of the twentieth century.

Dewey the functionalist saw humans and their musings as all embedded in a dense world of other musings and makings. Language is a tool like any other, its virtues being all instrumental, hence to be assayed not by a mapping between reified units of discourse and meaning but by its effectiveness in achieving desirable ends. Those ends and the means of their achievement being all interrelated, as are the communities pursuing them, holism is inevitable. Nor would there be reason to think that one set of tools would always prove best. There should be multiple means to our ends, and the choice among them should be, well, pragmatic.

4.3.3 Behavioral Semantics and the Indeterminacy of Radical Translation

Naturalism, holism, and underdetermination are ground from which flows the rest of Quine's philosophical project. Let us quickly survey the whole, keeping the focus, as far as possible, on the implications for a philosophy of science.

---

[11] Strictly speaking Popper's falsificationism is not algorithmic since what he termed "basis statements" are adopted as matters of convention (Popper 1935, pp. 51–67). Nonetheless, Neurath did famously accuse Popper of pseudorationalism (Neurath 1935).

We begin with the behavioral semantics that Quine, himself, regarded as the most important common heritage of pragmatism. Language, for Quine and Skinner, lives in the form of verbal behavioral dispositions. Quine simplifies in *Word & Object* by imagining a linguist who seeks to translate the discourse of a people with whom there is no history of cultural contact. That lack makes the translation "radical." Through full participation, the linguist notes loose correlations among vocalizings—such as "Gavagai"—and stimulus conditions like the sight of a rabbit. Assume for the sake of argument that the linguist can form queries in the native tongue, say by rising inflection, and can reliably distinguish assent and dissent to queries. He or she then does that, namely, queries, and notes the patterns of assent and dissent. What happens?

Anthropological linguistics is an empirical science. Its hypotheses, in this thought experiment, are guesses about the translation of native discourse. Its evidence comes in the form of verbal behavioral dispositions to assent and dissent to queries. As all evidence underdetermines theory choice, so, too, here. A manual of translation, built upon a set of "analytical hypotheses", is systematically underdetermined by empirical evidence in the form of native dispositions to assent and dissent to queries. The sentence, "Gavagai", might be translated as "Rabbit", or it might just as well be translated as "Lo, an undetached rabbit part", or "Aha, a temporal slice of a rabbit." A similar indeterminacy attends to the derivative translation of individual words, like "gavagai." Translation being, thus, systematically indeterminate, the postulate of determinate meanings has no place in such a behavioral semantics.

The device of "radical" translation is supposed to make the indeterminacy especially clear. Quine holds it to be there in all translation, even between, say, French and German, or between my discourse yesterday and my discourse today, only obscured by the history of cultural or autopsychological contact.

For Quine as for Neurath, language is the vehicle of science. Dewey would remind both that language is a form of doing, hence continuous with non-linguistic aspects of scientific and other practice. But there is no disagreement here. The Skinner whose behaviorism inspired Quine (and who, as noted, also belongs to the history of pragmatism), famously defined "verbal behavior" as any behavior in which the reinforcement was socially mediated, thereby emphasizing, like Dewey, the continuities among vocal and non-vocal behaviors.[12] Neurath's whole philosophy of science was based on a repudiation of the Cartesian distinction between practical and theoretical reason, Neurath arguing that rationally ungrounded auxiliary motives are central to both (Neurath 1913).

Language being the vehicle of science, and all translation being indeterminate, the Quinean version of a holist and underdeterminationist philosophy of science is more extreme than that of Duhem, certainly, and possibly also that of Neurath. One might hold that evidence in the form of empirically well-credentialed observation sentences underdetermines theory choice. Loosen the link of language to experience after the fashion of a naturalist and pragmatist like Quine, deny even the

[12] Skinner (1957). This point is discussed in Howard (2012).

possibility of determinate, hypostatized, empirical meanings attaching to observational primitives, and the hold of experience over theory becomes more tenuous still. Neurath already argued that the underdetermination goes all the way down to the level of physicalist protocol sentences, but Quine unpacked that point with care and theoretical subtlety.

Is there anything left of empiricism? Well, yes. There is at least the contingent, empirical fact of provisionally privileged sentences—call them "observation sentences" if one will—that history and sociology show to be less likely to be sacrificed when we minimize disturbances in the web of belief occasioned by newer experiences. This is not an empiricism that fetishizes experience as foundational, but it is an empiricism that respects both the fact that experience usually trumps dogma and the hope that it should and will continue to do so.

Is Quine's holism also more extreme than that of Neurath or Duhem also in countenancing an anything-goes form of relativism? That the underdetermination goes all the way down to the observation sentence surely suggests the question. I think that answer is, "no." But let's return to this question only in the conclusion.

4.3.4 Ontology and Ontological Relativity

Closely related to radical indeterminacy is the thesis of ontological relativity premiered in the 1968 Dewey Lecture with that title (Quine 1968), the lecture quoted above that opens with Quine's most expansive description of his debt to Dewey. Questions of ontology, for Quine, are questions about theories and the languages in which they are expressed. Quine was well known for the claim that "to be is to be the value of a bound variable" (Quine 1948), meaning that questions of ontology were nothing more than questions about the elements of our theories' models. But what are those elements? How do we identify them? Quine's answer is that we do that by translating a theory into a background language. What are the natural numbers? Translate Peano arithmetic into set theory. There is no other way of talking about that ontology.

That being so, ontology is, as Quine said, "doubly relative" (Quine 1968). It is relative, first, to the choice of a background language. I can translate Peano arithmetic into Zermelo-Fraenkel set theory, or I can translate it into von Neumann set theory. Which to choose is a question not of truth but of pragmatic utility. (There is that word, "pragmatic", again). Translation via Zermelo-Fraenkel set theory might better illuminate some problems, translation via von Neumann set theory might better illuminate others. But then, having chosen a background language, the translation is, itself, indeterminate, as are, for Quine, all matters of translation. Is the number 2 to be interpreted as the set of all two-membered sets, or is it to be interpreted as the set consisting of the null set and the unit set of the null set? There is the second way in which ontology is relative: It is relative, first, to a choice of a background language, and it is relative, second, to a choice of a manual of translation. Again, no truth of the matter, just instrumental or pragmatic utility.

Our topic being Quine as a pragmatist philosopher of science, we should pause here to consider a point not often made about Quine's philosophy of science. Questions about inter-theory and inter-level relations have long loomed large in the philosophy of science. In the 1930s, we argued about the kind physicalism that demanded the reducibility, in principle, of all science to physics.[13] Today we argue about emergence, taking that thesis sometimes to mean the denial of a reduction claim. Am I the only one who thinks that current debates about inter-theory and inter-level relations, from the philosophy of physics to the philosophy of mind, would be helped by a dose of Quine?

Quine would remind us that all such questions are, at least in part, questions about the translation of an object theory into a background theory. Stop right there for lesson one. Such questions are not well posed absent a clearly articulated object theory and a clearly articulated background theory translation between which is contemplated. Respecting just that requirement would and should shut down much of the debate. Does mind reduce to matter? That is not a well posed question absent clearly articulated theories of mind and matter, neither of which are on offer in the contemporary metaphysics literature. But suppose we do have reasonably clear object and background theories. Lesson two is that there will always be multiple background theories among which to choose. What is my matter theory? The standard model in particle physics? String theory? Some classical, macroscopic approximation to a microphysical theory? Does it make a difference? Surely. OK. Pick one. Lesson three. Translation from the object to the background theory will depend on a choice of a manual of translation. So even if the mind-matter reductionism question were well posed from the point of view of our having reasonably clear theories of mind and matter (which we do not), there would be no one right answer about how to connect the two. Are any of these ways of proceeding illuminating? It depends on our aims. Here, again, these are questions of pragmatics, not truth. A dose of Quine would have the salubrious effect, I think, of simply putting an end to a lot of philosophical silliness.

Alert readers will have noticed a fourth lesson. Not once was it asked whether translation were possible at all. Is mind reducible to mater? The discussion assumed that it was, the only questions being those about the constraints on translation. In the same way, Quine never suggested that his linguist might find impossible the translation of the native discourse, however alien or odd. The questions were all about the whereby. That should strike one as interesting, for when was the last time a philosopher told you that a problem was easy, not hard? But here there lurks another very deep point to which I want to return in the conclusion.

[13] Note that this was not Neurath's physicalism. For Neurath, physicalism was the thesis that the protocol or observation language for a unified science should be a physicalist one, meaning one in which the primitive terms putatively referred to medium-sized physical objects. Neurath condemned as "metaphysics" the reductionist version of physicalism. See Neurath (1931a).

4.3.5 Truth

Questions about ontology, for Quine, are not about truth but about the instrumental or pragmatic utility of this or that way of translating a theory into this or that background language. That claim will be an affront to many philosophical programs. Best to ask, then, about truth, itself, in Quine's philosophy.

Recall that Quine identified as the only common theme in pragmatism, after behavioral semantics, the view of "man" (the gratuitous sexism, by the way, is Quine's, not mine) as "truth-maker." Note that Quine did not borrow the cartoon characterization of pragmatism as the view that "truth" is to be defined as "pragmatic utility" or "success." He said, instead, that pragmatists view humans as "truth-makers." What does that mean, and what is Quine's theory of truth?

Truth is a topic that Quine discussed at length and often. But there is no better place to turn than to the opening chapter of *Word & Object* (Quine 1960, pp. 1–25), the chapter over which most of us skip to get right to the weird stuff about the gavagai. Quine explains there that his theory of truth is Tarski's. Lazy thinkers are puzzled. Tarski defended a correspondence theory of truth, did he not? "'Snow is white' is true iff snow is white." Right?.... Wrong.

Here is your crash course on Tarski on truth (see Tarski 1936). Tarski defines "truth" for a theory, T, in model-theoretic terms via the relationship of "satisfaction." A model structure "satisfies" a theory, roughly speaking, if an assignment of entities and relations in the model to appropriate terms in the language of the theory, L_T, makes the postulates of the theory true. A condition of adequacy on a Tarski truth definition for a theory, T, is that every sentence in the metatheory of the form "'X' is true iff X" is a consequence of the truth definition.

Right away one sees that one gives a Tarski truth definition only for specific theories, not for all theories in one fell swoop. Thus, "truth in T_1" means, in general, something different from "truth in T_2", if only for the trivial reason that T_1 is a different theory from T_2. But could not one just embed all of these different theories in, say, natural English? Sure. But Tarksi, himself, explained that one cannot give a Tarski truth definition for natural languages, because a necessary condition on the possibility of a Tarski truth definition is a recursive specification of sentencehood in the language and none such is available for natural languages. Indeed, the lack of a recursive specification for sentencehood is more or less definitive of the difference between natural and artificial languages.

Quine simply takes over all of this apparatus. For Quine, the pragmatist, there is no truth with a capital "T." There is only truth in theory T_1, truth in theory T_2, and so forth. In what sense, then, is "man" the "truth-maker"? Well, theory choice is underdetermined by logic and experience. When I choose my theory, I also choose what counts as truth, or better, when a community of inquirers chooses a theory, it chooses what counts as truth. Quine is a pragmatist about truth—in the sense in which he understood that claim—simply because he believes that theory choice is underdetermined by logic and evidence.

4.3.6 Realism

The truth question takes us close to the realism question. No debate dominated the philosophy of science literature of the late-twentieth century more than this. Quine chose not to mix in the debate, but he is nonetheless widely regarded as a defender of a view not too far removed from Hilary Putnam's "internal realism" (Putnam 1977) or Arthur Fine's "natural ontological attitude" (Fine 1984). It is not surprising that the Quine who held that to be is to be the value of a bound variable and added that ontology is doubly relative would be read as dissenting from scientific realism of the kind associated with Richard Boyd, or Karl Popper, or the Hilary Putnam of the pre- or post-internal realism phase. Underdetermination and ontological relativity certainly do not comport well with a realism like Popper's that postulates inquiry's long-term convergence on one final, consistent, complete, and true theory of the world that will fix the real ontology of science.[14]

It is curious, then, that when asked the question directly, Quine always denied that he was some species of anti-realist, insisting that he understood himself to be "as much of a realist as one can possibly be" (personal communication). What did he mean by that? I always took him to mean that doing science in an ontological mode was crucial, but that he recognized the obvious constraints on ontology: Which theory among the alternatives choice among which is underdetermined by logic and experience? Which background language? Which manual of translation? All of those questions receive pragmatic answers. Can we, then, characterize Quine's position as pragmatic realism?

4.4 Conclusion: Antifoundationalism Without Relativism

Dewey died one year after the publication of "Two Dogmas." By the time *Word & Object* appeared in 1960, as mentioned above, the Deweyan variant of American pragmatism had more or less disappeared from academic philosophy. It lingered in schools of education, but mainstream philosophers no longer took Dewey seriously. The political pressures of the Cold War and McCarthyism played a surprisingly large role in effecting this transformation.[15] One wonders how different might have been the reception and impact of *Word & Object* had John Dewey been Quine's chief interlocutor instead of Donald Davidson.

A dominant narrative tells the story of post-WWII philosophy of science, and philosophy more generally, as one in which neo-positivism and its progeny confront anti-foundationalism in forms ranging from the modest skepticism of Toulmin,

---

[14] Many have noted the similarity between Popper's convergent realism and that of the pragmatist, Peirce. But remember that Quine did not include either realism or antirealism in his short list of commitments common to all pragmatists.

[15] See McCumber (2001), Howard (2003), and Reisch (2005).

Hanson, and Kuhn, the antics of Feyerabend, and the wayward, Edinburgh, Marxism of strong programme sociology of scientific knowledge, to the excesses of post-structuralism and post-modernism. Defenders of science see the threat of radical relativism growing ever more ominous. Some try to salvage empiricist foundationalism, perhaps tricked out with a more powerful inductive logic. When that falters, a new transcendentalism appears. Sometimes it is implicit, as when proponents of the "no-miracles" argument hold the success of science to be impossible unless terms refer and theories converge in the long-run to the truth. Sometimes it is explicit, as with Michael Friedman's revival of the old relativized or contingent a priori program (Friedman 2001). Surely others sense, as I do, an air of desperation in much of this.

Too often lacking during most of my professional life have been philosophical projects that know how to be antifoundationalist without simply opening the door to radical relativism. But precisely that—antifoundationalism without relativism—defines what I have always found most valuable in the pragmatist legacy and in the philosophy of science of Quine.[16]

Many will be quick to object that Quine's project fails precisely because it cannot prevent the slide into radical relativism. After all, was it not it Quine, himself, who glossed his holism and underdeterminationism as showing that one can save any cherished belief as long as one is willing to make sufficiently drastic changes elsewhere in the web of belief? Yes, he did. And is that not tantamount to an anything-goes relativism? No, it is not. Let me explain.

Emphasize the first half of Quine's claim – "one can save any cherished belief"—and his view seems the epitome of radical relativism. But do not stop there. Read on to the end: "as long as one is willing to make sufficiently drastic changes elsewhere in the web of belief." Well, how easy is that, namely, to make sufficiently drastic changes elsewhere in the web of belief? What Quine the non-relativist would have us appreciate is that making those adjustments elsewhere is often a highly non-trivial task, and the more radical the nonsense we choose to cherish and preserve, the greater the price to be paid for the needed surgery elsewhere in the web. I can surely defend a geo-centric and geo-stationary model of the planetary system, but I would be a fool to try. I can cling to a belief in cold fusion, but I would have to make adjustments elsewhere in the web of belief so drastic that I should not expect an offer from Berkeley anytime soon.

[16] Here I part company with Brown, who sees important differences between Dewey's antifoundationalism and Quine's, about which he writes: "This view is unsatisfying in a variety of ways. Most importantly, it brushes over the obvious differences in scientific practice between data, evidence, and experiment on the one hand, and speculation, theory, and hypothesis on the other. These things don't play the same sort of role in actual inquiries (but neither do they play the role foundationalists attribute to them)" (Brown 2009, p. 54). In this he echoes one of Friedman's complaints about Quine (Friedman 2001). What might be fair in this criticism is that Quine did not write much, if at all, about scientific practice, which is a shortcoming that a Deweyan pragmatist should lament. But Quine was, nonetheless, attuned to the fact that, in practice, the scientist does approach differently the more speculative parts of the web of belief and those closer to the periphery.

Schlick tried this maneuver against Neurath back at the time of the protocol sentence debate. He charged Neurath with defending a coherence theory of truth and argued that a coherence theory of truth devolves into relativism. An old canard. But Neurath, like Quine, understood that the demand for coherence across the web of belief or within the total body of theory was, in fact, a very powerful one, and this precisely because of the highly interconnected or—if I may borrow a physics metaphor—highly entangled nature of that web. It is precisely because individual statements or beliefs do not stand alone, contrary to the phenomenalist reductionism that Quine dubbed one of the "two dogmas", that the Duhem-Neurath-Quine version of antifoundationalist empiricism works so well.

The quip above about not expecting an offer from Berkeley if one defends cold fusion points to another important feature of Quine's pragmatist naturalism. It is that, while there is no Archimedean point, no royal road to truth, no self-evident first principles, and no theory-choice algorithm, there are many contingent facts of history, sociology, psychology, and biology that make likely an agreement among a community of inquirers about how to respond to new discoveries. Spin it one way and it looks like the indoctrination or socialization that Kuhn said was necessary for the smooth functioning of normal science. Spin it another way and it looks like what Duhem called "bon sens", the collective, good, common sense evinced in the practice of a well-functioning community of well-trained scientists, a community of reasonable people reasoning together.

Dewey would have appreciated this aspect of Quine's philosophy of science. The naturalism that they shared would have led them to expect no revelation and no toehold in certainty. All is contingent, all is in flux, everything is up for grabs, more or less. But that same naturalism would have led them to expect a common nature and history that affords basis enough for getting on with the common task of making a better world.

What, then, about the indeterminacy of translation? How can we act in concert to make a better world if we cannot even understand one another? Is a gavagai a rabbit, a temporal slice of a rabbit, or a rabbit part? If we cannot sort that out, how can we hope to settle arguments over dark energy or the dangers of climate change? Does not the indeterminacy of radical translation prove Quine to be a radical relativist? Is this not proof of incommensurability far more extreme than ever imagined by Kuhn?

Well, no. Many of Quine's readers who remembered debates in linguistics from the 1930s through the 1950s used to say that Quine had just formulated a version of the Sapir-Whorf hypothesis, according to which different language groups had buried deep within their semantics and even their syntax different fundamental metaphysics, so different as to be incommunicable to speakers of other languages (see Hoijer 1954). Their best-known example was supposed to be the Hopi language, which was said to assume a process metaphysics so alien to the substance-accident metaphysics of Indo-European languages as to make the two untranslatable (Whorf 1936). When asked the question, as he often was, "Aren't you just asserting a version of the Whorf-Sapir hypothesis?", Quine would chuckle, and with a twinkle in his eye, and in a voice that sounded just like Hugh Downs, he would say, "Don't you find it curious that Sapir and Whorf tell us that Hopi has a deep metaphysics so different from ours as to be inexpressible in English, and then prove their point by telling us, in English, what that alien metaphysics is?" (personal communication).

It usually took a few seconds for the light to dawn. But Quine's point was really a very simple one. It was that the thesis of the indeterminacy of radical translation does not make translation harder; it makes it easier. The very fact that there will always be a multiplicity of manuals of translation that make equally good sense out of native dispositions to assent and dissent means that we should be optimistic about always being able to find a translation that works. How much so? Quine went so far as to dispute on these grounds the once widely discussed thesis of the existence of so-called "pre-logical" cultures, not because he could prove that such peoples were truly logical, but because one could always find a manual of translation that made it seem so while respecting the verbal behavioral evidence (Quine 1960, pp. 58–60).

I choose to end now on this optimistic note, because optimism about solving our problems should have been the third common commitment that Quine identified as uniting all of the thinkers in the pragmatist tradition. There is no denying that science is hard, let alone negotiating an end to the conflict in Ukraine, or securing the world against another pandemic. But American pragmatism voices the spirit of optimism that once characterized American culture, scientific and otherwise. Quine was like that. He was a science geek. He had a child's naïve enthusiasm for science (some would also say, perhaps with warrant, a child's understanding of science), and he expected science to be, on balance, a force for good. Enthusiasm must be tempered with an appreciation of the risks of applying science and deploying technology in ways that are blind to context. A pragmatist like Dewey would, however, be among the first to make that very point about understanding the context and tailoring the science to fit.

References

Angell, J. 1904. *Psychology: An Introductory Study of the Structure and Function of Human Consciousness*. New York: Henry Holt.

Brown, M. J. 2006. On What Quine Is. Review of Gibson 2004. *Mind, Culture, and Activity* 13:339–343.

Brown, M. J. 2009. *Science and Experience: A Deweyan Pragmatist Philosophy of Science*. Ph.D. Dissertation., University of California, San Diego.

Carnap, R. 1921. *Der Raum. Ein Beitrag zur Wissenschaftslehre*. Inaugural-Dissertation zur Erlangung der Doktorwürde der hohen philosophischen Fakultät der Universität Jena. Göttingen: Dieterichschen Univ.-Buchdruckerei, W. Fr. Kaestner. Reprinted as "Kant-Studien" Ergänzungshefte, no. 56. Berlin: Reuther & Reichard, 1922.

Carnap, R. 1928. *Der logische Aufbau der Welt. Berlin-Schlachtensee: Weltkreis-Verlag*. Translated as: *The Logical Structure of the World & Psuedoproblems in Philosophy*. Translated by R. George. Berkeley and Los Angeles: University of California Press, 1969.

Carnap, R. 1932a. Die physikalische Sprache als Universalsprache der Wissenschaft. *Erkenntnis* 2:432–465.

Carnap, R. 1932b. Über Protokollsätze. *Erkenntnis* 3:215–228.

Carnap, R. 1934. In *Logische Syntax der Sprache. Schriften zur wissenschaftlichen Weltauffassung*, ed. Philipp Frank and Moritz Schlick, vol. 8. Vienna: Julius Springer.

Carnap, R. 1950. Empiricism, Semantics, and Ontology. *Revue Internationale de Philosophie* 4 (11): 20–40.
Davidson, D., and J. Hintikka, eds. 1969. *Words and Objections: Essays on the Work of W.V. Quine*. Dordrecht and Boston: D. Reidel.
De Rosa, R., and E. Lepore. 2004. *Quine's Meaning Holisms*. In Gibson 2004, 65–90.
Dewey, J. 1896. The Reflex Arc Concept in Psychology. *Psychological Review* 3:357–370.
Dewey, J. 1925. *Experience and Nature*. Chicago: Open Court.
Dewey, J. 1929. *The Quest for Certainty: A Study of the Relation of Knowledge and Action*. New York: Minton, Balch.
Dewey, J. 1938. *Logic: The Theory of Inquiry*. New York: Holt, Rinehart and Winston.
Dewey, J., and A. Bentley. 1947. Definition. *The Journal of Philosophy* 44:281–306.
Dickstein, M., ed. 1998. *The Revival of Pragmatism: New Essays on Social Thought, Law, and Culture*. Durham: Duke University Press.
Duhem, P. 1906. *La Thèorie physique. Son objet et sa structure*. Paris: Chevalier & Rivière. [Originally published in the Revue de Philosophie 4 (1904), 387–402, 542–556, 643–671; 5 (1904), 121–160, 241–263, 536–569, 635–662, 712–737; 6 (1905), 25–43, 267–292, 377–399, 519–559, 619–641.] English translation of the 2nd. ed. (1914): *The Aim and Structure of Physical Theory*. Translated by P. Wiener. Princeton: Princeton University Press, 1954, rpt. New York: Athaneum, 1962.
Feyerabend, P. 1962. Explanation, Reduction, and Empiricism. In *Scientific Explanation, Space, and Time*, Minnesota Studies in Philosophy of Science, ed. H. Feigl and G. Maxwell, vol. 3, 28–97. Minneapolis: University of Minnesota Press.
Fine, A. 1984. The Natural Ontological Attitude. In *Scientific Realism*, ed. J. Leplin, 83–107. Berkeley: University of California Press.
Frank, P. 1957. *Philosophy of Science: The Link Between Science and Philosophy*. Englewood Cliffs: Prentice-Hall.
Friedman, M. 2001. *Dynamics of Reason*. Stanford: Center for the Study of Language and Information.
Gellner, E. 1975. The Last Pragmatist: The Philosophy of W.V. Quine. *Times Literary Supplement* 25 (1975): 848–853.
Gibson, R. 2009. W. V. Quine. In A Companion to Pragmatism, ed. J. Shook and J. Margolis, 101–107. Oxford: Blackwell.
Giere, R. 1996. A From Wissenschaftliche Philosophie to Philosophy of Science. In *Origins of Logical Empiricism*, ed. R. Giere and A. Richardson, 335–354. Minneapolis: University of Minnesota Press.
Goodman, N. 1955. The New Riddle of Induction. In *Fact, Fiction, and Forecast*, 59–83. Cambridge: Harvard University Press.
Hanson, N. R. 1958. *Patterns of Discovery: An Inquiry into the Conceptual Foundations of Science*. Cambridge: Cambridge University Press.
Hoijer, H. 1954. The Sapir-Whorf Hypothesis. In *Language in Culture: Conference on the Interrelations of Language and Other Aspects of Culture*, ed. H. Hoijer, 92–105. Chicago: University of Chicago Press.
Hothersall, D. 2004. *History of Psychology*. 4th ed. New York: McGraw-Hill.
Howard, D. 1996. Relativity, Eindeutigkeit, and Monomorphism: Rudolf Carnap and the Development of the Categoricity Concept in Formal Semantics. In *Origins of Logical Empiricism*, ed. R. Giere and A. Richardson, 115–164. Minneapolis: University of Minnesota Press.
Howard, D. 2003. Two Left Turns Make a Right: On the Curious Political Career of North American Philosophy of Science at Mid century. In *Logical Empiricism in North America*, ed. A. Richardson and G. Hardcastle, 25–93. Minneapolis: University of Minnesota Press.
Howard, D. 2009. Better Red than Dead: Putting an End to the Social Irrelevance of Postwar Philosophy of Science. *Science and Education* 18 (2009): 199–220.

Howard, D. 2012. Reference from a Behaviorist Point of View. In *Perception, Realism, and the Problem of Reference*, ed. A. Raftopoulos and P. Machamer, 183–211. Cambridge: Cambridge University Press.

Isaacson, D. 2004. *Quine and Logical Positivism*. Inbson 2004, 214–269.

Kim, J. 2003. The American Origins of Philosophical Naturalism. *Journal of Philosophical Research* 28 (Suppl): 83–98.

Kuhn, T. 1962. *The Structure of Scientific Revolutions*. Chicago: University of Chicago Press.

Legg, C. and C. Hookway. 2021. Pragmatism. In *The Stanford Encyclopedia of Philosophy* (Summer 2021 Edition), ed. E. Zalta, <https://plato.stanford.edu/archives/sum2021/entries/pragmatism/>.

McCumber, J. 2001. *Time in the Ditch: American Philosophy and the McCarthy Era*. Evanston: Northwestern University Press.

Menand, L. 2001. *The Metaphysical Club: A Story of Ideas in America*. New York: Farrar, Straus, and Giroux.

Misak, C., ed. 2007. *New Pragmatists*. Oxford: Clarendon Press.

Misak, C. 2013. *The American Pragmatists*. Oxford: Oxford University Press.

Morris, C. 1935a. Philosophy of Science and Science of Philosophy. *Philosophy of Science* 2:271–286. Reprinted in Morris 1937b, 6–21.

Morris, C. 1935b. The Relation of the Formal and Empirical Sciences within Scientific Philosophy. *Erkenntnis* 5:6–14. Reprinted in Morris 1937b, 46–55.

Morris, C. 1935c. Some Aspects of Recent American Scientific Philosophy. *Erkenntnis* 5:142–149.

Morris, C. 1936. Semiotic and Scientific Empiricism. In *Actes du Congrès International de Philosophie Scientifique. Actualitès Scientifique et Industrielles*, no. 388. Section 1, *Philosophie Scientifique et Empirisme Logique*. Paris: Hermann, 42–56. Reprinted in Morris 1937b, 56–71.

Morris, C. 1937a. The Concept of Meaning in Pragmatism and Logical Positivism. In *Logical Positivism, Pragmatism, and Scientific Empiricism*, Actualitès Scientifique et Industrielles, no. 449, ed. C. Morris, 22–30. Paris: Hermann.

Morris, C. 1937b. *Logical Positivism, Pragmatism, and Scientific Empiricism*, Actualitès Scientifique et Industrielles, no. 449. Paris: Hermann.

Moxley, R. 2004. B. F. Skinner's Adoption of Peirce's Pragmatic Meaning for Habits. *Transactions of the Charles S. Peirce Society* 40:743–769.

Nagel, E. 1950. Dewey's Theory of Natural Science. In *John Dewey: Philosopher of Science and Freedom*, ed. S. Hook, 231–248. New York: The Dial Press.

Neurath, O. 1913. Die Verirrten des Cartesius und das Auxiliarmotiv. Zur Psychologie des Entschlusses. *Jahrbuch der Philosophischen Gesellschaft an der Universität Wien*. Leipzig: Johann Ambrosius Barth. Translated as: The Lost Wanderers of Descartes and the Auxiliary Motive (On the Psychology of Decision). In Neurath, O. 1983, 1–12.

Neurath, O. 1931a. Physicalism: The Philosophy of the Vienna Circle. *The Monist* 41:618–623.

Neurath, O. 1931b. Soziologie im Phyikalismus. *Erkenntnis* 2:391–431. Translated as: Sociology and Physicalism. In Neurath 1983, 58–90.

Neurath, O. 1932. Protokollsätze. *Erkenntnis* 3:204–214. Translated as: Protocol Statements. In Neurath 1983, 91–99.

Neurath, O. 1935. Pseudorationalismus der Falsifikation. *Erkenntnis* 5:353–365.

Neurath, O. 1936. Physikalismus und Erkenntnisforschung. *Theoria* 2: 97–105, 23–236. Translated as: Physikalism and the Investigation of Knowledge. In Neurath 1983, 159–171.

Popper, K. 1935. *Logik der Forschung. Zur Erkenntnistheorie der modernen Naturwissenschaften*. Vienna: Julius Springer.

Putnam, H. 1977. Realism and Reason. *Proceedings and Addresses of the American Philosophical Association* 50:483–498.

Quine, W. V. O. 1936. Truth by Convention. In *Philosophical Essays for A. N. Whitehead*, ed. O. Lee, 90–124. New York: Longmans.

Quine, W. V. O. 1948. On What There Is. *Review of Metaphysics* 2:21–38.

Quine, W. V. O. 1951. Two Dogmas of Empiricism. *Philosophical Review* 60:20–43. Reprinted in: *From a Logical Point of View*. Cambridge: Harvard University Press, 1953, 20–46.
Quine, W. V. O. 1960. *Word & Object*. Cambridge: MIT Press.
Quine, W. V. O. 1968. Ontological Relativity. *The Journal of Philosophy* 65:185–212. Reprinted in Quine 1969b, 26–68.
Quine, W. V. O. 1969a. Epistemology Naturalized. In *Ontological Relativity and Other Essays*, ed. W. V. O. Quine, 69–90. New York and London: Columbia University Press.
Quine, W. V. O. 1969b. Replies. In *Words and Objections: Essays on the Work of W.V. Quine*, ed. D. Davidson and J. Hintikka. Dordrecht and Boston: D. Reidel.
Quine, W. V. O. 1981a. The Pragmatists' Place in Empiricism. In *Pragmatism: Its Sources and Prospects*, ed. R. Mulvaney and P. Zeltner, 21–39. Columbia: University of South Carolina Press.
Quine, W. V. O. 1981b. *Theories and Things*. Cambridge: Harvard University Press.
Quine, W. V. O. 1990. Comment on Parsons. In *Perspectives on Quine*, ed. R. Barrett and R. Gibson, 291–293. Oxford: Blackwell.
Quine, W. V. O. 1991. Two Dogmas in Retrospect. *Canadian Journal of Philosophy* 21:265–274.
Reichenbach, H. 1939. Dewey's Theory of Science. In *The Philosophy of John Dewey*, The Library of Living Philosophers, ed. P. Schilpp, vol. 1, 159–192. Evanston and Chicago: Northwestern University.
Reisch, G. 2005. *How the Cold War Transformed Philosophy of Science: To the Icy Slopes of Logic*. Cambridge: Cambridge University Press.
Rorty, R. 1979. *Philosophy and the Mirror of Nature*. Princeton: Princeton University Press.
Rorty, R. 1982. *Consequences of Pragmatism*. Minneapolis: University of Minnesota Press.
Ryan, A. 1995. *John Dewey and the High Tide of American Liberalism*. New York: W. W. Norton.
Skinner, B. F. 1957. *Verbal Behavior*. New York: Appleton Century Crofts.
Smart, J. J. C. 1969. Quine's Philosophy of Science. In *Words and Objections: Essays on the Work of W.V. Quine*, ed. D. Davidson and J. Hintikka, 3–13. Dordrecht and Boston: D. Reidel.
Smith, L. 1986. *Behaviorism and Logical Positivism: A Reassessment of the Alliance*. Stanford: Stanford University Press.
Strawson, P. 1982. What Can Be Done. *London Review of Books* 4:3.
Tarski, A. 1936. Der Wahrheitsbegriff in den formalisierten Sprachen. *Studia Philosophica* 1:261–405. Translated as: The Concept of Truth in Formalized Languages. In Alfred Tarski. *Logic, Semantics, and Metamathematics: Papers from 1923 to 1938*. Translated by J. Woodger, 152–278. Oxford: Clarendon, 1956.
Toulmin, S. 1953. *The Philosophy of Science: An Introduction*. New York: Harper & Row.
Toulmin, S. 1961. *Foresight and Understanding: An Enquiry into the Aims of Science*. New York: Harper & Row.
Tsou, J. Y. (forthcoming). Philosophical Naturalism and Empirical Approaches to Philosophy. In The Cambridge Handbook of Analytic Philosophy, ed. M. Rossberg. Cambridge: Cambridge University Press.
Whorf, B. 1936. The Punctal and Segmentative Aspects of Verbs in Hopi. *Language* 12:127–131. Reprinted in: *Benjamin Lee Whorf. Language, Thought & Reality*. Cambridge: MIT Press, 1956, 51–56.

Chapter 5
Science in Dewey's Great Community

Paul Howatt

Abstract Science and technology have become powerful forces in the modern world, continually expanding their influence over social and political life. But citizens have so far had little say in their development, or in how they get inserted into our lives. And this is crucial for us to correct, since, as John Dewey says, "any state of affairs which tends to render the things of the environment unknown and incommunicable by human beings in terms of their own activities and sufferings would be deplored as a disaster" (Dewey, J., The Public and its Problems: An Essay in Democratic Inquiry. Pennsylvania State University Press, University Park, [1927] 2012, p. 195), and the influence of science and technology upon social and political life has so far done this. But Dewey has offered us a way to solve this problem, so that modern citizens can understand and control the science and technology that is increasingly shaping their fates. By digging through Dewey's later works, we can uncover a model of democratic inquiry that gives citizens more control over science and technology, and which helps them overcome their cognitive limitations as individuals.

Keywords John Dewey · Pragmatism · Democracy · Democratic inquiry · Planning science and technology · Planning democracy

5.1 The Indeterminate Situation

Science and technology have become powerful forces in the modern world, continually expanding their influence over social and political life. But citizens have had little say in their development, or in how they get inserted into our lives. And this is crucial for us to correct, since, as John Dewey says, "any state of affairs which tends

P. Howatt (✉)
Department of Philosophy, Indiana University, Bloomington, IN, USA
e-mail: pjhowatt@iu.edu

J. Y. Tsou et al. (eds.), *Values, Pluralism, and Pragmatism: Themes from the Work of Matthew J. Brown*, Boston Studies in the Philosophy and History of Science 347, https://doi.org/10.1007/978-3-031-92958-8_5

to render the things of the environment unknown and incommunicable by human beings in terms of their own activities and sufferings would be deplored as a disaster" (Dewey [1927] 2012, p. 195). The influence of science and technology upon social and political life has so far done this. But Dewey has offered us a way to solve this problem, so that modern citizens can understand and control the science and technology that is increasingly shaping their fates. By digging through Dewey's later works, we can uncover a model of democratic inquiry that gives citizens more control over science and technology, and which helps them overcome their cognitive limitations as individuals.

A century ago, Walter Lippmann and several political scientists started questioning the value of democracy because of these cognitive limitations of citizens. Their observations of voter behavior led them to believe that we had to limit citizen involvement in politics in order to make government intelligent.[1] The reasons for this pessimism are well summarized by Lippmann:

> [T]he real environment is altogether too big, too complex, and too fleeting for direct acquaintance. We are not equipped to deal with so much subtlety, so much variety, so many permutations and combinations. And although we have to act in that environment, we have to reconstruct it on a simpler model before we can manage with it. To traverse the world men must have maps of the world. Their persistent difficulty is to secure maps on which their own need, or someone else's need, has not sketched in the coast of Bohemia. (Lippmann ([1922] 1997, p. 11)

Modern citizens live in an increasingly complex world, but rely on mental maps and habits developed for a much simpler world. To govern themselves wisely, citizens need to develop a good understanding of the complex world they inhabit, since governing themselves largely amounts to governing their shared environment. But citizens rely heavily on simplistic stereotypes to navigate their shared environment, forcing the complex features of this environment into simple categories. This stereotyping leads citizens to ignore many features of the environment, and to distort the features they don't ignore. So if we want citizens to govern themselves, it's hard to see how they could do this wisely; their typical ways of thinking generate narrow and distorted maps of the complex world they must govern together, making them more likely to steer the ship of state into the rocks than safely ashore.

So it became clear to many that individual citizens couldn't develop a good enough grasp of their world to govern it wisely, and social thinkers typically responded to this by trying to reduce the role of citizens in politics. But some, like John Dewey, responded by rethinking the nature of citizenship itself. In *The Public and Its Problems* (Dewey [1927] 2012; hereafter "*PP*"), Dewey argued that the main problem was that citizens were expected to know and do everything as *individuals*, rather than as members of a group. But if we think of citizenship as a group

[1] For more on the pessimistic turn in political science, see Purcell (2014), Ricci (1984) and Seidelman and Harpham (1985). For a detailed discussion of Dewey's engagement with this pessimistic turn, see Westbrook (1991).

effort, then each member only needs a small amount of knowledge and ability, and the division of labor between the members can combine their knowledge-sets and abilities into a rich store of knowledge and an arsenal of abilities. This culminates in what Dewey calls the "Great Community", where many small communities develop their own knowledge-stores and ability-arsenals, and each community can connect with others to generate a much larger knowledge-store and ability-arsenal.

There are two components to this Great Community, and when combined they serve as a promising response to Lippmann's pessimistic challenge. First, citizens affected by some shared problem form into publics to engage in organized inquiry, which aims at solving this shared problem. Within these publics citizens can work on their main task: "the perfecting of the means and ways of communication of meanings so that genuinely shared interest in the consequences of interdependent activities may inform desire and effort and thereby direct action" (Dewey [1927] 2012, p. 181). Publics provide citizens an organized forum for deliberative inquiry, which eases the intellectual burden they have to carry (citizens divide up their deliberative labor) and fosters the creation of more complex and accurate maps of their shared circumstances (through an intelligent communication of meanings).

But the other component of the Great Community is my focus in this chapter: democratic inquiry. Dewey called for a practical social science, which would rely on "methods for detecting the energies which are at work and tracing them through an intricate network of interactions to their consequences" (p. 199). Experts would perform these practical inquiries, and work with journalists and artists to convey their findings to publics, who could use this knowledge to solve their political problems: "The highest and most difficult kind of inquiry and a subtle, delicate, vivid and responsive art of communication must take possession of the physical machinery of transmission and circulation and breathe life into it" (p. 204). Like publics, this democratic inquiry could go a long way toward resolving the worry Lippmann raised: it gives citizens a more accurate and comprehensive picture of their political world, and eases the intellectual load they have to carry. Scientists have the time, training, and tools to develop great maps of the world, which citizens can use to more easily navigate their environment.

If we can develop the Great Community, then we can overcome the problems Lippmann raised for citizens; individual citizens only need a modest amount of knowledge and ability to play their roles in publics and democratic inquiry, working in a division of intellectual labor that organizes their knowledge-sets and skills (and those of experts) to create sophisticated and accessible maps. But there's a problem with Dewey's depiction of democratic inquiry in *PP*, which needs to be tackled for this inquiry to be democratic in the strong sense Dewey cared about. While Dewey gives a strong and relatively clear role to experts in democratic inquiry, he doesn't do the same for citizens. This risks undermining the democratic nature of this inquiry, since it empowers everyone except citizens in it. But we can resolve this problem by looking at some of Dewey's other works, especially those written in the 1930s, where he gives us resources that we can use to fill in the role of citizens in democratic inquiry, and so ensure that it empowers citizens as democracy demands.

And by filling in this role, we can also bolster Dewey's response to Lippmann, showing how one component of the Great Community makes it easier for citizens to govern themselves wisely in spite of their cognitive limitations.[2]

5.2 Observations

When he lays out his vision of democratic inquiry in *PP*, Dewey gives experts (scientists, journalists and artists) a strong and clear role in relation to citizens. Their role is *clearly* laid out in many places throughout *PP*:

> Systematic and continuous inquiry into all the conditions which affect association and their dissemination in print is a precondition of the creation of a true public. (pp. 232–233)
>
> Unless there are methods for detecting the energies which are at work and tracing them through an intricate network of interactions to their consequences, what passes as public opinion will be "opinion" in its derogatory sense rather than truly public… (p. 199)
>
> Inquiry, indeed, is a work which devolves upon experts. (p. 225)

Experts have two clearly defined roles here: (1) systematic inquiry and (2) relaying the results of inquiry to citizens. As the last quote shows, systematic inquiry is largely done by scientific experts, who use their resources and training to uncover the workings of social problems, work out the strengths and weaknesses of proposed solutions to those problems, and then hand off their results to those who relay them to citizens (pp. 198–201). This relay-work is done by journalists and artists, who use their communication skills to convey the results of expert inquiry to publics so that they can be easily understood and used (pp. 201–205). While Dewey doesn't provide many details on how experts can play these two roles, *what* the roles are seems fairly clear.[3]

The role experts play is *strong* as well; Dewey continually reminds us throughout *PP* that systematic inquiry and the delivery of its results plays a crucial role in the creation of public opinion:

> Application *in* life would signify that science was absorbed and distributed; that it was the instrumentality of that common understanding and thorough communication which is the precondition of the existence of a genuine public. (pp. 196–197)
>
> Communication of the results of social inquiry is the same thing as the formation of public opinion. (p. 198)

We can't have genuine public opinion unless we have systematic inquiry and delivery of its results to citizens. Genuine public opinion rests on a grasp of the public's problem and well-considered solutions to it. But such problems and solutions are

---

[2] For the sake of space, I won't address the role of publics in the Great Community in much detail. For a more detailed treatment, see Howatt (2023) on Dewey's response to Lippmann's challenge.

[3] I thank one of my anonymous reviewers for helping me distinguish these two claims.

discovered in inquiry, so if experts are the ones doing most of this inquiry and relaying the results, they will play a strong role in shaping public opinion.

But Dewey says little about what citizens could do to engage with experts as equal partners in inquiry. Even Dewey's clearest statements of citizens' role in democratic inquiry are *vague*:

> The final obstacle in the way of any aristocratic rule is that in the absence of an articulate voice on the part of the masses, the best do not and cannot remain the best, the wise cease to be wise. (p. 223)
>
> No government by experts in which the masses do not have the chance to inform the experts as to their needs can be anything but an oligarchy managed in the interests of the few. And the enlightenment must proceed in ways which force the administrative specialists to take account of the needs. (p. 225)

While both of these quotes support a strong role for citizens in democratic inquiry, they're vague about what that role will look like, or even *how* strong the role should be. Citizens must have some "articulate voice" in democratic inquiry, but it's unclear what this voice should look like; should citizens get to frame all democratic inquiry, or merely be able to comment on all inquiries, or…? There are many ways someone can develop an articulate voice, but to develop it, the speaker needs to know their speaking role: who they're talking to, what kind of speech acts they can use (such as demands, requests or comments), and the communication methods available to them.[4]

It's also unclear how strong citizens' voice should be; should it be a dominant voice, or an equal voice, or a minor voice in the background? Dewey's claim that experts should be forced to take citizens' needs into account suggests that citizens should have some authority over experts. But it isn't clear what counts as "taking needs into account", since this could range from those needs being the dominant standard to their being a minor consideration.

Some of the things Dewey says even seem to fit the model of expert rule supported by Lippmann and other pessimists, the very model Dewey is trying to reject:

> Communication of the results of social inquiry is the same thing as the formation of public opinion. (p. 198)
>
> Public opinion, even if it happens to be correct, is intermittent when it is not the product of methods of investigation and reporting constantly at work. (p. 199)

Dewey slips here into the language of the pessimists, and seems to claim that public opinion should be produced, packaged and delivered to citizens, who would then

---

[4] Some scientific practices that provide citizens such guidance have already been created, such as community-based public research (CBPR), participatory action research (PAR) and most recently, minipublics. Analyzing these practices could provide a large amount of context-specific guidance for citizens, showing them what options and challenges they have for engaging with scientists. For more on these practices, see Ottinger (2017), Fischer (2009), and Warren and Pearse (2008). Thanks to an anonymous referee for pointing out these practices and their importance.

seem to have little else to do but consume the results of expert-run inquiry.[5] If Dewey thinks public opinion is identical with the packaged results of expert-led inquiry, or the product of this inquiry, then he seems to agree with Lippmann after all.

5.3 Problem Institution

The clarity and strength of the role of experts in democratic inquiry, combined with the vagueness and relative weakness of the role of citizens, makes it seem as if Dewey's model of democratic inquiry leaves citizens on the sideline. This creates a situation ripe for the sort of expert rule that Lippmann and other pessimists about democracy wanted. But clearly Dewey doesn't want this, as it goes against his entire philosophical and democratic outlook:

> For what is the faith of democracy… except faith in the capacity of the intelligence of the common man to respond with commonsense to the free play of facts and ideas which are secured by effective guarantees of free inquiry, free assembly and free communication? (Dewey [1939] 1988b, p. 228)

In fact, Dewey is arguing against expert rule in *PP*, since he's trying to respond to the pessimistic turn in political thought that he found emerging in the 1920s (Westbrook 1991, Chap. 9). So the problem isn't Dewey's overall commitments, but rather the way he depicts democratic inquiry in *PP*. If Dewey can depict democratic inquiry in a way that better empowers citizens, the problem will dissolve.

The more fundamental problem is that Dewey leaves citizens without much guidance. Dewey ([1927] 2012) sketches several methods experts can use in democratic inquiry, such as practical, prospective and continuous inquiry (pp. 198–200), tracing the forces affecting social life to their sources (pp. 198–202), and using journalism and art to convey findings (pp. 200–204). These give experts pretty clear guidance on what their role is and some ideas for how they should play it.

But Dewey ([1927] 2012) doesn't lay out a similar set of methods for citizens, and without these citizens won't be able to freely engage in democratic inquiry:

> No man and no mind were ever emancipated merely by being left alone. Removal of formal limitations is but a negative condition; positive freedom is not a state but an act which involves methods and instrumentalities for control of conditions. (pp. 191)[6]

---

[5] We could read these two quotes another way, which I find more fitting: Expert-run inquiries should play a (but not the only) key role in how citizens form their opinions in publics. On this reading, expert-run inquiries are one of several key factors in how public opinion is formed. But even read this way, we still need to know what other factors should shape public opinion, and how these rival expert testimony in how strongly they shape public opinion, which Dewey doesn't do in *PP* (but as I will argue, he does elsewhere).

[6] For the sake of space I don't discuss the "instrumentalities" citizens could use in democratic inquiry. For some discussion of this, see James Bohman's (2010) discussion of "ecological rationality."

Citizens need some idea of what methods of inquiry they can follow and how they can relate to other inquirers in the organization of democratic inquiry. Without these, they won't be able to exercise control over the facts and ideas that this inquiry demands, and the experts who do have these methods and role-relations will be able to claim sole intellectual authority within democratic inquiry.

These two problems can be restated as a question, so that we can see where we should look to resolve the problem: Does Dewey ever give citizens a strong and clear role in democratic inquiry, so they can work with experts as partners instead of consumers? This can be broken down into two sub-questions:

SQ1: Does Dewey ever give citizens a strong role in democratic inquiry?
SQ2: Does Dewey ever give citizens a clear role in democratic inquiry?

If we can find other places where Dewey gives a stronger and clearer role in democratic inquiry, we can give citizens guidance on how to effectively engage in democratic inquiry (SQ2), and make it clear how Dewey's model of democratic inquiry fits Dewey's democratic vision (SQ1). With a strong, clear role in democratic inquiry, citizens would be partners in inquiry with experts, rather than experts' passive clients.

Of course, this will only serve as a (partial) response to Lippmann's concerns if there's reason to think that citizens can actually play this role well, which we may doubt because of citizens limited cognitive abilities and the complexity of their intellectual tasks. So we can add a third sub-question:

SQ3: Can citizens realistically develop the will and skill to play their role in democratic inquiry well?

If we can answer "yes" to each of these questions, we could say that Dewey offers a model of democratic inquiry that is both feasible and empowering, allowing them to develop the sophisticated and accurate maps they need to steer the ship of state.

5.4 Hypotheses

Dewey believed that the definition of a problem often goes halfway toward its solution, by offering up suggestions of how it could be solved (Dewey [1938] 1986b, p. 108), and each of our sub-questions does this. SQ1 points in the direction of Dewey's later political works, especially *Individualism Old and New* ([1933] 1986a), *Liberalism and Social Action* ([1935] 1987), and his political essays during the Great Depression. In these works, Dewey explored how economic institutions could be reconstructed so that they empower citizens within them, and how to make our crisis-ridden economic institutions more resilient and intelligent. This led him to create an empowering vision of citizens' role in democratic decision-making, which would give citizens a strong role in democratic inquiry. Dewey called this "social control" of industry and finance, which involved public co-operation of economic institutions to manage economic problems and conflicts. Dewey wanted social control to be intelligent, which for him means that it should follow the methods of scientific inquiry; social control should take the form of social inquiry. These observations suggest an answer to SQ1: Dewey's vision of social control offers citizens a strong role in democratic inquiry.

Since SQ2 concerns the clarity of some aspect of democratic inquiry, Dewey's conception of inquiry should provide a possible solution, especially his *Logic: The Theory of Inquiry* (Dewey [1938] 1986b). Two aspects of Dewey's view of inquiry ground a promising answer to SQ2: (1) its practice-based nature and (2) the pattern of inquiry. For Dewey, inquiry is practical, aiming at the solution to some problematic situation. This suggests that democratic inquiry could aim to solve the political problems that citizens in a democracy face. And Dewey's pattern of inquiry—a web of functional stages for inquiry to follow—suggests that the functional roles citizens should play in democratic inquiry could derive from the functions these stages perform. These two aspects of Dewey's view of inquiry can combine to answer SQ2: we can articulate citizens' role in democratic inquiry as aimed at resolving their political problems, and this role should stem from the functional stages Dewey lays out for effective inquiry.

But to address Lippmann's challenge, we need reasons for thinking that citizens are able and willing to wisely engage in democratic inquiry (SQ3). For this we can draw from Dewey's work on education, which develop Dewey's ideas on how to cultivate the scientific attitude (which provides the will to inquire) and intellectual abilities (through guided action), with the ultimate goal of producing citizens who are able and willing to engage in democratic inquiry. This suggests an answer to SQ3: citizens can realistically develop the will and skill for democratic inquiry by engaging in guided action and developing the scientific attitude.[7] If this claim is true, a key component (democratic inquiry) of Dewey's response to Lippmann (the Great Community) will give citizens a realistic way to map out their problems and solve them wisely.

5.5 Reasoning

In his later political writings, Dewey lays out a model of intelligent social control of industry and finance. These writings are heavily influenced by the Great Depression, and Dewey wants to respond to this industrial and financial breakdown by envisioning a more democratic economy:

> The only form of enduring social organization that is now possible is one in which the new forces of productivity are *cooperatively controlled* and used in the interest of the effective liberty and the cultural development of the individuals that constitute society… *Organized social planning*, put into effect for the creation of an order in which industry and finance are socially directed in behalf of institutions that provide the material basis for the cultural

[7] I thank one of my anonymous reviewers for both pointing out the problem posed in SQ3 and the direction a solution could be found.

> liberation and growth of individuals, is now the sole method of social action by which liberalism can realize its professed aims. (Dewey [1935] 1987, pp. 39–40, emphasis added)

Dewey typically called this cooperative social planning "social control", which we could characterize more fully by combining the two italicized phrases in the previous quote: social control is a form of *social planning* and action that involves *cooperation* between a range of actors, which is *organized* around a set of institutions that facilitate this cooperative planning. To exercise social control, actors cooperate in investigating social problems and experimentally testing solutions to them. Dewey calls this "the method of experimental and cooperative intelligence", and he sees science as the exemplar of this method in "subduing to human use the energies of physical nature" (p. 64). Social control is the application of this cooperative experimental method to social problems. This cooperative experimentation should follow the methods of scientific inquiry, since "[t]he crisis in democracy demands the substitution of the° intelligence that is exemplified in scientific procedure for the kind of intelligence that is now accepted" (p. 51; cf. Dewey [1933] 1986a, Chaps. 7 and 8). So we can say that social control is a form of social inquiry (which he usually calls "planning"), and so could give us the model of democratic inquiry we're looking for.

Social control also involves actors working together in an *organized* way; institutions facilitate organized cooperation by providing a structured forum within which actors and groups could coordinate their investigations of—and their experimental solutions to—social problems. For example, Dewey asks the hopeful reader to "envisage a permanent Economic Council finally growing out of the holding of a series of conferences, a council which shall take upon itself a planned coordination of industrial development" (Dewey [1933] 1986a, p. 98). He then says that within this council, "captains of industry and finance would meet with representatives of labor and public officials to plan the regulation of industrial activity" (p. 98). In this example, a number of actors engage in cooperative planning within a particular institution (the "permanent Economic Council"). Leaders of the major parties affected by an economic problem (like business owners, workers and consumers) would come together to engage in organized inquiry to solve it.

But social control isn't restricted to economic problems. Instead, it applies to all sorts of urgent social problems; since modern citizens live in a massive and heavily interconnected society, the "problem of democracy becomes the problem of that form of social organization, extending to all the areas and ways of living, in which the powers of individuals shall not be merely released from mechanical external constraint but shall be fed, sustained and directed" (Dewey [1935] 1987, p. 25). So we can say that social control should be established for any problem requiring organized cooperative inquiry, providing us with a general model of democratic inquiry that can guide citizens.

Unfortunately, Dewey doesn't make it clear what the role of workers and consumers would be on this vision, which can obscure the democratic nature of social control, since we can't say what role most of the people affected by the problem (a public on Dewey's definition) are supposed to play. But given his commitment to intellectually empowering workers and consumers (Dewey [1933] 1986a, pp. 104–106), it seems clear they should have a say in the cooperative planning that affects them. Here we could follow James Bohman, whose Deweyan model of "public control" connects citizens (in publics) to political institutions, where they can help shape how these institutions work.[8] On Bohman's (1996) model, "each such institution needs to develop a political public sphere around it", so that citizens can "ensure uptake by the agents who exercise power in institutions" (p. 190). This helps "increase the interchange between the public and institutions, thus harnessing the power of these institutions to the constraints of popular sovereignty" (p. 191). So we can embed workers, consumers and ordinary citizens into Dewey's model of cooperative institutional planning by placing them in the public that forms around this planning. On this interpretation, citizens' role in cooperative planning would be to engage with those who do cooperative planning in institutions (experts and representatives), providing them with information, steering their planning, and holding them accountable.

This seems to follow Mark Brown's (2009) view of "representative institutions", which categorizes "different institutions based on their distinctive modes of representation" (p. 204). On this view, workers, consumers and ordinary citizens within publics would be represented by members of cooperative planning institutions. Brown follows Iris Young's (2000) view of representation as "democratic discussion and decision-making…a mediated relation of constituents to one another and to a representative" (pp. 126–127). This fits how Dewey ([1927] 2012) thinks of representatives of publics; the public engages in "selecting officials and regulating their conduct as officials" (p. 121), and a "public articulated and operating through representative officers is the state" (p. 109). So we can see cooperative planning institutions like the permanent Economic Council as a state institution, whose officers serve as representatives of some public, who are selected and steered by citizens in a public. This would give citizens a strong role in the democratic inquiry that happens in these institutions, as they can choose and steer their representatives within them.

But even with this general model of democratic inquiry, it isn't clear what functions citizens are supposed to play. They're supposed to engage in organized democratic inquiry, aiming at mutually acceptable solutions to their shared (public) problems. But it isn't clear what this inquiry is supposed to look like, what citizens are supposed to do within it, and what methods they can use to play their role well. Fortunately, Dewey has extensive writings on the structure and functions of inquiry, which we can draw from to answer these questions.

For Dewey, every inquiry deals with a *problematic situation*: a problematic relation between certain actors and the context they engage with (Dewey [1938] 1986b, pp. 498–499). For democratic inquiry, this will be a problematic relation between citizens and the world they inhabit together (or between citizens themselves). But

[8] Dewey several times uses "public control" as equivalent to "social control" ([1935] 1987, p. 57; [1933] 1986a, Chap. 6), and talks frequently of public ownership and co-operation of the economy, so it seems plausible that publics would have a strong role in social control.

inquiry only starts when there is *recognition* of this problematic situation: those affected by a problem come to realize that something is wrong with their relationship to their world, and that some organized process for restoring a good relationship needs to occur.[9] In democratic inquiry, once some shared problem is recognized, citizens engage in *public formation*: they organize themselves into a deliberative body whose purpose is to solve this problem.[10] This process of public formation is clearly laid out in *PP*:

> When some social institution "conducts itself so as to affect large numbers outside of itself, those who are affected form a public which endeavors to act through suitable structures, and thus to organize itself for oversight and regulation. (Dewey ([1927] 2012), pp. 78–79)

Citizens form themselves into a public, and then work with each other and representatives in institutions to cooperatively plan a solution to the problem the public faces.

This cooperative planning should follow what Dewey ([1938] 1986b) calls the "pattern of inquiry", which is a structure of several distinct functional components, each of which works with the others to guide inquirers towards a solution to the problem (pp. 109–114).[11] These functions can clarify the role citizens could play in democratic inquiry, showing them what functions they could and should perform. And since each function in inquiry relies upon certain methods to achieve its end-in-view, citizens would also have methods to guide them through difficult inquiries. With a clear understanding of the ends-in-view and the methods for pursuing them, citizens would have a clearer idea of how to become co-operators of democratic inquiry.

Following Matthew Brown (2012), we can see the pattern of inquiry as composed of five functions:

1. **Observation** collects data or determines the facts of the case in order to take stock of the fixed conditions of the situation that shape the problem.
2. **Institution** of a problem specifies what is perplexing or indeterminate in the situation in order to formulate a problem statement.
3. **Suggestion** proposes hypotheses (or ideas) for solving the problem.
4. **Reasoning** refines hypotheses, brings hypotheses into relation with larger theories and conceptual frameworks, coordinates observations and hypothesis, suggests new observations, and proposes experimental tests.

[9] For some case studies on how stakeholders can address their shared problems through cooperative inquiries, see Busenberg (1999) and Ostrom (1990).

[10] Problem recognition can also stem from experts, whose expertise can allow them to see problems citizens don't recognize. In these cases, experts should publicize their findings and warn citizens of the danger, helping them to form a public around the issue. Thanks to Heather Douglas (personal correspondence) for this insight.

[11] Dewey often describes the problem as certain problematic consequences of activity, effects of interacting conditions that are perplexing or troubling to certain people, and in need of explaining or stopping, which is done by intervening in those conditions. I find this helpful in clarifying how observing conditions relates to instituting a problem, which is central to Dewey's view of inquiry.

5. **Experimentation** tests the hypothesis by tentative or limited application to the situation that involves intervening or acting in the situation. (p. 286)

As Brown rightly emphasizes, this isn't a linear process, but a set of "functionally specified, interlocking, and reciprocal phases" (p. 286). Each phase performs a role that guides inquiry toward solving the indeterminate situation, and each phase works in tandem with others in working towards this solution. This is especially true of the first three functions, as Dewey analyzes each of them in reference to the others. We can think of this as an *observation-institution-suggestion continuum*: inquirers bring forward a range of observations, chart out their relations to each other to institute the problem, which should then prompt suggestions for how the problem can be solved. By looking closer at this continuum, we can uncover most of the functions that citizens could perform in democratic inquiry and the methods that could guide them in their performance.

We can start with the *observation of facts*, where inquirers uncover all the information relevant to the problematic situation that they find themselves in (Dewey [1938] 1986b, pp. 112–113). These observations feed into the process of *problem-institution*, where inquirers organize the facts brought forth in order to chart out the problematic situation. Dewey ([1938] 1986b) says that "[a]ll of these observed conditions taken together constitute 'the facts of the case.' They constitute the terms of the problem, because they are conditions that must be reckoned with or taken account of in any relevant solution that is proposed" (p. 113). Inquirers use facts to institute a problem by organizing them in relation to each other to form a picture of the conditions that constitute their problematic situation. The "conditions that must be reckoned with" are located through observation, and then charted by finding structural connections between conditions and their relation to the problematic consequences, giving inquirers a diagnostic map of the problem.

This map prompts *suggestions* for how conditions can be altered so that the situation is no longer problematic. Here's how Dewey describes this process:

> What is meant by calling facts operational?… They are selected and described, as we have seen, for a purpose, namely statement of the problem involved in such a way that its material both indicates a meaning relevant to resolution of the difficulty and serves to test its worth and validity. In regulated inquiry facts are selected and arranged with the express intent of fulfilling this office. (Dewey [1938] 1986b, p. 116)

When facts are organized to reveal the nature and source of the problem, a number of suggestions for how to solve the problem should arise. These suggestions should be developed into ideas, where "[i]deas are anticipated consequences (forecasts) of what will happen when certain operations are executed under and with respect to observed conditions" (p. 113). These forecasts are used to develop a prospective solution to the problem, showing what changes can be made to the problematic conditions that would make them unproblematic. Ideas also assist observation and problem-institution, since "[i]deas are operational in that they instigate and direct further operations of observation; they are proposals and plans for acting upon existing conditions to bring new facts to light and to organize all the selected facts into a coherent whole" (p. 116). Ideas guide organized observation by posing questions relevant to their forecasts, and these questions are answered by bringing together observations.

So we have three functions of inquiry and methods for performing them: observation of conditions using a method of idea-guided questioning, institution of a problem using a method of organizing observed facts together based on their connection to the problematic consequences and guided by ideas for eliminating these consequences, and hypothesis-formation using a method of forecasting, projecting what might happen if the observed conditions were altered in particular ways. We can use these functions and methods of inquiry to clarify the role of citizens in democratic inquiry, providing a description of what they could do and how to do it.

Given that citizens will often have particular information about the conditions they face and the interests they have, there are many relevant facts they could bring to democratic inquiry that experts would otherwise be hard-pressed to find. This is especially true concerning facts about citizens themselves, and the ways the public problem impacts their lives. As Dewey says, the "man who wears the shoe knows best that it pinches and where it pinches, even if the expert shoemaker is the best judge of how the trouble is to be remedied" (Dewey [1927] 2012, p. 224). Since citizens will already be organized in a public that's addressing the public's problem, they will have a forum within which they could share those facts with each other and with cooperative planners, and where these planners could ask them what facts they have that might be of use.

Citizens could also do the same for instituting the problem; they could help develop ideas for dealing with the problematic consequences and use these ideas to guide a search for relevant facts about the conditions surrounding these consequences. They could then help in organizing these facts together and finding out how they connect to the problematic consequences. Citizens could work through their public and with the representative institutions that serve it to map the public's problematic conditions, proposing facts to depict on the map and ways to connect these facts to each other and the problematic consequences. And they could challenge expert diagnoses that conflict with facts that citizens have learned as those most embedded in the problematic situation. In fact, citizens must be involved if the institution of the public problem is to succeed, since they have the best access to a range of facts about what's wrong with their situation, especially facts about the nature of the problematic consequences (where the shoe pinches).

Last, citizens could make suggestions about how to resolve their public problem, using the problem-maps to chart possible paths out of the problematic conditions. Since citizens inhabit the space these maps depict, they should have plenty of ideas for how conditions could be changed and what sort of consequences these would have, stemming from their practical experience of how their institutions and practices work. And since citizens come from a range of backgrounds, they could bring forward novel suggestions from a variety of perspectives, using their diverse insider experiences of the problematic consequences and conditions to brainstorm solutions and to challenge expert solutions that don't mesh well with these experiences.

The solutions suggested need to be developed and refined, so that they can be tried out in practice in a controlled experimental fashion. This is the process of *reasoning*, which "refines hypotheses, brings hypotheses into relation with larger theories and conceptual frameworks, coordinates observations and hypothesis, suggests new observations, and proposes experimental tests" (Brown 2012, p. 286). This will be the hardest phase of democratic inquiry for citizens to engage in, as it involves a great deal of time, energy, training, and tools to do. So citizens will have to play a smaller role here, relying on experts to perform the technical investigations needed to develop and refine the hypotheses and the diagnosis of the problem. But, since the phases of inquiry aren't linear, citizens could still shape expert reasoning by continually bringing forward observations and hypotheses for experts to develop and refine, and co-controlling how the problem is framed.

Citizens could also constrain expert reasoning by helping *experimentally test* the solutions that experts have worked out. This involves the "execution of existential operations directed by an idea in which ratiocination terminates", which can then "bring about the re-ordering of environing conditions required to produce a settled and unified situation" (Dewey [1938] 1986b, p. 121). The results of these operations are then observed, and if the experiment fails, inquirers can return to the other phases of the inquiry to establish what went wrong and how to fix it, until the problem is finally solved. If the experiment works, then the problem is resolved, and the inquirers can preserve the lessons learned, so that they can be used in further inquiry.

Citizens should have a strong role here, since the public has the ultimate say over the (experimental) decisions that affect them, and citizens run their publics. They could deliberate in their public over how to implement the solution they have jointly created with experts and representatives. Once they're satisfied with the proposed solution and how it would be implemented, citizens could deliver a clear mandate to public officials to pass certain legislation and then carry it out. After the solution is set in motion, citizens could then observe how the solution pans out in practice, and engage in further inquiry if it fails, or note the lesson if it succeeds.

Of course, giving citizens all of these strong roles assumes that citizens can effectively engage in democratic inquiry, and that they have the will to do so. But if Lippmann's worries stem from reasonable doubts about citizens' intellectual abilities, this should make us skeptical that citizens have the ability to play the role democratic inquiry sets for them. Lippmann and other pessimists also doubt that citizens can muster the will to play this role, and political scientists have found plenty of evidence to feed this doubt.[12] So if we want a workable model of citizens' role in democratic inquiry, we need reason to think that citizens can develop the skill and the will to play their role.[13]

Fortunately, Dewey wrote a lot about cultivating the attitudes and abilities needed for effective inquiry. Two of his ideas are crucial here: (1) the scientific attitude and

[12] See Martinez (2012) and Verba et al. (1995).

[13] A much more thorough answer than given here could be found in Dewey's works cited here, in James Bohman's (2010) discussion of "ecological rationality", and in Carole Pateman's (1970) study of the impacts of worker participation on political attitudes and involvement.

(2) empowering education. Dewey defines the scientific attitude as "the will to inquire, to examine, to discriminate, to draw conclusions only on the basis of evidence after taking pains to gather all available evidence" (Dewey [1938] 1988a, p. 273). But this isn't a cold, abstract attitude. Instead, it is "the attitude which is rooted in the problems that are set and questions that are raised by the conditions of actuality" (ibid). So the scientific attitude is the will to thoroughly engage in practical inquiry. If citizens could develop this attitude, it seems plausible that they would have the will needed to engage in democratic inquiry. After all, this attitude just is "the will to inquire." The main question is whether citizens can develop this scientific attitude.

Along with the will to inquire, citizens need the skill to inquire, and we can find ways to develop this skill in citizens by looking at Dewey's educational philosophy. Most importantly, this philosophy sees learning as embedded in reflective practices, where knowledge and skills develop through a person's engagement with their context in their attempt to deal with problematic situations, and through adapting their methods based on the success or failure of their inquiry (Dewey [1916] 1980, Chaps. 8 and 10). In other words, we become better inquirers by experimentally engaging in inquiry: adaptive practice makes perfect. This view of learning fits perfectly with the scientific attitude as well, since both emphasize the practical and experimental nature of inquiry. Both the scientific attitude and active learning center around the drive to resolve problematic situations in an intelligent way, where the attitude reflects our commitment to this intelligent problem-solving and the active learning reflects the mechanism through which we improve our ability to inquire. They're both experimental as well, with the attitude reflecting a view of inquiry that treats ideas as possible solutions to a problem, and with active learning occurring through the method of (reflective) trial and error.

So if citizens want to develop the will and skill to engage in democratic inquiry, they need to develop the scientific attitude and a range of skills through active, reflective learning processes. But this attitude and these skills require a medium through which they can develop; some sort of structures that can inspire the scientific attitude and offer resources and guidance citizens can use to develop their skills as inquirers. This is where *groups* can help, as they can provide a structure that can inspire the scientific attitude and scaffold active learning. If the groups citizens inhabit are geared toward intelligent problem-solving, then they would provide a setting within which citizens can develop their love of inquiry. Dewey argues that "[b]ecause the free working of mind is one of the greatest joys open to man, the scientific attitude, incorporated in individual mind, is something which adds enormously to one's enjoyment of existence" (Dewey [1933] 1986a, p. 118). Inquiry can be exhilarating and engaging, which gives the scientific attitude a certain allure that can draw citizens to it, if they're exposed to it.

But this exposure is key, and the cooperative activities of groups provide a setting that exposes people to the scientific attitude. As Dewey says, "the office of the social medium is, as we have seen, to direct growth through putting powers to the best possible use" (Dewey [1916] 1980, p. 121). The groups people act within—such as unions, consumer cooperatives, and friend groups—all provide a structure within which group members can develop their intellectual interests and abilities. When people engage with others, they develop certain habits, since "habits are formed for the most part under the influence of the customs of a group" (Dewey ([1927] 2012), p. 184). If people want to form good habits of thought (such as the scientific attitude and effective methods of inquiry), well-organized groups are the key means for doing so:

> But in fact, knowledge is a function of association and communication; it depends upon tradition, upon tools and methods socially transmitted, developed and sanctioned. Faculties of effectual observation, reflection and desire are habits acquired under the influence of the culture and institutions of society, not ready-made inherent powers. (p. 183)

To develop the scientific attitude and effective methods of inquiry, citizens need the groups they act within to foster this attitude and develop these methods. This would provide a steady inspiration for engaging in inquiry, as well as a set of procedures citizens could experiment with to learn an effective style of inquiry. This would connect these learning processes to actual democratic inquiries, since these inquiries are largely conducted by groups (such as unions, expert communities, or consumer cooperatives). So citizens can learn the skills they need both inside and outside of democratic inquiry.[14]

But even if well-organized groups are necessary for developing the will and skill for democratic inquiry, we can still doubt their sufficiency, especially for developing skill. Why think that citizens can develop the ability demanded by democratic inquiry by inquiring within groups?[15] Ultimately, this will depend on the methods and tools the group offers its members, and the capacity of members to learn how to use them. Dewey does offer throughout his work certain ways of training the abilities needed for democratic inquiry, which groups can offer members for their intellectual development. For the sake of space, I will focus only on problem-institution and suggestion.

To teach members how to institute a problem, groups can facilitate an awareness of a wide range of facts relevant to the problem, while allowing each member to focus their study on a few facts, so they can learn to compare specific features of their conditions without losing sight of the whole (Dewey [1933] 1986a, p. 322). Specialization allows people to focus more closely on important details of the problem, allowing group members to develop analytical skills. They can then use these skills to institute problems, making detailed connections between the facts they

---

[14] For some detailed case studies of how scientific practices can be organized to foster trust between citizens and scientists, and a scientific attitude among citizens, see Wynne (1992) and Norton (2015).

[15] Dewey leaned heavily to the nurture end of the nature-nurture debate, while it seems many pessimists (such as Jason Brennan (2017), Christopher Achen and Larry Bartels (2017)) lean toward the nature side, given how they consistently make the leap from the fact that citizens in the past have failed to meet democracy's expectations to the claim that they will continue to do so, which excludes the obvious possibility that citizens can improve in the right (nurturing) conditions. Given how little we have invested in providing these conditions, it seems the best course is to try provide them and see if it works.

analyze and the problematic consequences the group is addressing. The members still have access to a wider body of facts, since these will be put into an observable medium: language. Group members can speak and write about the facts they uncover, giving each member access to all the facts that members uncover (pp. 304–305). This allows them to place their analytical skills into the context of a wider inquiry, so they can learn how to contribute to social inquiry by placing their contribution alongside others.

Groups can also support the ability to develop suggestions by fostering creative and reflective habits. Developing hypotheses requires imagination, since the inquirer must imagine what sort of changes would occur in their problematic conditions if they acted in certain ways upon those conditions. As Dewey says, "[t]he proper function of imagination is vision of realities and possibilities that cannot be exhibited under existing conditions of sense perception" (Dewey [1933] 1986a, p. 351). So if groups can provide guidance on how to imagine the possible trajectories of actions, they would be fostering the power of suggestion. Groups can provide this guidance, especially through creative play and the pursuit of novel activities that encourage creative thinking. Dewey often emphasized the value of play and creative action, as they can open up channels of intellectual exploration that can discover new realms of meaning. But as he also says, "our powers of observation, recollection, and imagination do not work spontaneously, but are set in motion by the demands set up by current social occupations" (Dewey [1916] 1980, p. 21). The groups citizens act within must nurture the creative play of imagination of their members, rather than dull them, as the social environments of workers and consumers often do.

5.6 Experimentation

To see if this model of democratic inquiry resolves the problems embodied in our three guiding questions, we need to try it out in practice, and see if citizens can gain control over their shared fate. So we return to our statement of the problem: Does Dewey give citizens and publics a strong and clear role in democratic inquiry (and can citizens develop the will and skill for this role?), so they can work with experts as partners instead of as consumers? I have argued that Dewey gives citizens a *strong* role; they help initiate and control democratic inquiry, working with representatives and experts to grasp and solve political problems. To test this, we would need to insert citizens into democratic inquiry and see if they can in fact act as partners in inquiry, or whether experts and politicians would still be able to ignore their voice.

I also argued that Dewey makes citizens' role *clearer*, as it offers several methods for citizens to use in democratic inquiry. Each stage of inquiry calls for a certain task to be done, and some methods to use to complete them, and this gives citizens a guide for democratic inquiry. We could test this claim by giving these tasks and methods to citizens embedded in democratic inquiry, and seeing if it provides them enough guidance to make significant contributions, or whether they would still be lost even with these tasks and methods. Last, I argued that Dewey gives citizens ways to develop the *will* and *skill* for democratic inquiry, cultivating the scientific attitude and scientific abilities through experimental engagement within groups. We could test this by seeing if ordinary people, when placed in inquiring groups, can develop a devotion to the scientific method and the skills to see it through. The fate of truly democratic inquiry, like the fate of any idea, must be decided in practice, for "while ideas are necessary to deal with facts, yet they are working hypotheses to be tested by the consequences they produce" (Dewey [1938] 1988a, p. 273).

References

Achen, C., and L. Bartels. 2017. *Democracy for Realists: Why Elections do not Produce Responsive Government*. Princeton: Princeton University Press.

Bohman, J. 2010. Participation Through Publics: Did Dewey Answer Lippmann? *Contemporary Pragmatism* 7 (1): 49–68.

Brennan, J. 2017. *Against Democracy*. Princeton: Princeton University Press.

Brown, M. B. 2009. *Science in Democracy: Expertise, Institutions, and Representation*. Cambridge: MIT Press.

Brown, M. J. 2012. John Dewey's Logic of Science. *HOPOS: The Journal of the International Society for the History of Philosophy of Science* 2 (2): 258–306.

Busenberg, J. 1999. Collaborative and Adversarial Analysis in Environmental Policy. *Policy Sciences* 32 (1): 1–11.

Dewey, J. 1916/1980. Democracy and Education. In *John Dewey. The Middle Works, 1899–1924, Volume 9: 1916*, ed. J. Boydston, 1–370. Carbondale: Southern Illinois University Press.

Dewey, J. 1933/1986a. How We Think: A Restatement of the Relation of Reflective Thinking to the Educative Process. In John Dewey. *The Later Works, 1925–1953, Volume 8: 1933*, ed. J. Boydston, 105–352. Carbondale: Southern Illinois University Press.

Dewey, J. 1938/1986b. Logic: The Theory of Inquiry. In *John Dewey. The Later Works, 1925–1953, Volume 12: 1938*, ed. J. Boydston, 1–527. Carbondale: Southern Illinois University Press.

Dewey, J. 1935/1987. Liberalism and Social Action. In *John Dewey. The Later Works, 1925–1953 Volume 11: 1935–1937*, ed. J. McDermott, 1–66. Carbondale: Southern Illinois University Press.

Dewey, J. 1938/1988a. Unity of Science as a Social Problem. In *John Dewey. The Later Works, 1925–1953, Volume 13: 1938–1939*, ed. J. Boydston, 271–280. Carbondale: Southern Illinois University Press.

Dewey, J. 1939/1988b. Creative Democracy: The Task Before Us. In *John Dewey. The Later Works, 1925–1953, Volume 14: 1939–1941*, ed. J. Boydston, 224–230. Carbondale: Southern Illinois University Press.

Dewey, J. 1927/2012. In *The Public and its Problems: An Essay in Democratic Inquiry*, ed. M. Rogers. University Park: Pennsylvania State University Press.

Fisher, F. 2009. *Democracy & Expertise: Reorienting Policy Inquiry*. Oxford University Press.

Howatt, P. 2023. Neither Wisdom Nor Folly. *Contemporary Pragmatism* 20 (4): 334–355.

Lippmann, W. 1922/1997. *Public Opinion*. New York: Free Press Paperbacks.

Martinez, M. 2012. Why Is American Turnout so Low, and Why Should we Care? In *The Oxford Handbook of American Elections and Political Behavior*, ed. J. Leighley, 107–124. Oxford: Oxford University Press.

Norton, B. 2015. *Sustainable Values, Sustainable Change: A Guide to Environmental Decision Making*. Chicago: University of Chicago Press.

Ostrom, E. 1990. *Governing the Commons: The Evolution of Institutions for Collective Action*. Cambridge: Cambridge University Press.
Ottinger, G. 2017. Reconstructing or Reproducing?: Scientific Authority and Models of Change in Two Traditions of Citizen Science. In *The Routledge Handbook of the Political Economy of Science*, ed. D. Tyfield, R. Lave, S. Randalls, and C. Thorpe, 351–363. New York: Routledge.
Pateman, C. 1970. *Participation and Democratic Theory*. Cambridge: Cambridge University Press.
Purcell, E. 2014. *The Crisis of Democratic Theory: Scientific Naturalism and the Problem of Value*. Lexington: University Press of Kentucky.
Ricci, D. 1984. *The Tragedy of Political Science: Politics, Scholarship, and Democracy*. New Haven: Yale University Press.
Seidelman, R., and E. Harpham. 1985. *Disenchanted Realists: Political Science and the American Crisis, 1884–1984*. Albany: State University of New York Press.
Verba, S., K. Schlozman, and H. Brady. 1995. *Voice and Equality: Civic Voluntarism in American Politics*. Cambridge: Harvard University Press.
Warren, M., and H. Pearse. 2008. *Designing Deliberative Democracy: The British Columbia Citizens' Assembly*. Cambridge: Cambridge University Press.
Westbrook, R. 1991. *John Dewey and American Democracy*. Ithaca: Cornell University Press.
Wynne, B. 1992. Misunderstood Misunderstanding: Social Identities and Public Uptake of Science. *Public Understanding of Science* 1:281–304.
Young, I. M. 2000. *Inclusion and Democracy*. New York: Oxford University Press.

Chapter 6
Dismantling the Deficit Model of Science Communication Using Ludwik Fleck's Theory of Thinking Collectives

Victoria Min-Yi Wang

Abstract Numerous societal issues, from climate change to pandemics, require public engagement with scientific research. Such engagement reveals challenges that can arise when experts communicate with laypeople. One of the most common frameworks for studying these communicative interactions is the deficit model of science communication, which holds that laypeople lack scientific knowledge and/or positive attitudes towards science, and that imparting knowledge will fill knowledge gaps, lead to desirable attitude/behavior changes, and increase trust in science. Section 6.2 introduces the deficit model in more detail and shows that adhering to this model often fails to achieve its aims, which motivates the main question of this chapter: how can Ludwik Fleck's theory of thinking collectives address the persistent problem of deficit approaches in science communication? I suggest that it can do so by exposing the deficit model's implicit assumption of an expert-lay divide. Accordingly, Sect. 6.3 lays out Fleck's theory and Sect. 6.4 contrasts it with contemporary debates about science communication. Following this descriptive work, Sect. 6.5 draws on Fleck's ideas to make four concrete suggestions for further questioning the expert-lay divide.

Keywords Science communication · Deficit model of science communication · Expertise · Ludwik Fleck · Science and values · Thinking collectives

V. M.-Y. Wang (✉)
School of Philosophy, Psychology and Language Sciences, University of Edinburgh, Edinburgh, Scotland, UK
e-mail: v.m.wang@sms.ed.ac.uk

J. Y. Tsou et al. (eds.), *Values, Pluralism, and Pragmatism: Themes from the Work of Matthew J. Brown*, Boston Studies in the Philosophy and History of Science 347, https://doi.org/10.1007/978-3-031-92958-8_6

6.1 Introduction

Numerous societal issues, from climate change to pandemics, require public engagement with scientific research. Such engagement reveals challenges that can arise when experts communicate with laypeople. One of the most common frameworks for studying these communicative interactions is the deficit model of science communication, which holds that laypeople lack scientific knowledge and/or positive attitudes towards science, and that imparting knowledge will fill knowledge gaps, lead to desirable attitude/behavior changes, and increase trust in science. In Sect. 6.2, I introduce the deficit model in more detail and show that adhering to this model often fails to achieve its aims, which motivates the main question of this chapter: how can Ludwik Fleck's theory of thinking collectives address the persistent problem of deficit approaches in science communication? I suggest that it can do so by exposing the deficit model's implicit assumption of an expert-lay divide. Accordingly, I lay out Fleck's theory in Sect. 6.3 and contrast it with contemporary debates about science communication in Sect. 6.4. Following this descriptive work, I draw on Fleck's ideas to make four concrete suggestions for further questioning the expert-lay divide in Sect. 6.5.

6.2 Introducing the Deficit Model

The classic version of the deficit model of science communication posits that one-way transfer of value-neutral knowledge—in the form of scientific facts—from scientific experts to laypeople will fill knowledge gaps, lead to desirable attitude/behavior changes, and increase trust in science. This *information* deficit model assumes that scientific "facts speak for themselves" (Seethaler et al. 2019, p. 379) in the sense that "the interpretation of these facts is assumed to be identical for all members of the public" and occurs in a "rational and objective manner" (Simis et al. 2016, p. 401). The alliance between the value-free ideal of science and information deficit communication strategies is worth emphasizing since, even if "purely" scientific statements turn out to be value-laden, this is often not recognized and/or acknowledged by the communicators, so that "lay publics are presented with a version of science that is stripped of non-epistemic values" (Branch 2022, p. 564). It is also noteworthy that most discussions of the information deficit model refer to the communication of research stemming from the natural rather than the social sciences, especially in the anglophone world (cf. Cassidy 2021; Lewis et al. 2023).

Since its inception in the 1960s, the information deficit model has evolved in response to various criticisms.[1] Originally, there was an emphasis on the publics'

[1] For the purposes of this chapter, I am focusing on science communication practices in liberal democracies in Western Europe and North America. For developments in this thriving area see Cheng et al. (2008), Jamieson et al. (2017), and Bucchi and Trench (2021).

lack of knowledge of scientific facts.[2] This prompted attempts to increase the publics' scientific literacy, with concomitant surveys to measure this literacy by building on the hitherto more common surveys measuring interest in and attitudes towards science (cf. Miller 1983, 1998). The second wave of science communication in the UK, motivated by The Royal Society's Bodmer report (1985), doubled down on the literacy rhetoric and also focused on the publics' attitudes towards science. Despite generally high levels of interest in science and technology, the report recommended pushing this further and assumed that more scientific knowledge would lead to more positive attitudes towards science.[3] More recently, following the UK House of Lords report (2000) and the COVID-19 pandemic, the *trust* deficit model has come to the forefront of science communication efforts in an attempt to avoid the pitfalls—outlined in the next paragraph—of pure information deficit models.[4] Given this trust deficit, the reasoning goes, public participation and engagement with science are needed to "re-enchant" publics. However, Martin Bauer, an expert in the public understanding of science, among others (e.g., Wynne 2006; Trench 2008), emphasizes that these different science communication discourses do not reflect "a narrative of progress, but one of multiplication of discourses … the latter [discourse] does not entirely supersede the former" (Bauer 2009, p. 222). Furthermore, scientific literacy is resurfacing as a test for how effective trust-building initiatives, such as citizen juries and science festivals, are (cf. Jensen 2014). For example, a 2017 UK House of Commons report on science communication and engagement acknowledges the recommended move towards two-way engagement laid out in the 2000 House of Lords report, only to state in the next sentence that their "predecessor Committee examined in 2013 the public's *understanding* of the science on climate change, including where people look for science information and how that influences climate change policy" (2017, p. 5, my italics). Despite gesturing at the importance of dialogue between scientific establishments and publics, the 2017 report continues to emphasize how science communication can foster scientific understanding and "complement science learning in formal education" (2017, p. 26), using the term "engagement" largely as a placeholder for top-down communication without discussing how said engagement could become bidirectional.

This proliferation of discourses notwithstanding, one enduring critique of information deficit approaches is that they do not achieve their aims. There are many instances in which relaying facts from places of scientific authority to people lacking institutional scientific expertise have been ineffective in changing people's

[2] I use "publics" in the plural to highlight the diversity of this group.

[3] Jon Turney's review (1998), commissioned by the Committee on the Public Understanding of Science (formed as a result of the Bodmer report) summarizes evidence that there is no clear correlation between increased science literacy and positive attitudes towards science, let alone a causal effect; for controversial topics there is no correlation between understanding and attitude. Recent work bolsters this conclusion (e.g., Lackner et al. 2023).

[4] See Goldenberg (2021, especially chs. 1 and 2) for an overview of the development of various deficit models.

behavior (and sometimes rightly so). For example, one prominent case report details how (institutionally credentialled) scientific experts tried to impose restrictions on hill sheep farming in Cumbria, England in the wake of the Chernobyl nuclear accident. While mitigating risks from the radiation, farmers however needed to continue to make their livelihoods and felt that "scientists ignored local variations in radioactive fallout effects and … were evidently ignorant of local farming realities and neglected local knowledge" (Wynne 1989, p. 14). More recent studies show that communicating the cancer risks associated with alcohol consumption can change beliefs but not drinking patterns (Martin et al. 2018; Young et al. 2018), and that relaying information about vaccine safety and efficacy does not necessarily increase vaccine uptake (Kitta and Goldberg 2017; Goldenberg 2021).

Despite efforts to move beyond the information deficit model, newer models that underscore the importance of public participation (e.g., trust deficit models), as well as some scholarly literature on science communication, often still rely on a dichotomy between experts and laypeople. As Michel Callon observes, models of public education and public debate "share a common obsession: that of demarcation. [Both models] deny lay people any competence for participating in the production of the only knowledge of any value: that which warrants the term 'scientific'" (Callon 1999, p. 89). Recent work in the UK and USA suggests that many scientists themselves continue to think of science communication as predominantly one-way communication and education (e.g., Davies 2008; Besley and Nisbet 2013; Dudo and Besley 2016; Choi et al. 2023).[5] For example, a 2014 survey of members of the American Association for the Advancement of Science showed that 84% of the respondents think that the public not knowing much about science is a major problem for science in general, with 40% suggesting that too few scientists communicating their findings is a major reason for this lack of knowledge (Pew Research Center 2015). Thus, although communication specialists know that information deficit models do not work, pressing problems such as the COVID-19 pandemic can lead scientists and policy-makers to think that if "people just had *accurate* information … they would follow science on policies like mask wearing, vaccines, and other policy recommendations" (Scheufele 2022, p. 299). Despite its well-documented shortcomings, the deficit model is still prevalent, and exemplifies that the dominant way of thinking about knowledge involves a dichotomy between experts and laypeople.

The thesis of this chapter is that Ludwik Fleck's ([1935] 1979) concept of "thinking collectives"[6] and his understanding of science communication ([1936] 1986c), which have not received enough attention in ongoing discussions of the deficit

[5] Of course, not *all* scientists think of communication with publics as unidirectional; there may well be differences between scientists from different cultures, disciplines and research structures (e.g., universities, pharmaceutical industry, military research).

[6] The published translations of the German *Denkkollektiv* and *Denkstil* are rendered as "thought collective" and "thought style", respectively. However, "thinking collective" and "thinking style" might be more accurate, emphasizing the active nature of these concepts. Fleck himself translated them as "community of thinking" and "style of thinking" (Fleck [1960] 1986a, p. 154). Thanks to Jossi Berkowitz for pointing this out.

model, can address the persistent problem of deficit thinking in science communication by exposing the assumption of deficit model approaches that there is such an expert-lay divide, and by troubling the distinction between experts and laypeople. To do this, I first explicate Fleck's theory of thinking collectives (Sect. 6.3), before exploring how his concepts can be used to describe science communication practices, and how they contrast with contemporary deficit model approaches (Sect. 6.4). Following this analysis, my prescriptive aim is to suggest improvements in science communication by further problematizing the divide between experts and laypeople (Sect. 6.5).

6.3 Fleck's Theory of Thinking Collectives and Styles

Unlike the relatively strict separation of people into "experts" and "laypeople"—often presumed by scientists and science communicators from the mid-twentieth century until now—Fleck, writing in the 1930s, describes groups of people, including experts and laypeople, bound together by particular thinking styles as thinking collectives. A collective is defined as "*a community of persons mutually exchanging ideas or maintaining intellectual interaction*" (Fleck [1935] 1979, p. 39). Within such a thinking collective, e.g., the population of Canada in 2020, there are, with respect to any particular topic, so-called esoteric and exoteric circles. The former are comprised of specialists who have gone through an educational process. Depending on the topic in question, this might mean a graduate degree or an apprenticeship. In the first instance, members of exoteric circles are defined by the absence of "initiation" into a specialty. Generally, esoteric circles are smaller than exoteric circles. For example, during the COVID-19 pandemic there are multiple esoteric circles tackling the pandemic from various angles (e.g., virologists, epidemiologists, public health experts, social scientists, respiratory and critical care doctors and nurses). At the same time there are multiple exoteric circles whose members are more heterogeneous; they include everyone who is not a pandemic expert but also cannot be lumped together into a single "public." For instance, exoteric circles might be loosely characterized by educational, socio-economic status, religious affiliation, voting preferences, urban-rural lifestyles, allegiance to a particular sporting team, whether there are children in a given household, and so on. Fleck's description of these circles therefore avoids monolithic depictions of "Science" and the "Public." Importantly, these circles are comprised of individuals who "may belong to several exoteric circles but probably only to a few, if any, esoteric circles" (Fleck [1935] 1979, p. 105). Moreover, the status of circles changes depending on the issue at hand. Different people constitute the more expert, esoteric circles pertaining to the climate crisis compared to the COVID-19 pandemic. With respect to questions about greenhouse gas emissions or the efficiency of electric cars, the expert epidemiologist returns, so to speak, to an exoteric circle. According to Fleck, although the composition of these circles changes depending on the subject matter, the intersecting circles are connected because individuals carry thoughts, ideas, and moods[7] with them. So, although the epidemiologist is a member of an exoteric circle when it comes to climate change, she brings with her a certain scientifically-informed way of thinking, for example. Simultaneously, as a parent she may have different concerns and thoughts compared to other (exoteric) members who do not have children, but share a mood with other parents in the collective. Thus, Fleck's model emphasizes the fact that experts have a narrow range of expertise and are therefore laypeople with respect to the vast majority of issues.[8] This stands in contrast with much of the current rhetoric, which pitches experts writ large *versus* laypeople.

Thinking collectives form around so-called thinking styles, which arise out of particular moods that have "two closely connected aspects: readiness both for selective feeling and for correspondingly directed action" (Fleck [1935] 1979, p. 99). Though undoubtedly important in Fleck's philosophy, the concept of mood is a difficult one. Jarnicki (2022) gives a comprehensive overview of Fleck's use of "mood" (*Stimmung* in German, *nastrój* in Polish) and shows how the musical metaphor is not captured by the English word. Importantly, when rendered as "mood", there is a temptation to equate it with a transitory, personal psychological state, whereas the original meaning, Jarnicki argues, is closer to a "social mood" at the interpersonal level. Individuals can become attuned to or achieve a social mood. These moods are thus inculcated in people through "social forces", interactions with others, different types of education, and the surrounding society and culture in general. A thinking style can then be defined as:

> [the readiness for] *directed perception, with corresponding mental and objective assimilation of what has been so perceived.* It is characterized by common features in the problems of interest to a [thinking] collective, by the judgment which the [thinking] collective considers evident, and by the methods which it applies as a means of cognition. The [thinking] style may also be accompanied by a technical and literary style characteristic of the given system of knowledge. (Fleck [1935] 1979, p. 99)

The thinking style of a large thinking collective, such as the population of Canada in 2020, is heterogeneous, defying straightforward characterization, but there might be widespread rejection of practices such as witchcraft, for example, or shared awareness of, if not agreement on, science reporting related to climate change through various media channels.[9] The thinking styles of a collective's scientific esoteric circles will include things such as particular factual background knowledge

[7] More on moods below. It is possible that Fleck was influenced by Heidegger's concept of *Stimmung*, which is now often translated as "attunement" rather than "mood." Thanks to Marie-Eve Morin for suggesting this possibility. See also Janik (2006, p. 188 fn. 11).

[8] Collins and Evans make a similar point: with respect to a specific technical decision-making process "the wider scientific community *should* be seen as indistinguishable from the citizenry as a whole; the idea that scientists have special authority purely in virtue of their scientific qualifications and training has often been misleading and damaging" (Collins and Evans 2002, p. 250).

[9] Are these coarse-grained common features—e.g., rejection of witchcraft, awareness of climate change reporting—enough to define a thinking style and mood holding the population of Canada together as a thinking collective, rather than merely as a national demographic, given its heterogeneity e.g., in terms of ethnicities and religious affiliations? This is a valid question that requires

and assumptions, know-how regarding scientific instrumentation, as well as following certain communicative conventions. Thinking styles can be more or less stable, with the thinking style of contemporary science being relatively stable. In fact, "as an entity [the thinking collective] is even more stable and consistent than the so-called individual, who always consists of contradictory drives" (p. 44). This stability gives rise to the "intrinsic harmony of [a thinking] style" (p. 87), which induces a certain way of thinking and "looking" at things. Like Thomas Kuhn later, Fleck was influenced by Gestalt psychology. According to him, prevailing social moods influence which gestalt can be perceived, leading to "mood-conforming gestalt-seeing" (p. 179). Changes in mood help to bring about changes in what can be perceived, and seeing a new gestalt can amount to a scientific discovery. The "directed perception" afforded by thinking styles thus often restricts how and what can be thought. However, these styles also enable discoveries when members of the collective work together.

6.4 Fleck's Science Communication and the Deficit Model

With this outline of Fleck's theory of thinking styles, collectives and their ever-shifting esoteric and exoteric circles in place, we can now examine Fleck's conceptions of communication within such collectives. Depending on the participants and purposes of a conversation, there are three types of intra-collective communication: information exchange, legitimization, and popularization.[10] Regardless of the type of communication, for Fleck the "circulation of thought is always related, in principle, to its transformation" (Fleck [1936] 1986c, p. 85). Even between two people in the same esoteric circle "[t]houghts pass from one individual to another, each time a little transformed, for each individual can attach to them somewhat different associations. Strictly speaking, the receiver never understands the thought exactly in the way that the transmitter intended it to be understood" (Fleck [1935] 1979, p. 42). There is no communication, not even between experts, that does not entail some information gain, loss and/or change, though these changes tend to be larger the bigger the differences in thinking style are.

Fleck's notion of popularization—thoughts moving from esoteric to exoteric circles—is most akin to what we call science communication today. He says, "[p]opular science in the strict sense is science for nonexperts, that is, for the large circle of adult, generally educated amateurs" (p. 112). Popularization, especially through science journalism and popular science books, is characterized by the omission of technical details and statements that are contested by members of esoteric

further unpacking. In Sect. 6.5 I discuss how thinking collectives may be viewed as pragmatic constructs. Thanks to Michela Massimi and an anonymous reviewer for pressing me on this.

[10] Fleck also discusses propaganda, which operates between radically different thinking collectives (Fleck [1936] 1986c, pp. 85–86). Due to space restrictions and its relevance to modern science communication, I focus only on popularization.

circles, thus knowledge becomes "simplified, lucid, and apodictic" (ibid). Concomitant with this mode of communication are "specific social forces" between members of esoteric and exoteric circles:

> On the one hand it is the specific trust of laymen in the 'initiates' or specialists, and on the other it is the specific dependence of the latter on so-called public opinion and so-called common sense. The effect of both these forces is identical: they strengthen, intensify and carry into effect each thought circulating in the collective. (Fleck [1936] 1986c, p. 102)

Today, some of the science communication literature that has moved on from information deficit models recognizes and discusses exoteric circles' trust deficits at length. For example, Maya Goldenberg stresses that it is frequently not a lack of knowledge that makes parents hesitant to have their children vaccinated, but rather an often-understandable lack of trust (Goldenberg 2021). Given the now-infamous Tuskegee syphilis experiment, she argues, it is no wonder that some Black Americans today harbor mistrust towards governmental and scientific authorities. I do not dispute that a lack of trust in scientific institutions can be problematic, but I want to highlight that Fleck's insight about the two-way relation is often overlooked in today's discourses, with little focus on the ways in which experts also depend on "public opinion and common sense." Scientists and scientific institutions sometimes take for granted that "Science" is owed trust by the publics. For many scientists, this may be partly because of their self-conception as the "thinking" partner to the "feeling" public, where rationality somehow automatically garners trust (Cook et al. 2004).

Although Fleck does not spell out explicitly what he means by public opinion and common sense, one interpretation is that of values and "moods" held more or less generally in a given thinking collective. Fleck's main case study of the development of the study of syphilis and the eventual discovery of a diagnostic blood test in 1906 can serve as an illustration of this point (Fleck [1935] 1979, chs. 1 and 3). By tracing historical notions of diseases, and venereal diseases in particular, Fleck isolates multiple strands of popular thinking that, according to him, feed into esoteric thinking and practices. In particular, he traces the idea of a blood test for syphilis back to the notion of "noxious, foul mixtures of humors" that, over time, took hold in the popular imagination as a "change in the blood" (p. 11). (The popular metaphor of "bad/foul blood" was still in use during the Tuskegee experiment). This, in turn, led to several failed attempts at developing a blood test. Fleck suggests that one of the reasons serologists, microbiologists and physicians continued to conduct these unsuccessful experiments was because of the "persistent idea—demanding justification—of change in syphilitic blood" (p. 77). Additionally, he claims that since syphilis was considered a "carnal scourge" in fifteenth century Europe, and this association of the disease with "punishment for fornication" (p. 41) continued to hold sway in exoteric circles "[t]he attention, importance, and power of development that this research gained from the special moral emphasis on syphilis cannot be overestimated" (p. 77). He contrasts this with a relative lack of research interest in tuberculosis—despite its greater prevalence—because it was viewed as "romantic" rather than "accursed and disgraceful" (ibid). Fleck refrains from evaluating these exoteric influences on esoteric circles, but uses this example to show how prevailing moods drive certain avenues of research while precluding others.

Moreover, unlike the information deficit model, which holds that "popularization" is the dissemination of facts, Fleck recognizes that popularization is not merely an attempt at informing laypeople, but that the act of popularizing esoteric knowledge also "exert[s] a mental influence" (p. 43) on expert thinking in return. Popular science "furnishes the major portion of every person's knowledge. Even the most specialized expert owes to it many concepts, many comparisons, and even his general viewpoint" (p. 112). For example, he suggests that contemporary (i.e., from the 1920s) histories of syphilis are written in such a way as to obscure the way in which the term "syphilis" changed its meaning over time, and "the general conviction follows that there is no development of thought. This is a conviction that in turn also influences the expert" (p. 116). Fleck also cites examples of specialists importing terminology from their "fund of popular knowledge" (p. 112), such as when economists use the metaphor of an organism to refer to the economy, or when biologists refer to certain cellular arrangements as "cell states." Or as a more contemporary case: when molecular biologists explain to laypeople that DNA is (like) an "instruction manual" for "building" an organism, this popular notion can feed back into the workings of scientists themselves. This can be detrimental if, consequently, molecular biologists adopt reductionist views of development, potentially precluding cooperation and communication with biologists who hold more holist views (cf. Pigliucci and Boudry 2010).[11]

Fleck also acknowledges the difficulties attendant upon communication when thinking styles are very different. At the extreme he asks us to imagine the exchange between scientists from the 1930s and thinkers from the Middle Ages, "[w]hat if we could present our symbols—the potential, or physical constants, or the gene of heredity, etc.—to thinkers of the Middle Ages? Could we expect them to be delighted with the "correctness" of these symbols and instantly listen to reason?" (Fleck [1935] 1979, p. 115). Although the differences in thinking style between members of esoteric and exoteric circles in twenty-first century Canada are presumably smaller than between the people in his thought experiment, Fleck is pointing out one of the main flaws of information deficit thinking, namely the assumption that simply providing facts is enough for people to "listen to reason", an assumption that many scientists themselves hold (cf. Simis et al. 2016).

Fleck clearly does recognize knowledge differentials between specialists and members of exoteric circles, but only with respect to suitably well-defined scientific questions. There is no doubt that years of working as a physician or laboratory researcher equip the practicing doctor and scientist with knowledge, know-how, and

[11] I agree with an anonymous reviewer that causal claims should be made cautiously: does the simplification during popularization *cause* biologists to adopt reductionist views, or did they already hold these views before? Fleck might respond by suggesting that textbooks, which are used to "initiate" specialists, already contain many popular elements, thus "clos[ing] the circle of intra-collective dependence in knowledge" (Fleck [1935] 1979, p. 113). More empirical work may shed light on this question.

intuitions that the "uninitiated" lack.[12] However, Fleck casts a critical eye even on interactions that laypeople would lump together as "expert" interactions. For example, although both bacteriologist and doctor are part of esoteric circles, even they cannot achieve mere information exchange without thought transformation. Accordingly, the bacteriological report "is specially written to suit the general practitioner, but it does not represent the knowledge of the expert. It is vivid, simplified, and apodictic" (Fleck [1935] 1979, p. 113). Fleck suggests that the ideal of an informing statement is "merely a certain boundary possibility" that could occur only when the receiver of information has both confidence in the sender's abilities and the "full possibility of checking the contents" (Fleck [1936] 1986c, p. 86). Otherwise—and this applies to the vast majority of communicative situations even within esoteric circles—one specialist's "attitude is colored either with an excess of confidence and a lack of the possibility of checking (and thus appears… *to a certain extent as a layman*), or else with an excessive desire of testing" (p. 87, my italics). Even among esoteric circles there are these two-way relations otherwise associated with popularization. The account thus rightly emphasizes the fact that any individual expert has only a narrow range of expertise. However, because of the trust of laypeople in specialists "the layman has a tendency to overestimate the capabilities of the specialist and to underestimate his limitations" (p. 102).[13] Additionally, all specialists are laypeople with respect to the vast majority of societally important issues and may underestimate their own limitations outside their expertise.[14] Overall, Fleck's account makes distinctions between individuals in a collective much blurrier by highlighting the many two-way interactions in which thought transformations take place.

Others have also compared Fleck's theory with present-day science communication practices (e.g., Bucchi 2008; Bauer 2009; Mößner 2016). Some of these comparisons, however, seem to misconstrue Fleck's insights regarding the inextricable links between exoteric and esoteric circles. For example, Bauer, after discussing iterations of the deficit model, closes his account with a description of Fleck's

[12] It would be a stretch to interpret Fleck as recognizing "lay expertise" or knowledge co-production between "experts" and "laypeople." However, Fleck's insistence on the context-dependence of who counts as a member of an esoteric circle suggests that, if scientific questions were reframed to make the societal consequences at stake more explicit, he might assent to the idea that members of exoteric circles could be as well equipped to contribute to inquiries as "experts." Thanks to Maya Goldenberg for raising this point.

[13] Of course, not all laypeople trust experts across the board. For example, the Tuskegee experiment helps explain why Black Americans today distrust scientific institutions more than members of other demographic groups. Moreover, public trust in scientists has fallen since the start of the COVID pandemic in the USA, especially among Republicans, while trust in science has increased in the UK. Fleck's statements may thus not apply wholesale to the contemporary USA and UK, but despite fluctuating levels of public trust, scientists remain trusted by a majority of people (Fonseca et al. 2023; Pew Research Center 2023).

[14] The phenomenon of experts asserting knowledge in areas outside their expertise has recently received the label "epistemic trespassing" in analytic philosophy (e.g., Ballantyne 2019).

theory.[15] Bauer interprets Fleck as "suggest[ing] an image of concentric spheres of science, similar to a planetary system of a centre and circulating peripheries" (Bauer 2009, p. 236; Fig. 6.1). The monolithic core of experts in the middle is surrounded by an exoteric circle, which tentatively includes consumers, students, and politicians. Based on this visualization, Bauer claims that "the problem of [public understanding of science] research for the future will be to map the esoteric-exoteric distance of the public conversation from science" (p. 237).

However, I believe that this interpretation mistakenly reifies the expert-lay dichotomy and misinterprets Fleck, who is clear that thinking collectives consist of "many such mutually *intersecting* esoteric and exoteric circles" (Fleck [1936] 1986c, p. 101, my italics). We are invited to imagine the above diagram much more like a complex Venn diagram, in which members of each circle shift when different questions arise in the collective. Aren't all experts also always consumers and at least residents of a country? In Fig. 6.2, I sketch a diagram, which hints at the

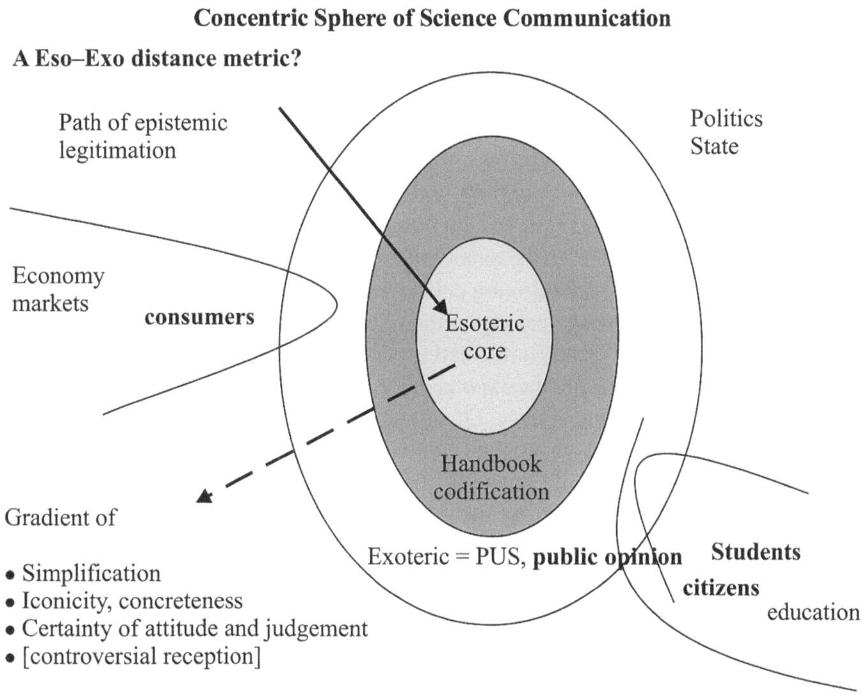

Fig. 6.1 Bauer's representation of Fleck's ideas about science communication. *PUS* public understanding of science. (Copied from Bauer (2009))

[15] I focus on Bauer's work here partly because of space constraints and partly because of his prominence in the public understanding of science field.

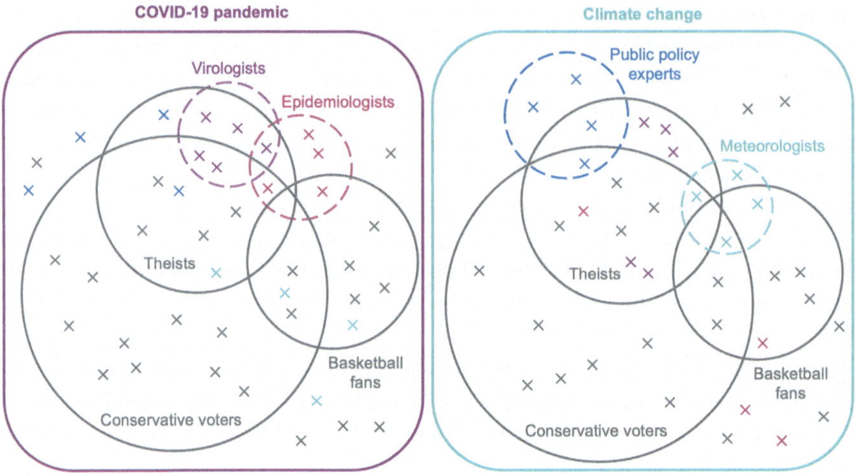

Fig. 6.2 Fleck's intersecting esoteric and exoteric circles. Left: thinking collective addresses the COVID-19 pandemic; right: same collective addresses climate change. Each X represents an individual. Circles represent a limited number of porous esoteric (colored) and exoteric (grey) circles

diversity of esoteric and exoteric circles, their interconnectedness, and how members of each shift with respect to two different societal questions. On the left, the "pink thinking collective" is grappling with the COVID-19 pandemic: the dashed pink circles represent esoteric circles and the pink crosses represent individuals who count as experts with respect to the pandemic, including virologists and epidemiologists.[16] On the right, the *same* thinking collective is addressing the climate crisis: the "pink experts" now no longer belong to the relevant esoteric circles, although they need not lose their scientifically-informed way of thinking (their "scientific mood"). Instead, "blue individuals" are the experts—e.g., meteorologists, public policy makers—with respect to climate change. Of course, this diagram is simplified; I am not, for example, claiming that virologists cannot be basketball fans. Moreover, given the socio-political structure of "Western" thinking collectives, it would be plausible to equate the rectangles delimiting the collective with consumer society. To avoid cluttering the diagram, I have also not included concentric esoteric circles to represent the "various *degrees* of initiation and numerous links between them" (p. 102), which would distinguish between, for example, experts on the coat proteins of coronaviruses versus virologists more broadly. Finally, although only esoteric circles are represented with dashed lines—indicating the porosity of these circles to the moods of the collective—exoteric circles are also porous to moods. Despite these simplifications, I believe this representation more accurately captures Fleck's intentions. Unlike Bauer, who wants to measure the distance between esoteric and exoteric

[16] Thank you to an anonymous reviewer for alerting me to my own biases as a natural scientist: there is certainly also a need for considering esoteric circles of social scientists during a pandemic (e.g., Lohse and Canali 2021; Pickersgill and Smith 2021).

circles, the figure below highlights that distance might not be the problem. Rather, the focus should be on studying the solidity or porousness of these circles. I expand on the porousness metaphor in the next section. Taken together, Fleck's theory successfully problematizes the distinction between experts and laypeople, and also shows how members of a thinking collective are bound to each other by pervasive moods.

6.5 Making the Expert-Lay Divide More Porous

Before making suggestions for how to understand the porousness of the boundaries between experts and laypeople, and how this porousness could be fostered, I address a worry a critic might have regarding Fleck's concept of thinking collectives, namely that it is too loose or amorphous. If some of the examples above were confusing this might be because "we could agree with anybody who calls the [thinking] collective fictitious" (Fleck [1935] 1979, p. 44). Although it plays such a central role for Fleck, the concept of a thinking collective can seem ambiguous or vague in the sense that there might be multiple interpretations of what such a collective is. At times collectives are large and disparate, held together by "Christian ideas" (Fleck [1936] 1986c, p. 86). At other times, thinking collectives exist when "two or more people are actually exchanging thoughts... A special mood arises" (Fleck [1935] 1979, p. 44). This vagueness can initially cause confusion, but if we remember that collectives are characterized by their moods and thinking styles, and that esoteric/exoteric circles are defined with respect to a particular question, then it becomes clear that Fleck does not intend his descriptions to have some metaphysical or ontological grounding. Thinking collectives are pragmatic constructs rather than static or fixed social structures, and Fleck remains aware of this fact. As we will see, this distinguishes him from others who are involved in constructing boundaries and drawing distinctions, but do so without this critical awareness. For contemporary discussions of science communication, it might, by turns, be useful to apply the "fictitious" concept of a thinking collective to a population's subsections that have been historically marginalized, or to certain nation states when the questions concern a specific country, or the European Union when questions are of bloc-wide interest. At other times it can be productive to attribute certain thinking styles, and their "literary styles", to particular academic subdisciplines addressing similar research questions in different ways.[17] In sum, a thinking collective is "not to be understood as a fixed group or social class. It is functional ... rather than substantial" (Fleck [1935] 1979, p. 102). Nonetheless, Fleck's theory fruitfully problematizes the expert-lay divide. How do we take the idea of thinking collectives seriously *as constructs*? How else can we

[17] Think, for example, of the different literary and thinking styles associated with philosophy of science and science and technology studies.

question the expert-lay dichotomy? How can the boundaries between esoteric and exoteric circles be made porous?

The clearest answer Fleck gives to the question of porousness between esoteric and exoteric circles is the idea of two-way dependence between these circles. Above I explicated the proposal that popularization, popular knowledge and thinking styles (tacitly) influence the thinking and acting of experts. Given that eliminating thought transformations is impossible, what would promote transformations that foster genuine two-way communication? As the communications scholar Dietram Scheufele notes, "every single issue emerging at the science–society interface is at least *partially* a science communication challenge" (Scheufele 2022, p. 298). Exactly how to manage thought transformations will be context-dependent, but important components might include communicating uncertainties surrounding scientific findings and highlighting that the fact/value distinction is frequently artificial. However, I want to resist the temptation—which Scheufele yields to by referring to the binary "science-society interface"—of presupposing a stable separation between experts and laypeople. As the philosopher Matthew Sample points out, the imaginaries of philosophers of science influence how they theorize about science and society, affecting both their descriptions of and prescriptions for communicative practices. When philosophers presuppose or idealize a particular science-society interface, their theories will yield certain conclusions while precluding others. At the very least, these theorists ought to hold themselves responsible for their imaginaries (Sample 2022). With these points in mind, I turn to making four suggestions for making the expert-lay divide more porous, and although these suggestions have largely been made before, here I use Fleck's conceptual resources in the context of science communication to provide new and additional justification for them.

6.5.1 Esoteric Diversity

First, we ought to increase the diversity of members of esoteric circles by drawing on many more and different exoteric circles. This point has been argued by many feminist philosophers of science, but maybe most prominently by Helen Longino. For her, the goal of diversity is to increase objectivity and improve scientific knowledge production. Accordingly, "the exclusion of women and members of certain racial minorities from… scientific professions constitutes not only a social injustice but a cognitive failing" (Longino 2002, p. 132). More recently, Goldenberg advocates for increased diversity among scientific experts so that trust deficits between experts and lay communities can be repaired, claiming that "the publics need more than academic credentials. They need some assurance that the experts understand their values" (Goldenberg 2021, p. 55). So, although this idea is not new, Fleck's framework provides a different rationale for the purposes of science communication. Instead of drawing from a limited number of exoteric circles, research ought to be conducted by people who represent a wider variety and range of "moods" of a thinking collective. Recall that individuals belong to multiple exoteric circles and carry their moods with them when they become experts. One might therefore imagine that diversifying esoteric circles will reduce the barriers for communication between esoteric and exoteric circles, leading, perhaps, to smaller, or at least less distorting, thought transformations. With respect to this first suggestion, then, the porousness metaphor refers to the ability of people to move more easily between different esoteric and exoteric circles.

6.5.2 Science Journalism

Even if esoteric circles are diversified, not everyone will join an esoteric circle through "initiation" at the graduate level, and those that do are limited in expertise to their particular esoteric circles. A second way to conceptualize and increase the porousness between esoteric and exoteric circles is to promote the role of science journalism as a means to poke holes into the boundaries between experts and laypeople, such that the latter can "see" into the workings of esoteric circles. For Fleck, oversimplified reporting by journalists can lead to distorting thought transformations. This certainly happens and is a particular worry for science journalism that operates under time pressure or aims at entertainment.

After dismissing this distorting kind of science journalism, Fleck discusses the roles of art critics at some length, which, I argue, can be analogous to those of investigative science journalists. With respect to fine art, Fleck contends that there are two esoteric circles: one comprised of artists and one of art critics. The critics mediate between artists and wider exoteric circles, thus "popularizing" art to some extent. But the critics also influence the artists and have "a specific knowledge of art in which both the artists and the public form the exoteric circle" (Fleck [1936] 1986c, p. 104). Similarly, investigative journalists with time and resources to conduct in-depth reporting can mediate between esoteric and exoteric circles, influencing both.[18] For example, building on critiques of the value-free ideal of science, Kevin Elliott suggests that "[s]cience journalists have valuable opportunities to contribute... by identifying and explaining major value judgments in scientific research... on behalf of non-specialists" (Elliott 2019, p. 4). This move could help laypeople appreciate that disagreements between experts can arise not because of disagreements about the empirical evidence itself, but also because of interpretive differences, which often, and sometimes tacitly, depend on a variety of (non-epistemic) value judgements. Revealing these value judgements through journalism

[18] Science journalism is underexplored in philosophy of science but garnering interest. Previous work includes Gerken's "justification reporting": in addition to reporting purported facts, "science reporters should, whenever feasible, report aspects of the nature and strength of scientific justification or lack thereof for a reported scientific hypothesis" (Gerken 2020, p. 95). However, this proposal still follows deficit thinking because justification is given in terms of more information. Figdor (2023) discusses the epistemic virtues of good science reporting and Schipani (2024) shows that science journalists, like scientists, face inductive risk in their reporting.

is distinct from the classical information deficit approach discussed in Sect. 6.2 since laypeople are learning about scientists' decision-making at a different level and not merely acquiring more "scientific facts." Moreover, journalists' reporting on value judgements has the benefit of potentially making these values explicit to scientists themselves. I would add to Elliott's proposal that it might be useful not just to reveal important value judgements at crucial decision-making points, but that some investigative science journalism should report on the "nature of science." That is, journalists could show both that climate scientists disagree about potential mitigating measures because of differences in their risk tolerances, for example, but also that the expertise to create and interpret models of e.g., rising sea levels is spread across a large number of scientists who come from neighboring esoteric circles (e.g., statistics, oceanography). This kind of reporting could also help scientists recognize their interdependence.[19]

6.5.3 Science Education: K-12

A third, related suggestion for moving beyond the information deficit model is to help laypeople and scientists understand that science is not a monolithic, value-free enterprise of knowledge creation. The idea is to change science education starting in kindergarten (cf. Lederman et al. 2013; Rudolph 2023). As Heather Douglas says, "[w]hat is *not* needed is a passing grade on fact-based literacy tests. What *is* needed is an understanding of the process and practice of science" (Douglas 2021, p. 137). Although there is increasing attention to the "nature of science" in school curricula, these still tend to focus on contemporary science, leaving out its history and sociology (Berkovitz 2017). In Fleck's terminology, this education would aim to make members of exoteric circles—especially those that will not join an esoteric circle—more aware of the practices of esoteric circles. Importantly, and this aspect is underemphasized by Douglas, this kind of education should reveal the interconnectedness of esoteric and exoteric circles. This suggestion is similar to the idea of investigative science journalism above, but intended for children passing through the educational system. Importantly, this is likewise not to be thought of as an information deficit approach precisely because the education should go beyond the "scientific facts." Going through such an education will help laypeople come to the "science-society interface" suspecting that this interface is partly constructed, and better equipped to communicate with experts so that there are fewer detrimental thought transformations.

[19] It might be apt to describe this as a double deficit model since members of both exoteric and esoteric circles can be unaware of these value judgements, and the "nature of science" more broadly. Thanks to Denis Walsh and Rebecca Korf for (separately) pointing this out.

6.5.4 Science Education: At Esoteric Initiation

Finally, while the above suggestion seems right, there are two problems with it, one practical, the other conceptual. The practical worry is that overhauling school curricula, (re-)educating teachers, and rolling out this kind of education is time-consuming. If this approach succeeds it will, of course, also affect how experts understand their roles within a thinking collective, since they start their education with everyone else. The conceptual worry is that scientists themselves often create and maintain the boundaries between "science" and "society", which affects how they communicate. When experts are "initiated" into their specialties, the thinking styles of these esoteric circles are often such that members' exoteric moods are suppressed in favor of the prevailing "scientific mood." If this is true, then diversifying esoteric circles will not automatically lead to communicative changes.

This scientific "boundary-work" has been characterized by the sociologist Thomas Gieryn (1983). He describes how scientists, often through their communication efforts, demarcate the projects, questions, and practices they deem scientific, so that they can keep other actors, including journalists, politicians and other members of the public, separate. Gieryn illustrates this with an example from the Cold War during which the US government tried to control scientific knowledge circulation to prevent the Soviet military from gaining information. In response, the National Academy of Sciences issued a report recommending that the vast majority of scientific knowledge not be restricted, in part because "basic science" should not be censored or held responsible for downstream military uses. In this case, Gieryn contends that boundary-work aims to "protect professional autonomy: public scientists construct a boundary between the production of scientific knowledge and its consumption by non-scientists... The goal is immunity from blame for undesirable consequences of non-scientists' consumption of scientific knowledge" (Gieryn 1983, p. 789). As mentioned previously, more recent empirical studies of scientists' predominant attitude towards science communication—as one-way knowledge transfer—suggest that this boundary-work is ongoing (e.g., Cook et al. 2004; Davies 2008; Pew Research Center 2015; Dudo and Besley 2016). Therefore, educating people who are becoming members of esoteric circles about the social situatedness of science could raise awareness of this boundary-work. For example, undergraduate and graduate science programs could include compulsory courses in the history, philosophy, and sociology of science. Making scientists aware of their social situation, how intricately bound they are to their exoteric circles, and the many ways in which values legitimately enter scientific practice could change how they then communicate science. Fleck recognized that scientists, in part because they "grow out of" exoteric circles, often lack awareness of the workings of science. He says, "[i]t is highly interesting to establish to what extent scientists who devote the whole of their life to the problem of distinguishing illusions from reality are unable to distinguish their own dreams about science from the true form of science" (Fleck [1946] 1986b, p. 113). Sample argues that "we philosophers have the opportunity to hold ourselves to a higher standard [than scientists, policy-makers, and publics], reflecting on our choice of idealization and its relationship (if any) to messy human practices" (Sample 2022, p. 13). I see no reason why scientists should not also reflect on and take responsibility for their imaginaries (cf. Brown 2020).

6.6 Conclusion

Scientists, and experts more generally, are engaged in boundary-work; merely telling them that they are constructing boundaries and thereby closing off avenues of two-way communication will not necessarily disrupt these boundaries. However, Fleck's notion of collective moods could suggest a more productive way forward. According to him, these moods bind people together in a collective, bridging esoteric and exoteric circles. So, to question or undermine the partly constructed boundaries between experts and laypeople we might appeal to shared feelings/affects/emotions, rather than relying only on shared information. In the context of science communication specifically, the science and technology studies scholar Sarah Davies suggests that affect has been under-theorized in work on public engagement, and that we should "become attentive to the role of materialities, affects, and place in encounters between publics and science" (Davies 2014, p. 98). To be sure, there is no single shared feeling that people rally around to support vaccine or climate change policies. However, it is often affect-laden events or experiences—such as art-science collaborations or protest movements—that prompt people to reflect on their assumptions, values, and practices (e.g., Salmon et al. 2017).[20] If Fleck is correct in thinking that moods are the basis for holding people together in a collective, then this offers us a different perspective through which to view the expert-lay dichotomy: paying attention to sharing and shared moods may be a step towards making boundaries more porous.

Acknowledgements Thanks to Jossi Berkovitz for feedback and for introducing me to Fleck in his social epistemology seminar. I presented an earlier version of this project at the IHPST in Toronto, the philosophy graduate student conference at the University of Alberta and at the 11th Values in Medicine, Science, and Technology Conference in Dallas; many thanks to participants at those conferences for their feedback, especially: Alison Wylie, Heather Douglas, and Holly VandeWall. Thank you to Maya Goldenberg, Joyce Havstad, Michela Massimi, and three anonymous reviewers for their written feedback.

[20] See also Furman (2024) for a discussion of how information/beliefs, values, and emotions interact to shape people's (dis)trust in science and public health.

References

Ballantyne, N. 2019. Epistemic Trespassing. *Mind* 128 (510): 367–395.

Bauer, M. 2009. The Evolution of Public Understanding of Science—Discourse and Comparative Evidence. *Science, Technology & Society* 14 (2): 221–240.

Berkovitz, J. 2017. Some Reflections on "Going Beyond the Consensus View" of the Nature of Science in K–12 Science Education. *Canadian Journal of Science, Mathematics and Technology Education* 17 (1): 37–45.

Besley, J., and M. Nisbet. 2013. How Scientists View the Public, the Media and the Political Process. *Public Understanding of Science* 22 (6): 644–659.

Branch, T. Y. 2022. Enhanced Epistemic Trust and the Value-Free Ideal as a Social Indicator of Trust. *Social Epistemology* 36 (5): 561–575.

Brown, M. 2020. *Science and Moral Imagination: A New Ideal for Values in Science*. University Of Pittsburgh Press.

Bucchi, M. 2008. Of Deficits, Deviations and Dialogues: Theories of Public Communication of Science. In *Handbook of Public Communication of Science and Technology*, Routledge International Handbooks, ed. M. Bucchi and B. Trench, 1st ed., 57–76. London: Routledge.

Bucchi, M., and B. Trench, eds. 2021. *Routledge Handbook of Public Communication of Science and Technology*, Routledge International Handbooks. 3rd ed. London: Taylor & Francis.

Callon, M. 1999. The Role of Lay People in the Production and Dissemination of Scientific Knowledge. *Science, Technology and Society* 4 (1): 81–94.

Cassidy, A. 2021. Communicating the Social Sciences and Humanities: Challenges and Insights for Research Communication. In *Routledge Handbook of Public Communication of Science and Technology*, Routledge International Handbooks, ed. M. Bucchi and B. Trench, 3rd ed., 198–213. London: Taylor & Francis.

Cheng, D., M. Claessens, T. Gascoigne, J. Metcalfe, B. Schiele, and S. Shi, eds. 2008. *Communicating Science in Social Contexts: New Models, New Practices*. Dordrecht: Springer.

Choi, S., A. Anderson, S. Cagle, M. Long, and N. Kelp. 2023. Scientists' Deficit Perception of the Public Impedes Their Behavioral Intentions to Correct Misinformation. *PLoS One* 18 (8): e0287870.

Collins, H., and R. Evans. 2002. The Third Wave of Science Studies: Studies of Expertise and Experience. *Social Studies of Science* 32 (2): 235–296.

Cook, G., E. Pieri, and P. Robbins. 2004. 'The Scientists Think and the Public Feels': Expert Perceptions of the Discourse of GM Food. *Discourse & Society* 15 (4): 433–449.

Davies, S. 2008. Constructing Communication: Talking to Scientists About Talking to the Public. *Science Communication* 29 (4): 413–434.

Davies, S. 2014. Knowing and Loving: Public Engagement Beyond Discourse. *Science & Technology Studies* 27 (3): 90–110.

Douglas, H. 2021. Science, Values, and Democracy: The 2016 Descartes Lectures. In *Consortium for Science, Policy & Outcomes*, ed. T. Richards. Arizona State University.

Dudo, A., and J. Besley. 2016. Scientists' Prioritization of Communication Objectives for Public Engagement. *PLoS One* 11 (2): e0148867.

Elliott, K. 2019. Science Journalism, Value Judgments, and the Open Science Movement. *Frontiers in Communication* 4:1–10.

Figdor, C. 2023. Science Journalism and Epistemic Virtues in Science Communication: A Defense of Sincerity, Transparency, and Honesty. *Episteme* 21:1–12.

Fleck, L. 1935/1979. *Genesis and Development of a Scientific Fact*. Chicago: University of Chicago Press.

Fleck, L. 1960/1986a. Crisis in Science. In *Cognition and Fact: Materials on Ludwik Fleck*, Boston Studies in the Philosophy of Science, vol. 87, eds. R. Cohen and T. Schnelle, 153–158. Dordrecht: D. Reidel Pub. Co.

Fleck, L. 1946/1986b. Problems of the Science of Science. In *Cognition and Fact: Materials on Ludwik Fleck*, Boston Studies in the Philosophy of Science, vol. 87, eds. R. Cohen and T. Schnelle, 113–127. Dordrecht: D. Reidel Pub. Co.

Fleck, L. 1936/1986c. The Problem of Epistemology. In *Cognition and Fact: Materials on Ludwik Fleck*, Boston Studies in the Philosophy of Science, vol. 87, eds. R. and T. Schnelle, 79–112. Dordrecht: D. Reidel Pub. Co.

Fonseca, C., A. Woollard, L. Hurst, and J. Pettitt. 2023. *Covid-19 and the Public Perception of Genetics*. London: The Genetics Society. https://genetics.org.uk/public-perception-of-genetics/.

Furman, K. 2024. Beliefs, Values and Emotions: An Interactive Approach to Distrust in Science. *Philosophical Psychology* 37 (1): 240–257.

Gerken, M. 2020. Public Scientific Testimony in the Scientific Image. *Studies in History and Philosophy of Science Part A* 80:90–101.

Gieryn, T. 1983. Boundary-Work and the Demarcation of Science from Non-Science: Strains and Interests in Professional Ideologies of Scientists. *American Sociological Review* 48 (6): 781–795.

Goldenberg, M. 2021. *Vaccine Hesitancy: Public Trust, Expertise, and the War on Science*. Pittsburgh: University of Pittsburgh Press.

Jamieson, K., D. Kahan, and D. Scheufele, eds. 2017. *The Oxford Handbook of the Science of Science Communication*. New York: Oxford University Press.

Janik, A. 2006. Notes on the Origins of Fleck's Concept of "Denkstil". In *Cambridge and Vienna: Frank P. Ramsey and the Vienna Circle*, ed. M. Galavotti, 179–188. Dordrecht: Springer.

Jarnicki, P. 2022. Stimmung/Nastrój as Content of Modern Science: On Musical Metaphors in Ludwik Fleck's Theory of Thought Styles and Thought Collectives. *Foundations of Science* 27 (3): 1207–1228.

Jensen, E. 2014. The Problems with Science Communication Evaluation. *Journal of Science Communication* 13 (1): C04.

Kitta, A., and D. Goldberg. 2017. The Significance of Folklore for Vaccine Policy: Discarding the Deficit Model. *Critical Public Health* 27 (4): 506–514.

Lackner, S., F. Francisco, C. Mendonça, A. Mata, and J. Gonçalves-Sá. 2023. Intermediate Levels of Scientific Knowledge are Associated with Overconfidence and Negative Attitudes Towards Science. *Nature Human Behaviour* 7 (9): 1490–1501.

Lederman, N., J. Lederman, and A. Antink. 2013. Nature of Science and Scientific Inquiry as Contexts for the Learning of Science and Achievement of Scientific Literacy. *International Journal of Education in Mathematics, Science and Technology* 1 (3): 138–147.

Lewis, J., A. Bartlett, H. Riesch, and N. Stephens. 2023. Why We Need a Public Understanding of Social Science. *Public Understanding of Science* 32 (5): 658–672.

Lohse, S., and S. Canali. 2021. Follow *the* Science? On the Marginal Role of the Social Sciences in the COVID-19 Pandemic. *European Journal for Philosophy of Science* 11 (4): 99.

Longino, H. 2002. *The Fate of Knowledge*. Princeton: Princeton University Press.

Martin, N., P. Buykx, C. Shevills, C. Sullivan, L. Clark, and D. Newbury-Birch. 2018. Population Level Effects of a Mass Media Alcohol and Breast Cancer Campaign: A Cross-Sectional Pre-Intervention and Post-Intervention Evaluation. *Alcohol and Alcoholism* 53 (1): 31–38.

Miller, J. 1983. Scientific Literacy: A Conceptual and Empirical Review. *Daedalus* 112 (2): 29–48.

Miller, J. 1998. The Measurement of Civic Scientific Literacy. *Public Understanding of Science* 7 (3): 203–223.

Mößner, N. 2016. Scientific Images as Circulating Ideas: An Application of Ludwik Fleck's Theory of Thought Styles. *Journal for General Philosophy of Science* 47 (2): 307–329.

Pew Research Center. 2015. Public and Scientists' Views on Science and Society. https://www.pewresearch.org/internet/wp-content/uploads/sites/9/2015/01/PI_ScienceandSociety_Report_012915.pdf.

Pew Research Center. 2023. Americans' Trust in Scientists, Positive Views of Science Continue to Decline. https://www.pewresearch.org/science/2023/11/14/americans-trust-in-scientists-positive-views-of-science-continue-to-decline/.

Pickersgill, M., and M. Smith. 2021. Expertise from the Humanities and Social Sciences is Essential for Governmental Responses to COVID-19. *Journal of Global Health* 11:03081.

Pigliucci, M., and M. Boudry. 2010. Why Machine-Information Metaphors are Bad for Science and Science Education. *Science & Education* 20 (May): 453–471.

Rudolph, J. 2023. *Why We Teach Science (and Why We Should)*. Oxford: Oxford University Press.

Salmon, R., R. Priestley, and J. Goven. 2017. The Reflexive Scientist: An Approach to Transforming Public Engagement. *Journal of Environmental Studies and Sciences* 7 (1): 53–68.

Sample, M. 2022. Science, Responsibility, and the Philosophical Imagination. *Synthese* 200 (2): 79.

Scheufele, D. 2022. Thirty Years of Science–Society Interfaces: What's Next? *Public Understanding of Science* 31 (3): 297–304.

Schipani, V. 2024. Journalism and Public Trust in Science. *Synthese* 204:56.

Seethaler, S., J. Evans, C. Gere, and R. Rajagopalan. 2019. Science, Values, and Science Communication: Competencies for Pushing Beyond the Deficit Model. *Science Communication* 41 (3): 378–388.

Simis, M., H. Madden, M. Cacciatore, and S. Yeo. 2016. The Lure of Rationality: Why Does the Deficit Model Persist in Science Communication? *Public Understanding of Science* 25 (4): 400–414.

The Royal Society. 1985. *The Public Understanding of Science*. London: The Royal Society.

Trench, B. 2008. Towards an Analytical Framework of Science Communication Models. In *Communicating Science in Social Contexts: New Models, New Practices*, ed. D. Cheng, M. Claessens, T. Gascoigne, J. Metcalfe, B. Schiele, and S. Shi, 119–135. Dordrecht: Springer.

Turney, J. 1998. *To Know Science is to Love it? Observations from Public Understanding of Science Research*. London: Committee on the Public Understanding of Science (COPUS).

UK House of Commons, Science and Technology Committee. 2017. *Science Communication and Engagement. Eleventh Report*. London: UK House of Commons. https://publications.parliament.uk/pa/cm201617/cmselect/cmsctech/162/162.pdf.

UK House of Lords, Select Committee on Science and Technology. 2000. *Science and Society. Third Report*. London: UK House of Lords. https://publications.parliament.uk/pa/ld199900/ldselect/ldsctech/38/3802.htm.

Wynne, B. 1989. Sheepfarming after Chernobyl: A Case Study in Communicating Scientific Information. *Environment: Science and Policy for Sustainable Development* 31 (2): 10–39.

Wynne, B. 2006. Public Engagement as a Means of Restoring Public Trust in Science—Hitting the Notes, but Missing the Music? *Community Genetics* 9 (3): 211–220.

Young, B., S. Lewis, S. Katikireddi, L. Bauld, M. Stead, K. Angus, M. Campbell, et al. 2018. Effectiveness of Mass Media Campaigns to Reduce Alcohol Consumption and Harm: A Systematic Review. *Alcohol and Alcoholism* 53 (3): 302–316.

Part III
Values in Socially Relevant Contexts

Chapter 7
Who the Computer Sees: Race, Gender, and AI

Carla Fehr

Abstract In this chapter, I develop what I call the Gender Shades Case Study (GSCS), which focuses on Joy Buolamwini's master's research on racial and gender bias in facial recognition and classification systems. I situate that research in its demographic and cultural context and use it to demonstrate the significance of diversity in computing and technology, arguing, as Buolamwini puts it, that "[w]ho codes matters" (2017). Buolamwini's research is relevant to many topics in Philosophy and AI, including discussions of algorithmic fairness, privacy, and surveillance. However, I leave those important discussions for other scholars to investigate. In this chapter, I develop this case to provide multifaceted evidence of the impact of diversity in computer science and technology, including AI, mainly as it plays out concerning biases and values around gender and race. I develop this case an example of situated knowledge and Helen Longino's Critical Contextual Empiricism.

Keywords Artificial intelligence · Diversity in computer science · Values and bias · Gender · Race · Joy Buolamwini · Helen Longino

7.1 Introduction

Dr. Joy Buolamwini was named one of the 100 most influential people in AI by Time Magazine in 2023. Her Ted talk, "How I'm fighting bias in algorithms," has more than 1.7 million views, and her work is featured in a documentary, "Coded Bias," which premiered at the Sundance Film Festival (Buolamwini 2016b; Kendall

C. Fehr (✉)
Department of Philosophy, University of Waterloo, Waterloo, ON, Canada
e-mail: carla.fehr@uwaterloo.ca

J. Y. Tsou et al. (eds.), *Values, Pluralism, and Pragmatism: Themes from the Work of Matthew J. Brown*, Boston Studies in the Philosophy and History of Science 347, https://doi.org/10.1007/978-3-031-92958-8_7

2020). Dr. Buolamwini testified about algorithmic bias before the US Congress in 2019, and her research and activism have been showcased in major news outlets and magazines, ranging from The New York Times (Buolamwini 2018; Lohr 2018) to The Rolling Stone (O'Neil 2023). She is a founder of the Algorithmic Justice League, whose mission is to

> raise awareness about the impacts of AI, equip advocates with empirical research, build the voice and choice of the most impacted communities, and galvanize researchers, policy makers, and industry practitioners to mitigate AI harms and biases. We're building a movement to shift the AI ecosystem towards equitable and accountable AI. (Algorithmic Justice League 2022)

In this chapter, I develop what I call the Gender Shades Case Study (GSCS), which focuses on Dr. Buolamwini's master's research on racial and gender bias in facial recognition and classification systems.[1] I situate that research in its demographic and cultural context and use it to demonstrate the significance of diversity in computing and technology, arguing, as Buolamwini puts it, that "[w]ho codes matters" (2017). Buolamwini's research is relevant to many topics in Philosophy and AI, including discussions of algorithmic fairness, privacy, and surveillance. Those discussions are significant; however, in this chapter, I develop GSCS to explore the impact of diversity in computer science and technology, including AI, on gender and racial biases and values. I develop this case study to provide a thick, detailed, contextualized example of situated knowledge and Helen Longino's Critical Contextual Empiricism.

The Gender Shades Case Study is important for many reasons. First, GSCS provides reasons to be cautious about facial recognition systems, which are used in applications with the potential to exacerbate racism and other kinds of significant social harms. Consider for example, Mr. Robert Williams, an African American man, arrested at his home in front of his wife and children, or Ms. Porcha Woodruff, an African American woman, arrested in front of her home while 8 months pregnant (Hill 2020, 2023). Both arrests were based on errors in facial recognition systems. Facial recognition is increasingly being used by Canadian police forces, with two Toronto area police services announcing the use of this technology in Spring 2024 (CBC 2024). Facial recognition technology can contribute to racism and other social injustices, which make it important to attend to the weaknesses in this technology.

Second, this case provides reason to increase diversity in computing and technology, which is essential for economic and social justice. Overall, computing and technology have been slow to respond to efforts to improve gender and racial diversity. This leads to the exclusion of members of equity-deserving groups from highly lucrative careers in a fast-growing sector.

Third, GSCS advances the philosophy of science by adding a case in computer science, technology, and AI to the literature supporting CCE and other views on the consequences of the lack of diversity within epistemic communities. Buolamwini's public presence and stature make the Gender Shades Case Study evidentially significant. She provides ample, publicly available first-person testimony about her

[1] This case study focuses on Dr. Buolamwini's MA work. She has continued her research and activism and opportunities remain for examining this additional work.

experiences of being unrecognized and unseen by facial recognition systems and being misgendered by gender classification systems while her white classmates and colleagues used these technologies without problems. She explicitly described how those experiences motivated her research and influenced her methodological choices. Philosophers are not left to infer how her social-material location influenced her knowledge-producing projects and practices. GSCS is socially, scientifically and technologically, and philosophically significant.

7.2 Gender Shades Research

As an undergraduate student, Joy Buolamwini worked on a project called UpBeat Walls, an immersive, interactive art installation that used commercially available automated facial analysis algorithms (AFAAs), what some people refer to as facial recognition systems, to responded to users' faces to create patterns on the walls. Although Dr. Buolamwini could work on the coding end of the project, she could not demonstrate or use it herself because the AFAAs her group employed failed to detect her face. It was not that the AFAA didn't recognize her as Joy Buolamwini, it didn't 'see' her face—didn't recognize her face as a human face at all. This is just one of many stories Dr. Buolamwini has shared of AFAAs failing to recognize her face or miscategorising her (Buolamwini 2023, p. 16).

Dr. Buolamwini is a dark-skinned Black woman who had to wear a white Halloween-style mask for some of her work with AFAAs because, as she reports, her "bare face was not consistently detected by facial recognition software" (Buolamwini 2016a). This decidedly inhuman looking mask was more reliably recognized as a face than her own very human countenance. Shocking images of her holding this stark white mask in front of her face are widely available in the media and a drawing of that image is on the cover of her 2023 book, *Unmasking AI: My Mission to Protect What is Human in a World of Machines*, in which she tells the story of her work on AFAAs and social justice.

As an undergraduate, Dr. Buolamwini assumed that software companies would quickly identify and fix this problem. However, in graduate school she discovered that the problem remained. She ran an image of her face through facial analysis systems produced by several companies and found that they often failed to recognize her face and those that succeeded at the recognition task misgendered her (Buolamwini 2023, pp. 66–67).

At this point, she turned her focus from AFAAs that detected faces to those that classified the gender of the face in an image. She decided to conduct a systematic investigation to determine whether failures to classify images of her face were idiosyncratic or were symptomatic of a larger pattern of bias. This experiment formed the basis of her MIT Master's Thesis, *Gender Shades: Intersectional Phenotypic and Demographic Evaluation of Face Datasets and Gender Classifiers* (Buolamwini 2017), and a subsequent scientific publication, "Gender Shades: Intersectional Accuracy Disparities in Commercial Gender Classification," coauthored with Timnit Gebru (2018).

7.2.1 The Research

Automated facial analysis algorithms (AFAAs) are machine learning systems that are trained and then evaluated using large datasets, in this case comprised of labelled pictures of faces, called benchmarks. AFAAs can accomplish a range of tasks including determining whether there is a face in a picture, identifying a particular face, or classifying a face in terms of things like age, gender, or emotion. Commercial gender classifiers (CGCs) are AFAAs that classify faces according to perceived gender/sex. CGCs typically assume a gender/sex binary and, as we will see, are trained on images of faces labeled as male or female. Similarly to Dr. Buolamwini, I highlight the idea of 'perceived gender/sex' because someone or something labelled the training images according to a male-female binary, and we often don't know the criteria used to develop and apply those labels. Presumably these criteria could be related to sex or to gender or more likely to some combination of sex and gender.[2]

In her MA thesis, Dr. Buolamwini was interested to explore whether the errors CGCs made classifying her face were part of a larger pattern of race and gender bias (Gender Shades 2022). She describes the focus of her thesis as an "intersectional evaluation of face datasets and gender classifiers with respect to gender and skin type (Buolamwini 2017, p. 18). And she asks questions about the gender and skin type composition of benchmarks as well as about the accuracy of her target CGCs with respect to different demographic groups. In her 2023 book she characterizes

[2] Sari van Anders' term 'gender/sex' refers to "[w]hole people/identities and/or aspects of women, men, and people that relate to identity and/or cannot really be sourced specifically to sex or gender" (2015, Table 2). I am using this term loosely because of the lack of information about how training and testing images were labeled as male or female. In order to be consistent with the technology and the literature about that technology I primarily use the term 'gender,' however in some cases 'gender/sex' is more apt and in others we don't know whether 'gender/sex' or 'gender' is more apt.

her research question slightly differently, asking "[d]o AI systems that classify gender based on an image of a face perform differently based on the skin type of a face" (Buolamwini 2023, p. 72).

Although the social, ethical, and political framing and implications of Dr. Buolamwini's research focus on race, with particular attention to gendered anti-Black racism, her research methodology relies on phenotypic traits based on skin type while maintaining a focus on skin types of people from Africa and Northern Europe. She reports that her decision to shift from race to skin type arose because of the ethical and methodological complexity she faced trying to classify images of people based on race. How we should understand racial categories and their relationship to biological features is a complex matter where there are a lot of competing views with profound social and ethical implications at stake.[3] Buolamwini writes that

> After a few weeks of trying to disentangle and arrange this evolving social construct of race, I tried another perspective. Inasmuch as I was looking at gender classification, my aim was not to see if I could come up with a better race or ethnic classification system, but to see if someone's appearance made a difference in the accuracy of gender classification. I decided instead of looking at race I wanted to look for a more objective measure, which is when I started to focus on not just demographic attributes like gender and race, but also phenotypic attributes, namely the color of someone's skin. (Buolamwini 2023, p. 95)

To do this she used the Fitzpatrick Classification of Skin Types, which is a measure used by dermatologists to determine susceptibility to skin cancer based on skin colour and other phenotypic features, not race itself. But notably here, different cultures use different racial categorizations, and skin colour has no simple correlation with these. So, skin type is an imperfect indicator and is only one of many potentially relevant features that could be used to sort images of people. Nonetheless, she identifies the Fitzpatrick Classification of Skin Types as the best methodological option she could find to assess diversity in existing benchmarks, and, as we will see, to develop her own benchmark that she could then use to test CGCs. Even given her careful consideration of the power and politics of classification she reports "dreading explicitly labeling strangers by the appearance of their skin" (Buolamwini 2023, p. 96). It is deeply troubling to see that systemic and individual racism makes it difficult to conduct research demonstrating the manifestation of that racism in AFAAs.

Using the Fitzpatrick Classification of Skin Types, Buolamwini assessed two commonly used benchmarks, IJB-A and Adience. IJB-A is a US government benchmark released by the National Institute of Standards and Technology and constructed to include geographic variation among its subjects (Buolamwini and Gebru 2018, p. 4). It includes 89 subjects from Asia, 7 from Oceania, 135 from North America, 50 from South America, 149 from Europe, 29 from the Middle East, and 41 from Africa (Klare et al. 2015, Table 2). Adience is a benchmark specifically developed to classify faces by gender and age group. While these two benchmarks might seem to promise relevant diversity, Buolamwini and Gebru found that

[3] For an introduction to these complexities related to racial categories see Alcoff (2023).

subjects with lighter skin types were overrepresented, comprising 79.6 of the subjects in IJB-A and 86.2% of the subjects in Adience (Buolamwini and Gebru 2018, Table 3).

Given the absence of benchmarks with diverse representations of skin type and gender Buolamwini created what she called the Pilot Parliaments Benchmark (PPB). To create this benchmark, Buolamwini curated a collection of 1270 pictures of subjects selected from the members of parliament of three African countries (Senegal, Rwanda, and South Africa) and three Northern European countries (Sweden, Finland, and Iceland). She used parliamentarians for ethical and practical reasons: parliamentarians are public figures who would likely not have an expectation of privacy and they include people with a range of ages and body types. This data set includes a balanced representation of images of people who she perceived to be men and women, and who have a range of skin types.

In addition to the challenges labelling racial and phenotypic variation, Buolamwini also faced challenges classifying images with respect to gender while creating the PPB. She came to use the phrase "perceived as men or women" because the people in the images she labelled did not report their own identity. Buolamwini used criteria including her perceptions of people's faces, hair, and clothing, as well as their titles (ex. Mr., Ms., or Mrs.) and pronouns to label images (Buolamwini 2023, pp. 117–118). Throughout her description of her work in her book *Unmasking AI*, she is explicit about the power that accompanies labelling images by race or skin type and by sex.

With PPB in hand, Buolamwini was able to assess Microsoft's, IBM's, and Face++'s CGCs. These companies did not provide a detailed description of the methodology they used to create their CGCs and did not mention what data sets they used to train these CGCs (Buolamwini and Gebru 2018, p. 8). Overall, these commercial gender classifiers performed well. However, the variations in their error rates when considering different categories of people are significant. All systems worked better on perceived male than female subjects and on subjects identified as having light skin types than dark skin types. When Buolamwini conducted an intersectional analysis, in which she looked at combined effects of skin type and perceived gender, she found that these systems had the highest error rates for faces labeled as female and having dark skin types. Microsoft and IBM systems had error rates of 0.0% and 0.3% for images labeled as male and having light skin types and error rates of 20.8% and 34.7% for images labeled as female and having dark skin types respectively (Buolamwini and Gebru 2018, Table 4). Face++'s system had error rates of 0.7% for images labeled as male and having dark skin types, 0.8% for images labeled as male and having light skin types, and 34.5% for images labeled as female and having dark skin types (Buolamwini and Gebru 2018, Table 4). Buolamwini and Gebru conclude that

> The substantial disparities in the accuracy of classifying darker females, lighter females, darker males, and lighter males in gender classifications systems require urgent attention if commercial companies are to build genuinely fair, transparent, and accountable facial analysis algorithms (p. 1).

I want to highlight two elements of Dr. Buolamwini's methodology. First, when the data used to train AFAA's and the data used to test AFAA's have the same flaws, for example having a substantial overrepresentation of images of people with light skin types, these systems will not learn to detect faces with dark skin types *and* that flaw will be masked—it will appear that these systems make fewer errors than they will actually make because they are tested using a benchmark that rarely presents them with subjects for whom they tend to fail. In this case the tests will not show that the AFAA systematically fails to classify images perceived to be of women with dark skin types. This is one reason why Buolamwini's PPB is important. PPB facilitates research that unmasks this bias in the CGCs Buolamwini tested.

A second significant feature of the Gender Shades methodology is that the authors conducted an intersectional analysis, which in this case means that they considered the interaction of skin type and gender.[4] Buolamwini and Gebru's analysis is intersectional because they do not limit their attention to differences in AFAA's ability to classify people with light and dark skin types, and to classify people who they perceived as male or as female. Additionally, they attend to four groups of people: men with light skin types, women with light skin types, men with dark skin types, and women with dark skin types.

Using PPB, rather than a benchmark in which light skin types are overrepresented, allows researchers to assess the error rates of CGCs more accurately. However, if Buolamwini and Gebru failed to conduct an intersectional analysis of skin type and perceived gender, even while using the PPB, the gender classifiers they tested would still have looked moderately effective, they would not have revealed the error rates of 34.5% and 34.7% that Face++ and IBM's systems exhibited when classifying people perceived as women with dark skin types. The classifiers performed exceptionally well on men with light skin types, and moderately well on men with dark skin types and women with light skin types. To unmask the significant weakness of these classifiers, researchers needed to attend to the category of women with dark skin types. Because PPB is balanced by skin type and perceived gender using it makes it possible to detect these classifiers' weaknesses classifying Black women's faces. But researchers need to use this tool *and* conduct an intersectional analysis for this possibility to be actualized.

These two features of Buolamwini and Gebru's methodology, developing a benchmark with a balanced representation of skin type and gender, and choosing to conduct an intersectional analysis support at least two scientific and epistemic values, increased objectivity and creativity, in terms of providing methodologies that can be used in a much wider range of contexts than this experiment. PPB is publicly available and is now part of the repertoire of benchmarks that researchers and companies can use to train and assess additional AFAAs. However, as I discuss later it also facilitates the collection of data that can be used to argue for limitations on the development and use of AFAAs.

---

[4] Kimberli Crenshaw (1991) developed to concept of intersectionality to capture the interaction of multiple axes of oppression.

7.2.2 *Gender Shades Demographic and Institutional/ Cultural Context*

The demographic context of computer science and the tech industry is significant for the Gender Shades case study because it reveals a large epistemic community in which women of colour are underrepresented. West et al. (2019, p. 7) quote machine learning researcher Stephen Merity saying that, "[b]ias is not just in our data sets, it's in our conferences and community." Even after years of work to increase diversity in STEM, the industry that produces AFAAs, and the academic disciplines that educate people who work in this industry and conduct this research are disproportionately white and male. Cave and Dihal (2020) use an analysis of popular media to demonstrate that AI is largely portrayed as white and male and postulate that one of the reasons for this is the overrepresentation of white men among those who create AI.

The representation of women students in computer science reached an all-time high in the late 1980s and early 1990s, after which time it dropped precipitously and has only, in the last 10 or 15 years, started to slowly increase again. And while the number of women computer science is slowly increasing, it is nowhere near what it once was. The representation of women of colour is much lower than the representation of white women. We see a similarly low representation of white women and a lower representation of women of colour in the private technology sector.

McAlear et al. in their 2018 report "Data Brief: Women of Colour in Computing" provide an overview of the demographic terrain in the United States. They reveal that Black women comprise 3% of all computing degree earners, make up less than 2% of the Silicon Valley professional workforce and hold less than 0.5% of Silicon Valley leadership positions (McAlear et al. 2018, pp. 4–5). They also note that 82% of bachelor's degrees in computer science go to men of all races, 8% go to white women, 3% go to Black women, 2% go to both Asian and Latinx women, and less than 1% go to Native America/ Alaskan Native and Native Hawaiian/ Pacific Islander Women (McAlear et al. 2018, Fig. 6). It is important to note that this depressed representation of women of colour exists in a rapidly growing and highly lucrative sector. As a result, it exacerbates patterns of economic injustice associated with gender and race.

McAlear et al. (2018, p. 5) also point out the low representation of women in computing jobs over the last several decades and note that the underrepresentation of women of colour is especially significant, particularly in Silicon Valley. They show that women comprise 35% of the US computer and information science workforce, and of those women, 56% are white, 32% are Asian, 7% are Black, and 5% are Latinx. In Silicon Valley, Asian and white women hold a combined 26% of the workforce and Black, Latinx, and Native American/Alaskan Native women each constitute 2% or less. These disparities are larger in leadership positions. While Asian and White women are represented in leadership positions in numbers proportionate to their representation in the Silicon Valley workforce, less than 1% of Silicon Valley's leadership positions are held by Latinx women and less than 0.5%

by Black women. In 2016 Google reported that 19% of its tech workforce were women and 1% were Black, Microsoft reported that 17.5% of its tech workforce were women and 2.7% Black, and Facebook reported that 17% of its tech staff were women and 1% Black (Kleinman 2017).[5] None of the data reported are intersectional and so we don't know the representation of Black women in these companies.

The overrepresentation of white men in computing and technology is accompanied by a work culture in these fields that many have described, in general, as hostile to women and particularly to women of colour. For example, in a 2023 survey conducted by Women Who Tech, 42% of women and 24% of men who worked in tech and were surveyed reported experiencing harassment. Although many types of harassment were reported, sexism was the most common type of harassment reported by women, and harassment based on race or religion were the most common types of harassment reported by men.[6] West et al. (2019) document allegations of discrimination and harassment at leading tech companies including Microsoft, Uber, Apple, Google, Facebook, and Tesla (West et al. 2019, p. 9).

Sarah West Meyers has extensively studied the culture of computer science and concludes that it tends to be inhospitable for women writing of a culture of

> geek masculinity, bolstering the status of male computer nerds, in which individual artistic genius, eccentricity, and anti-authoritarian behaviour, and dislike of activities involving human interactions acted to bolster male software programmers personal and professional authority. (West 2020, see also Varma 2007)

West et al. (2019) cite Buolamwini's coauthor Timnit Gebru, AI researcher and co-founder of the organization Black in AI, saying "[w]e are in a diversity crisis for AI." Gebru goes on to say that "[i]n addition to having technical conversations, conversations about law, conversations about ethics, we need to have conversations about diversity in AI. This needs to be treated as something that's extremely urgent" (Snow 2018, in West et al. 2019).

7.2.3 Harms of Seeing and Not Seeing; Gendering and Misgendering

Ethically and socially fraught issues of sight, surveillance, and classification course through the Gender Shades Case Study. It is a clear example of gendered anti-Black racism. Buolamwini experienced being unseen by AFAAs and misgendered by CGCs and she demonstrated that the problem of being misgendered was significant

[5] These numbers are important because they were published near the same time as Gender Shades and hence describe the state of the field when Dr. Buolamwini was conducting this research. There have been some small improvements since these data were published. For example, in 2021 Microsoft reported that women represent 29.7% of their global workforce and Black people 5.7% of their U.S. workforce (McIntyre 2021).

[6] See also National Academy of Sciences (2018) and U.S. Equal Employment Opportunity Commission.

for women with dark skin types. Black women are underrepresented in the industry that created these systems and the university departments that produced the workers who created these systems. Before exploring the significance of the GSCS for feminist philosophy of science in Sect. 7.3, I briefly acknowledge and explore a complicated set of harms that arise from both the failures Buolamwini pointed out in AFAAs and CGCs, and also harms that can arise from the successes of these systems, of them working well and doing the job they are intended to do.

The technical failures of these systems to recognize and categorize women with dark skin types is a technological twist on dehumanization that pre-exists AI. AFAAs and CGCs that exclude women with dark skin types reproduces a harm perpetrated by flesh and blood people and social structures, and that feminists of colour and critical race scholars have long studied. Racialized feminist scholars, often speaking to their white-Anglo feminist colleagues, have produced voluminous scholarship on the harms associated with being unseen, unheard, unread, unstudied, unacknowledged, uncited. Maria Lugones, in a paper exploring how white people can be in good relationship and can come to know or befriend people who are members of racialized groups, describes this phenomenon when she writes that she is interested in those many cases in which "White/Anglo women do one or more of the following to women of colour: they ignore us, ostracize us, render us invisible, stereotype us, leave us completely alone, interpret us as crazy. All of this *while we are in their midst*" (Lugones, 1987, p. 7). Ortega (2006) laments the phenomenon of white feminists studying women of colour without taking sufficient care to check that their analyses are correct. Charles Mills describes the white supremacy inherent in the structure of social relations as an epistemology of ignorance:

> On matters related to race, [Whites hold] an inverted epistemology, an epistemology of ignorance, a particular pattern of localized and global cognitive dysfunctions (which are psychologically and socially functional), producing the ironic outcome that Whites will in general be unable to understand the world they themselves have made. (Mills 1997, p. 18)

The AFAAs that required Buolamwini to wear a white mask and the CGCs she tested in Gender Shades are dysfunctional in a manner similar to the white disfunction of not noticing or attending to or seeing women of colour. It is an exercise of power to classify, to recognize, and in this instance to determine who is seen and who isn't. Safiya Noble in her authoritative critique of Google search engines points out that a group's identity development is guided by

> social experiences and interactions typically organized around race, gender, education, and other social factors that are ideological in nature. These issues are at the ready to a 'politics of recognition,' which is an essential form of redistributive justice for marginalized groups that have been traditionally maligned, ignored or rendered invisible by means of disinformation on the part of the dominant cultures. (Noble 2018, p. 84)

While the Gender Shades case is not about producing disinformation about Black women, it documents a literal absence of recognition and misclassification by technologies produced by leading technology companies.

I briefly point to this literature on the erasure and systematic ignorance of women of colour and Black people by white people to contextualize the impact of creating, selling, buying, and using AI systems that systematically fail see and misclassify people in some racial categories. This is not a new phenomenon. It is taking an old set of biases that are built into many white people's ways of living in the world and expressing those biases in a new medium and a new venue with a problematic veneer of objectivity that comes along with being rendered in code. Safiya Noble (2018) notes that the denial of racism in Silicon Valley is often framed in terms of people "not seeing colour," or being "colourblind." Gender Shades highlights the harm that arises in the case in which that platitude is actually true.

However, although Buolamwini's work exposed biases in AFAAs and CGCs, it does not automatically support the idea that these systems should be 'fixed' or continue to be developed and deployed. There are general and significant concerns about the impact of AFAAs on privacy, over-surveillance, and over-policing Black communities and members of other racialized groups. This is not the primary focus of this chapter, but I encourage readers to attend to Robyn Maynard's 2017 book *Policing Black Lives* and Ruha Benjamin's 2019 *Race after Technology*.

In addition to concerns about racism, it is important to attend to the impact of CGCs on sex and gender minorities—in this case on members of trans communities. Os Keyes (2018) demonstrates that automatic gender recognition systems (AGR), which include CGCs, operationalize gender in a way that assumes a binary, that gender is fixed, and that gender is based on physiological characteristics. This set of assumptions "fundamentally ignore[s] the existence of transgender people, with dangerous results" (Keyes 2018, p. 2). These systems threaten the privacy and autonomy of transgender people and can put them in physical danger. Hamidi, Scheuerman, and Branham (2018) found that transgender people have "overwhelmingly negative attitudes towards AGR" including concerns about privacy and about being misgendered. These matters are significant when considering the role that race plays in transgender people's experiences of violence. Keyes writes that,

> the situation is likely to be worst for trans feminine people of colour; not only are they already subject to a disproportionate amount of discrimination and violence from the police…, AGR has already been shown to be inaccurate at classifying feminine people with dark skin. (Keyes 2018, p. 11)

It may not be possible to find technical solutions to the dangers faced by transgender people being misclassified by CGCs. In an ideal world, this sort of technology might not ever exist. And in a world better than the one we currently live in it would be more carefully regulated and its users held accountable for ameliorating the harms associated with its development and implementation. AFAAs and CGCs, even when they function well, can contribute to the existing hyper vigilance of people of colour and the both the erasure and invasion of privacy of transgender and nonbinary people.

So, there is a range of responses to the bias that Buolamwini's methodology revealed. On one hand, it can provide motivation for developing CGCs that are less biased, that function more closely to the way they are advertised to function, which is supported by using benchmarks such as Buolamwini's PPB and by calling for more representation of Black people and other people of colour in other data sets.

On the other hand, given systemic racism and the harms faced by trans and non binary people, the bias revealed by this methodology can motivate concerns about the development and use of AFAA's in general and support arguments for limiting the development and implementation of this technology, or for the implementation and enforcement of careful regulatory and accountability measures. For example, she writes that "I was clear that I did not want to be in the business of building better facial recognition technologies that could then be used for increased surveillance" (Buolamwini 2023, p. 176).

7.3 Critical Contextual Empiricism and Gender Shades

7.3.1 Critical Contextual Empiricism

Buolamwini's and Gebru's perspectives and experiences lead to scientific research that revealed gender and racial bias. Their work exemplifies the benefits that can attend diversity in research communities. Drawing on the work of Helen Longino (1990, 2002), I have argued that epistemic benefits can occur when diverse practitioners join relatively homogeneous epistemic communities (Fehr 2007, 2011; Fehr and Jones 2022). Increasing diversity can increase rigour and creativity in a community's knowledge-producing practices.

Helen Longino's Critical Contextual Empiricism (CCE) diagnoses some cases in which communities of well-intentioned researchers follow their field's best practices and fail to recognize the impact of social biases on their knowledge-producing practices and the knowledge they produce. CCE also provides a mechanism to help communities identify and evaluate their biases and produce more objective knowledge. Longino points out that assumptions will always play a role in the relationship between evidence and the hypothesis or view that evidence putatively supports. These assumptions can include social values and biases. However, individual people lack the resources to know that they have succeeded in identifying all their assumptions and the role those assumptions play in evidential reasoning. To address this problem, Longino moves our attention from individual to social knowledge-producing processes, arguing that one way to identify more of our assumptions is to engage in critical democratic dialogue with people who hold different assumptions. She recommends that communities foster diversity, writing that many instances of racism and sexism in scientific knowledge arose from excluding women and people of colour from scientific communities (1990). According to Longino, an ideal epistemic community includes diverse perspectives and has a culture that facilitates social interactions that can reveal assumptions informing individual members' evidential reasoning. The closer a community comes to this ideal, the more objective or well justified the knowledge it produces. I argue that in addition to increased rigour (objectivity or justification), a diverse community that engages in critical democratic dialogue can also enjoy benefits in terms of creativity or innovation (Fehr 2011).

Situated knowledge refers to the notion that what one knows is at least partly conditioned on one's situation and experience. A person's first-person experiences, what is interesting or salient to them, and their attitudes toward what they know will at least partly depend on their social-material location or their identity (Haraway 1988; Code 2008; Anderson 2020). It is not a coincidence that feminist philosophers often identify as women, and philosophers of race are often people of colour. Katherine Phillips (2017) points out that the mere presence of diversity can improve some epistemic practices in some communities. For example, white people tend to reason more carefully about race when a Black person is in the group. However, the mere presence of diverse perspectives is insufficient for CCE. CCE requires that diversity be integrated into a community's social interactions. This integration is essential because critical democratic dialogue among people with different perspectives, who have differently situated knowledge, is a route to recognizing and evaluating biases.

Longino identifies four features of an ideal epistemic community that support critical democratic dialogue. Two of these features are structural: a community should have shared standards for assessing knowledge claims and public venues for those knowledge claims. The other two features describe norms and practices of community members that help the full range of perspectives within a community encounter one another in critical dialogue. The first of these features is that community members must give each other's views uptake, meaning that they must be willing and able to hear each other's views and be open to the possibility of changing their minds in response to those alternative views. The second feature is tempered equality of intellectual authority, which is intended to block power differences from negatively impacting the reception of a community member's ideas. Research in various disciplines shows that it takes skill and care to develop a community culture that instantiates uptake and tempered equality of intellectual authority.

Already in 1990, Longino hypothesized that the lack of women and people of colour in many scientific communities explains the production of racist and sexist research by leading scientists and scientific institutions using the current best practices in their fields. Years later, Longino writes that.

> the exclusion of women and members of certain racial minorities from scientific education and the scientific professions constitutes not only a social injustice but a cognitive failing. (Longino 2002, p. 132)

Nearly 25 years later, we find another example of this in terms of the correlation between the drastic underrepresentation of Black women in technology with the production of sexist, racist facial recognition and gender classification systems.

7.3.2 *Gender Shades Through a CCE Lens*

The Gender Shades Case Study is a good illustration of the importance of diverse perspectives. At first glance the problems that motivated Gender Shades are puzzling. The huge overrepresentation of faces with light skin types in the Adience and

IJB-A benchmarks, the public distribution of facial detection systems that failed to detect faces like Dr. Buolamwini's, the creation and distribution by top technology companies of gender classification systems with stark differences in error rates between women with dark skin types and men with light skin types are all states of affairs that seem surprizing and beg explanation.

One reason for these failures could be the dramatic underrepresentation of women with dark skin types in computer science and tech. It is not a coincidence that Dr. Buolamwini is a Black woman with a dark skin type. She repeatedly describes the frustration of engaging systems that worked for her light skinned colleagues, but not for her. And she was concerned about the broad use of AFAAs. She reports that this friction and concern motivated her to conduct the research that has come to be known as Gender Shades.

In the opening remarks of her congressional testimony on facial recognition technology she said

> Today, I speak to you as both a researcher and someone who has personally experienced erasure and bias from machines that attempt to analyze human faces. (May 22, 2019. Hearing on Facial Recognition Technology (Part 1): Its Impact on our Civil Rights and Liberties)

Joy Buolamwini brought an underrepresented perspective to the AFAA technology. She used that perspective to uncover sexist and racist background assumptions in those products. She identified a problem and developed methods for addressing that problem. She found and addressed bias and she created a new benchmark that is available for other researchers and developers to use.

The following is an exchange between Joy Buolamwini (JB) and Representative Alexandria Ocasio-Cortez (AOC) during Buolamwini's congressional testimony.

> AOC: "Ms. Buolamwini, I heard your opening statement… We saw that these algorithms are effective to different degrees. Are they most effective on women?"
>
> JB: "No"
>
> AOC: "Are they most effective on people of color?"
>
> JB: "No, in fact it excludes them."
>
> AOC: "Which demographic is it mostly effective on?"
>
> JB: "White men."
>
> AOC: "And who are the primary engineers and designers of these algorithms?"
>
> JB: "Definitely, white men." (ibid)

As Joy Buolamwini has said before, "who codes matters." It matters to philosophy as well as technology and social justice.

We can consider GSCS in terms of Longino's criteria of shared standards and public venues. Buolamwini's research, was presented in a Master's thesis successfully defended at MIT and published in a respectable venue, *Proceedings of Machine Learning Research*. Both research products are publicly available and are easy to find online. Which communities gave her research uptake and whether Buolamwini has been treated with tempered equality of intellectual authority are more complicated issues. Buolamwini went to significant efforts to ensure that the biases she documented, as well as other kinds of algorithmic bias, are publicized. Her Ted talks, work with the Algorithmic Justice League, media presence and congressional testimony name only a portion of her activism around this research and algorithmic bias more generally. Given this activism and media attention it is difficult to imagine that Buolamwini's research could fail to get uptake and that the communities of which Buolamwini is a member could fail to grant her equality of intellectual authority. And it will be difficult to determine whether or how much of the uptake she received was due to the striking results of the research itself or due to social pressure arising from the publicity and activism connected with this research—two things that are intimated interconnected.

Buolamwini's exchange with Alexandria Ocasio-Cortez continued, highlighting the broader social significance of Buolamwini's work:

> AOC: "And so we have a technology that was created and designed by one demographic, that is only mostly effective on that one demographic, and they are trying to sell it and impose it on the entirety of the country?"
>
> "So we have the pale male dataset being used as something that's universal when that isn't actually the case when it comes to representing the full sepia of humanity?
>
> "And do you think it could exacerbate the already egregious inequalities in our criminal justice system?"
>
> JB: "It already is." (ibid)

The Gender Shades Case Study provides an example of diversity supporting creativity and objectivity. As part of her research, Buolamwini created the Pilot Parliaments Benchmark, a tool she used to show that two commonly used benchmarks included an overrepresentation of subjects with light skin types, making them ineffective for training and testing automated facial analysis algorithms. Further, she used the Pilot Parliaments Benchmark to show that her experiences with automated facial analysis algorithms and commercial gender classification systems were not idiosyncratic but rather that these products had previously undocumented and remarkably high error rates when classifying images of women with dark skin types. I don't find it surprising that this research was conducted by a Black woman with a dark skin type, which places her in a tiny minority in a field with a remarkably chilly climate for women and, in particular, for women of colour. Gender Shades research revealed a cognitive failure in terms of the creation and use of benchmarks and the design and testing of CGCs. This pattern of localized and global cognitive dysfunction led to the creation of a technology that rendered Black women unseen and misclassified.

Acknowledgements Thank you to those who offered generous feedback on earlier versions of this paper including Drs Kathryn Plaisance and Patrica Marino, and audiences at the Canadian Society for Women in Philosophy and the Consortium for Responsible Innovation in Technology conferences both in 2019, The Feminism, Social Justice, and AI Workshop in 2021, the Center for Values in Medicine, Science, and Technology Annual Conference in 2023, and audiences at Trent University and The University of Saskatchewan.

References

Alcoff, L. 2023. Critical Philosophy of Race. In *The Stanford Encyclopedia of Philosophy*, ed. E. Zalta and U. Nodelman. https://plato.stanford.edu/archives/fall2023/entries/critical-phil-race/.

Algorithmic Justice League. https://www.ajl.org/. Accessed 28 February 2022.

Anderson, E. 2020. Feminist Epistemology and Philosophy of Science. In *The Stanford Encyclopedia of Philosophy*, ed. Edward N. Zalta. https://plato.stanford.edu/archives/spr2020/entries/feminism-epistemology/.

Benjamin, R. 2019. *Race After Technology: Abolitionist Tools for the New Jim Code*. Cambridge: Polity.

Buolamwini, J. 2016a. The Algorithmic Justice League: Unmasking Bias. *MIT Media Lab*. https://medium.com/mit-media-lab/the-algorithmic-justice-league-3cc4131c5148.

Buolamwini, J. 2016b. How I'm Fighting Bias in Algorithms. https://www.ted.com/talks/joy_buolamwini_how_i_m_fighting_bias_in_algorithms?language=en. Accessed 28 February 2022.

Buolamwini, J. 2017. *Gender Shades: Intersectional Phenotypic and Demographic Evaluation of Face Data Sets and Gender Classifiers*. MIT Master's thesis.

Buolamwini, J. 2018. When the Robot Doesn't See Dark Skin. *New York Times*. https://www.nytimes.com/2018/06/21/opinion/facial-analysis-technology-bias.html. Accessed 28 February 2022.

Buolamwini, J. 2023. *Unmasking AI: My Mission to Protect What is Human in a World of Machines*. New York: Random House.

Buolamwini, J., and T. Gebru. 2018. Gender Shades: Intersectional Accuracy Disparities in Commercial Gender Classification. *Proceedings of Machine Learning Research* 81:1–15.

Cave, S., and K. Dihal. 2020. The Whiteness of AI. *Philosophy and Technology.* 33:685–703.

CBC. 2024. York, Peel Police now Using Facial Recognition Technology. https://www.cbc.ca/news/canada/toronto/police-facial-recognition-software-1.7216242.

Code, L. 2008. Taking Subjectivity into Account. In *The Feminist Philosophy Reader*, ed. A. Bailey and C. Cuomo, 718–741. New York: McGraw-Hill.

Crenshaw, K. 1991. Mapping the Margins: Intersectionality Identity Politics, and Violence Against Women of Colour. *Stanford Law Review* 43 (6): 1241–1299.

Fehr, C. 2007. Are Smart Men Smarter Than Smart Women? The Epistemology of Ignorance, Women and The Production of Knowledge. In *The 'Woman Question' and Higher Education: Perspectives on Gender and Knowledge Production in America*, ed. A. M. May, 102–116. Northhampton and Cheltenham: Edward Elgar.

Fehr, C. 2011. What Is in It for Me? The Benefits of Diversity in Scientific Communities. In *Feminist Epistemology and Philosophy of Science: Power in Knowledge*, ed. H. Grasswick, 133–156. Dordrecht: Springer.

Fehr, C., and J. Jones. 2022. Culture, Exploitation, and Epistemic Approaches to Diversity. *Synthese* 200 (6): 1–25.

Gender Shades. http://gendershades.org/. Accessed 28 February 2022.

Hamidi, F., M. Scheuerman, and S. Branham. 2018. Gender Recognition or Gender Reductionism? *The Social Implications of Embedded Gender Recognition Systems*. 1–13. https://doi.org/10.1145/3173574.3173582.

Haraway, D. 1988. Situated Knowledges: The Science Question in Feminism and the Privilege of Partial Perspective. *Feminist Studies* 14 (3): 575–599.

Hill, K. 2020. Wrongfully Accused by an Algorithm. *New York Times*. https://www.nytimes.com/2020/06/24/technology/facial-recognition-arrest.html?unlocked_article_code=1.vE0.FIej.LIkM8Jgtm8ru&smid=url-share. Accessed March 2025.

Hill, K. 2023. Eight Months Pregnant and Arrested After False Facial Recognition Match. *New York Times*. Accessed March 2025.

Kendall, D. 2020. Now Streaming: 'Coded Bias' Exposes Tech Made without Women and BIPOC in Mind. www.sundance.org. Retrieved 9 August 2024.

Keyes, O. 2018. The Misgendering Machines: Trans/HCI Implications of Automatic Gender Recognition. *Proceedings of the ACM on Human-Computer Interaction* 2 (CSCW): 88.

Klare, B., B. Klein, E. Taborsky, A. Blanton, J. Cheney, K. Allen, P. Grother, A. Mah, and A. Jain. 2015. Pushing the Frontiers of Unconstrained Face Detection and Recognition: Larpa Janus Benchmark A. In *Proceedings of the IEEE Conference on Computer Vision and Pattern Recognition*, 1931–1939.

Kleinman, Z. 2017. Artificial Intelligence: How to Avoid Racist Algorithms. *BBC News*. https://www.bbc.com/news/technology-39533308. Accessed 22 February 2022.

Lohr, S. 2018. Facial Recognition Is Accurate, if You're a White Guy. *New York Times*. https://www.nytimes.com/2018/02/09/technology/facial-recognition-race-artificial-intelligence.html. Accessed 28 February 2022.

Longino, H. 1990. *Science as Social Knowledge: Values and Objectivity in Scientific Inquiry*. Princeton: Princeton University Press.

Longino, H. 2002. *The Fate of Knowledge*. Princeton: Princeton University Press.

Lugones, M. 1987. Playfulness, "World"-Travelling, and Loving Perception. *Hypatia* 2 (2): 3–19.

Maynard, R. 2017. *Policing Black Lives: State Violence in Canada from Slavery to the Present*. Halifax: Fernwood Publishing.

McAlear, F., A. Scott, K. Scott, and S. Weiss. 2018. Data Brief: Women of Colour in Computing. *Kapor Center/ASU CGEST*.

McIntyre, L.-R.. 2021. Microsoft's 2021 Diversity & Inclusion report: Demonstrating Progress and Remaining Accountable to our Commitments. https://blogs.microsoft.com/blog/2021/10/20/microsofts-2021-diversity-inclusion-report-demonstrating-progress-and-remaining-accountable-to-our-commitments/. Accessed 24 March 2022.

Mills, C. 1997. *The Racial Contract*. Ithaca: Cornell University Press.

National Academy of Sciences. 2018. *Engineering, and Medicine: Sexual Harassment of Women: Climate, Culture, and Consequences in Academic Sciences, Engineering, and Medicine*. Washington, DC: The National Academies Press.

Noble, S. 2018. *Algorithms of Oppression: How Search Engines Reinforce Racism*. New York: New York University Press.

O'Neil, Lorena. 2023. These Women Tried to Warn Us About AI. *Rolling Stone*.. https://www.rollingstone.com/culture/culture-features/women-warnings-ai-danger-risk-before-chatgpt-1234804367/. Accessed 9 August 2024.

Ortega, M. 2006. Being Lovingly, Knowingly Ignorant: White Feminism and Women of Colour. *Hypatia* 21 (3): 56–74.

Phillips, K. 2017. What is the Real Value of Diversity in Organizations? Questioning our Assumptions. In *The Diversity Bonus: How Great Teams Pay off in the Knowledge Economy*, ed. S. Page, 223–245. Princeton: Princeton University Press.

Snow, J. 2018. "We're in a Diversity Crisis": Cofounder of Black in AI on What's Poisoning Algorithms in Our Lives. *MIT Technology Review*. https://www.technologyreview.com/s/610192/were-in-a-diversity-crisis-black-in-ais-founder-on-whats-poisoning-the-algorithms-in-our/.

Time Magazine. 2023, October 9. The 100 Most Influential People in Artificial Intelligence: Time 100/AI. https://time.com/collection/time100-ai/.

Van Anders, S. 2015. Beyond Sexual Orientation: Integrating Gender/Sex and Diverse Sexualities via Sexual Configurations Theory. *Archives of Sexual Behavior* 44 (5): 1177–1213. https://doi.org/10.1007/s10508-015-0490-8.

Varma, R. 2007. Women in Computing: The Role of Geek Culture. *Science as Culture* 16 (4): 359–376.

West, S. M. 2020. Redistribution and Rekognition. A Feminist Critique of Algorithmic Fairness. *Catalyst* 6:2. https://doi.org/10.28968/cftt.v6i2.33043.

West, S. M., M. Whittaker, and K. Crawford. 2019. *Discriminating Systems: Gender, Race and Power in AI*. AI Now Institute. https://ainowinstitute.org/discriminatingsystems.html.

Women who Tech. 2023. Women Who Tech Startup and Tech Culture Survey. https://womenwhotech.org/data-and-resources/state-women-tech-and-startups-2023. Accessed 20 September 2024.

Chapter 8
Conceptions of Machine Learning: Limitations and Weaknesses from the Viewpoint of Ethics

Brett Bolander

Abstract This chapter examines questions concerning the ethical oversight of Machine Learning (ML), focusing on approaches that emphasize the principles of transparency, fairness, and explainability. I begin in Sect. 8.2 by describing the most dominant approach to ML ethics, which focuses on finding technical means of implementing abstract ethical principles. In Sect. 8.3, I probe the dominant cognitive conception for understanding ML. The primary contention here is that the cognitive framing is useful for developers in creating these programs and politically useful for the organizations these developers operate within, but that it obfuscates the relevant factors at play when the conception is exported into ethical or regulatory discussions. Sections 8.4 and 8.5 delve into two alternative conceptions of this class of programs, one from the sociologist Elena Esposito that attempts to frame algorithms as communicative agents, which allow various systems and individuals to interface with another through constructing what she calls a 'virtual double contingency', and another framing from political geographer Louise Amoore, which aims to center the discussion of algorithms and machine learning within the always socio-political context the operate within and upon which they are empowered to act. In Sect. 8.6 I apply these framings to a toy case and describe where they fall short in illuminating the ethical landscape. Finally, I conclude in Sect. 8.7 by pointing to areas I believe require more attention from philosophers if a coherent understanding of ML is to be reached.

Keywords Artificial intelligence · Machine learning · Ethics of machine learning · Cognitive understandings of machine learning · Elena Esposito · Louis Amoore

B. Bolander (✉)
Department of Philosophy, University of Pennsylvania, Philadelphia, PA, USA
e-mail: brittabo@sas.upenn.edu

J. Y. Tsou et al. (eds.), *Values, Pluralism, and Pragmatism: Themes from the Work of Matthew J. Brown*, Boston Studies in the Philosophy and History of Science 347, https://doi.org/10.1007/978-3-031-92958-8_8

8.1 Introduction

Current debates surrounding the role of machine learning (ML) in social institutions primarily operate within a shared paradigm for understanding precisely what these software programs are, how they operate, and what it is that they do when they are implemented in a given context. The paradigm for this understanding can broadly be understood as 'cognitive' in nature; the program 'learns' representations which allow it to perform tasks within a given social context. Issues with anthropomorphizing computer programs have been raised even by professionals working in the field (Shanahan 2023), especially insofar as these programs are conceived as "knowing", "believing", and "thinking". This framing is also dominant in the ethical literature, which primarily has focused on how to intervene in this "learning" or "training" process in order to ensure that it makes fairer or unbiased "decisions". The most extreme manifestations of this framing can be seen in discussion of whether or not large language models have a "theory of mind" (Kosinski 2023) or whether some forms of computer programs have claim to certain rights (Schwitzgebel and Garza 2015).

There are myriad facets of the discussion around the ethical oversight of ML, but the focus of this chapter will primarily be on approaches which emphasize the principles of transparency, fairness, and explainability. These principles can all be understood as aiming to better understand and describe the "how" and "what" that correspond to the machine's "learning", and especially how what it has learned leads to specific outcomes. However, this focus on the "behaviors" exhibited by software programs has led to a myopic focus on technical solutions to non-technical problems. As a consequence, much of the focus has been on developing something akin to a business or professional ethics, which I argue is inadequate for fully grappling with the consequences of implementing novel methods of automating social processes, especially those with the power to shape the trajectories of individual's lives. This approach also presupposes and reinforces an attitude of inevitability and acquiescence when the starting point for ethics in experimental technologies should be at the exact opposite end of the spectrum, something closer to the precautionary principle.

In discussing this topic, I am not so much concerned with whether any given conception is right or wrong in how it describes the subject in question. I also set the question of whether the conceptions are themselves responsible for the adoption of specific ethical principles and oversight approaches to the side. Rather, I only seek to expound where these conceptions nudge discourses into specific enclaves, where they are limited in making the subject itself comprehensible, and what sorts of interests and frameworks are made more obvious under specific ways of understanding this technology. Additionally, for my purposes here I am not concerned with any specific instance or usage of the technology, either large language models which power the currently hyped chatbots, ML programs used in "loss-prevention" in retail contexts (Matsakis 2020), or those which influence policing and parole decisions (Biddle 2022).

I begin in Sect. 8.2 by describing the most dominant approach to ML ethics which focuses on finding technical means of implementing abstract ethical principles. In Sect. 8.3, I then probe the cognitive conception itself, showing where, why, and for whom it is helpful as well as where it fails to provide fruitful insight or actively confuses. The primary contention here is that the cognitive framing is useful for developers in creating these programs and politically useful for the organizations these developers operate within, but that it obfuscates the relevant factors at play when the conception is exported into ethical or regulatory discussions. Sections 8.4 and 8.5 delve into two alternative conceptions of this class of programs, one from the sociologist Elena Esposito which attempts to frame algorithms as communicative agents which allow various systems and individuals to interface with another through constructing what she calls a 'virtual double contingency', and another framing from political geographer Louise Amoore which aims to center the discussion of algorithms and machine learning within the always socio-political context they operate within and which they are empowered to modulate. In Sect. 8.6 I apply these framings to a toy case and describe where they fall short in illuminating the ethical landscape. Finally, I conclude in Sect. 8.7 by pointing to areas I believe require more attention from philosophers if a coherent understanding of ML is to be reached.

8.2 Dominant Ethical Paradigm

Thilo Hagendorff (2020) analyzed 22 major "ethical AI" guidelines across public, private, and professional organizations, over 80% of which included "accountability", "privacy" and "fairness" in their principles, with "transparency" also appearing at a similar rate. Notably, "explainability" is present in nearly every set of guidelines published by a government[1] or professional organization, while IBM is the only

[1] The only exception, at the time of Hagendorff's publication as well now under the second Trump Administration, was and is the United States. However, the Biden administration released a "Blueprint for an AI Bill of Rights" in October 2022 which included "explainability" (Office of Science and Technology Policy 2022). The creation of an 'AI Bill of Rights' was then mandated under Executive Order 14110, signed on October 30th 2023. The blueprint itself is no longer accessible online, but the author has a partially saved image of the original html file. The United States' policy has since changed under the second Trump Administration, with Executive Order 14110 being revoked on January 20th 2025. Executive Order 14179 was signed on January 23rd 2025 (Removing Barriers to American Leadership in Artificial Intelligence 2025). This order directed Michael Kratsios, Assistant to the President for Science and Technology, Director of the Office of Science and Technology Policy, and Science Advisor to the President; David Sacks, Special Advisor for AI and Crypto; and Marco Rubio, Assistant to the President for National Security Affairs; to develop an Artificial Intelligence Action Plan as well as to "immediately review, in coordination with the heads of all agencies as they deem relevant, all policies, directives, regulations, orders, and other actions taken pursuant to the revoked Executive Order 14110" (EO 14179, Secs. 4 & 5). The AI Action Plan is intended to "achieve the policy set forth in section 2" of Executive order 14179 (Sec. 4). Section 2 reads as follows: "It is the policy of the United States to sustain and enhance America's global AI dominance in order to promote human flourishing, economic competitiveness, and national security" (EO 14179, Sec. 2). Both Michael Kratsios and David Sacks, as well as 'special government employee' Elon Musk and Vice President J.D. Vance, have intimate ties with the co-founder and chairman of Palantir, Peter Thiel. Palantir surpasses the market capitalization all 5 defense primes (Raytheon, Boeing, Lockheed Martin, General Dynamics, and Northrop Grumman) by at least 100 billion dollars (as of market close on 5/17/25) and has "won more than $2.7bn in US contracts since 2009", with just under half of this figure being from Pentagon contracts (O'Murchu and Kinder 2025).

corporation that included it in their principles. The governments analyzed included the United States, the European Union, and Beijing, and the industry organizations included the Information Technology Industry Council and the Institute of Electrical and Electronics Engineers. The corporations included Google, Microsoft, Deepmind, OpenAI, and IBM. The definitions given of these terms are often fairly loose and sometimes painfully abstract. For my purposes, "accountability" will be understood to be the primary oversight principle at work here, with "privacy" and "fairness" providing the core ethical principles that developers are being held accountable for maintaining. "Privacy" is less relevant for my purposes because I am primarily concerned with the consequences the program itself has within a given social system, something which would remain regardless of whether the program can be made a perfect protector of privacy. "Fairness" generally refers to ensuring that the outputs of the ML program do not effect discriminatory outcomes against any given individual or demographic. The more nebulous "transparency" and "explainability" will be taken to refer to the specific means by which accountability can be implemented within an institution. It is generally thought that if the institutional users of ML programs can explain why a specific output is what it is and they transparently disclose the data and operations which underlie that explanation, then accountability for user privacy and output fairness can be enforced by a governing body, whether internal or external. Importantly for our purposes, this constellation of ethical oversight focuses primarily on the policing of specific *behaviors* attributed to the devices, rather than the devices *themselves*.

Greene et al. (2019) also analyzed a battery of values statements using frame analysis to draw out the common themes which constitute what they call, following the sociologist Gabriel Abend (2014), the 'moral background' of the discourse around the ethical oversight of this technology. They identify seven specific elements of this moral background within the values statements they analyze: (1) universal concerns, objectively measured; (2) expert oversight; (3) values-driven determinism; (4) design as locus of ethical scrutiny; (5) better building; (6) stakeholder-driven legitimacy; (7) machine translation. Elements 1–4 are particularly relevant for our discussion here as the technical and design driven focus is reflected in the principles mentioned above.

The biggest concern in the discussion around ethical use of ML has been in how the programs themselves disparately impact specific groups of people, especially already disadvantaged minority groups, because of biases in the data or training process which end up encoding discriminatory outcomes (O'Neil 2016; Noble 2018; Benjamin 2019). Many approaches focus on the development of methods to explain why a certain outcome came to pass instead of a different one in the form of a counter-factual explanation (Guidotti 2022), while others focus on ensuring that the outcomes produced by these programs adhere to some procedural metric of fairness such as the 'conditional demographic disparity' measure proposed by Wachter et al. (2020). In all cases that proceed along these lines, technical intervention within the design process itself is taken to be the means for securing the ethical values of fairness, transparency, and explainability.

The decision to take this design-based approach to ethics is not a trivial one, and it is certainly not obviously superior to any other approach. An approach such as this has the tendency to become hyper-focused on the particularities of single applications and their measurable consequences at the expense of broader political and sociological considerations. This presupposition that design is the primary area of intervention for ethical oversight also presupposes the permissability of applying these computational methods to the management of a given system. As I will develop in the next section, the cognitive conception of this class of software makes these assumptions more comfortable to adopt. In other words, there is an affinity between conceiving of these programs as engaging in task-completing behavior in a way similar to humans and focusing on technical interventions for achieving supposedly ethical outcomes.

8.3 Cognitive Conception

Russel & Norvig's influential textbook *Artificial Intelligence: A Modern Approach* describes the field itself as being "focused on the study and construction of agents that do the right thing," as defined by whatever objective is provided (Russell and Norvig 2021, p. 4). This is also glossed by them as creating agents that "act rationally". The technical frameworks underlying ML approaches are themselves inextricable from a cognitive framing, as the name 'machine learning' itself and its theoretical foundations in 'computational learning theory' and 'statistical learning theory' should make evident. The tendency among those who create these programs is to frame them as cognitive in nature, as 'learning' the most useful distinctions between given data points in order to complete a given task relating to them. The task to which the program is put is often something like categorization or prediction. This conception of ML programs is instrumental for developers as they attempt to frame the problem in such a way that the program can consistently produce satisfactory results. In application, an ML program is generally intended to be used as a substitute for a human process of some sort, from something as simple for us as deciphering the addresses on envelopes to something as complex as interpreting radiological images for signs of disease. Conceiving of the software program as itself exhibiting a cognitive process of some sort makes the translation of the human task into a data processing problem much easier. If one wants the program to "behave" in some particular way, one then only has to consider how a human would produce behavior of that sort and find some means for creating analogous conditions within the program.

We can see this cognitive framing reflected in how experts talk about these technologies. Sebastien Bubeck, a Microsoft researcher who worked on integrating OpenAI's large language model GPT-4 into Bing, gave a talk at MIT in April of 2023 which demonstrates this. The title of the talk was "*Sparks of AGI: Early experiments with GPT-4*" (Sebastien Bubeck 2023) and it has an associated paper (Bubeck et al. 2023). For those who have managed to avoid the term until now, "AGI" stands for "Artificial General Intelligence", the philosopher's stone of the field. Close to the beginning of the talk, Bubeck primes the audience to think about machine learning technology in the following way in order to best understand the rest of the presentation:

> From my perspective, we shouldn't think about those neural networks as learning simple concepts like 'Paris is the capitol of France'. It's doing much more like learning operators, it's learning algorithms. So inside it, it's not just retrieving information, not at all, it has built internal representations that allows it to reproduce the data it has seen succinctly.

This description will be taken as generally representative of how ML programs are understood within the cognitive framing by those who create them, using mostly terms which normally require a rational-intentional agent as their subject.

The understanding of machine learning programs that the cognitive framing provides is easily mapped onto the previously mentioned ethical paradigm: transparency aims to make explicit what datasets it has "seen"; fairness is primarily concerned with ensuring that the "representations it builds" do not encode or effect discriminatory practices; and finally, explainability seeks to show how what it has "learned" relates to what it produces. When the focus is on the behaviors of a hypothetical agent, it makes sense to be concerned with the kind of information it receives, the representations it constructs, and how it uses those to produce a determination or output. This projection of intentional agency allows developers to skip ahead to the very practical matter of controlling and taming the "behavior" of the program before they've even constructed a clear notion of what this "behavior" is, or whether that label is itself even appropriate. In this way, the conception provides engineers with certain affordances for practical action and thus serves an essential instrumental role in the design and deployment process.

However, this attempt to assimilate the operations of software programs and the workings of the human mind does far more to obscure our picture of what is going on than it does to enlighten us. The assimilation is incapable of thinking about what is inherently *different* between the processes that a program effects and those that a human effects, and it obscures the extent to which the replacement of a human with a program alters the sociological landscape of any given social system. In conceiving of these programs as cognitive in their very nature, discourses are nudged into considerations that are akin to constructing a professional ethics for software programs, defining the boundaries of acceptable actions for these programs to take through technical restrictions and auditing procedures. The affinity between the cognitive conception and the design-based approach to ethics is seen most concretely here, in the attempt to translate the ethical considerations and actions that the human would ideally adhere to into standards and metrics which can be met through technical intervention.

The most important failure of this conception, in my view, lies in the incapacity to appropriately consider these differences. This conception treats the integration of an ML program into a specific social system as though it were simply changing what entity is carrying out a specific role in the institution, either a human or an ML program. A conflation between *roles* and *functions* occurs here; while the human and the program may be able to fulfill one of the same *functions* within a social system or institution, it is highly unlikely that they will ever play the same *role*. The assumption that the cognitive conception makes most obvious is that there is no real difference between having the program fulfill a function and having a human fulfill the function, except that the program is likely to be faster and more accurate than the human. For those in the business of selling these things, this makes it quite easy to explain the benefits that a customer would gain (greater efficiency, accuracy, etc.) by integrating these methods into their organization. However, the mere fact that a function is fulfilled by a machine rather than by a human necessarily changes the role itself since some subordinate function of the role is delegated to an optimization regime. The effects of this replacement could even mean that the entire institution or system itself has been fundamentally changed or reoriented. These larger pictures of the political and sociological effects are lost when one is focused on designing well-behaved programs.

8.4 "Artificial Communication"

The first alternative to be explored comes from the field of sociology, specifically from the tradition of Luhmannian social systems theory. Elena Esposito has introduced the concept of "artificial communication" (Esposito 2017, 2022) which aims to reframe algorithms as non-cognitive communicative agents. Esposito's understanding of communication begins from the receiver of a message rather than the sender, so that communication can only be said to have occurred when a cognitive agent registers an utterance as conveying meaningful information. Under this framing, Eposito argues that algorithms—such as those behind social media and advertising recommendations which focus on the construction, assignment, and networking of certain users into neighborhoods of "profiles"—can be understood as communicative partners that users interact with in order to receive information relevant to their interests and desires. She argues that the algorithm constructs what she calls a "virtual double contingency", which is her way of saying that it represents and reproduces the communicatively informative aspects of other user's behavior in its own output, thereby meeting the criteria for communication under her framework. For Esposito, these devices can be understood as taking on a role similar to that of a concierge, a role which no doubt involves communication of some sort. That the algorithms reproduce the *communicatively* informative actions of many users is central to the framing of these use cases as communicative. As Esposito clarifies, something like the layers of rock in an area do not "communicate" with a geologist despite containing information of some kind because this information can only be *communicatively* informative once they are deciphered by that geologist, who then puts them into words and utterances that convey said information about the rock layers (Esposito 2017, p. 254).

We can take the example of ChatGPT as an easy way to demonstrate precisely what Esposito trying to get at. The program itself is created using vast amounts of text data scraped from the internet which provide the foundation for all outputs of the program. The act of communication comes into play when the user enters a prompt and receives a response; Esposito would argue that this is where the program is acting as a communicative agent as it uses past communications that it has access to in order to provide a response to a user inquiry. This act counts as communication for Esposito precisely because the program is not merely reflecting the user themselves nor simply reproducing the utterances it was trained on. Rather, the program refracts past communications in order to produce a response to a specific user inquiry. This reaction to the user who enters the prompt is what constitutes the "double"; the program does not simply reflect the data or the user, but mediates between them in order to produce communicative information which did not necessarily exist before. Esposito applies this idea to many kinds of algorithms, even those which do not fall under the umbrella of ML such as Google's original PageRank algorithm.

What is admirable about Esposito's work is how she translates the insights of this sociological framework of communication into the domain of algorithms, and it could be especially helpful in providing an understanding of information flows in the use cases similar to those she focuses on. It is important to note, however, that the only reason the communication framing has this explanatory power is because the contextual usages of algorithms that she explores can be feasibly modeled as being intermediaries—or more accurately in my view, mediums—for communication. The conception deteriorates when applied to use cases which cannot be so neatly framed as communication, such as algorithms used in cancer screenings, automated chemical formulations, and computer vision applications like face and object recognition. As she herself explained in the geology example, the fact that information can be retrieved is insufficient for determining that communication has occurred, and the same holds true in these non-communicative use cases which must also be deciphered by an expert. She even indulges in the same exact cognitive language of the program "knowing" and "learning" in various sections of the book (see Esposito 2022, p. 133, as just one example).

While Esposito does attempt to shift the locus of agency and intentionality from the algorithm itself and onto the designers and data used to train them, nothing much has really changed. In fact, Esposito's communication framing retains the same problematic projections as the cognitive one, an unsurprising fact considering that "building" a representation which "reproduces" the data they have "seen" is essentially the same thing as "building" a virtual contingency which "re-elaborates" the "observations" of other users. Even the ethical paradigm maps onto the understanding in the same way: transparency now corresponds to the "reflection" of perspectives; fairness to the "construction" of an unbiased and accurate virtual-contingency; and explainability to the method of "re-elaboration" which produces the output the user observes. It just is the cognitive conception, albeit with a coat of communicative paint, in the sense that it has also constructed a false intentional agent in order to explain the model's "behavior". This is not only entirely unnecessary, but it actively confuses and occludes many of the ethical issues inherent to the production and deployment of any large scale technical device because it allows us to ignore that it even is such a thing. Any framing which similarly posits or implies a rational-intentional agent enables this short circuiting of thought into the shallow domain of managing the unruly "behaviors" of fictional agents.

What this tells us is less that algorithms *are* communicative agents, but more that they can act as communicative agents. This is certainly helpful for better understanding something about the social functions that these programs are able to fulfill within the world, but it does not necessarily tell us much about the programs themselves. This is, however, a step in the right direction in the sense that it allows for alternative conceptions to come on the scene when communication may not be an adequate framework for understanding a particular program or application thereof. The cognitive conception is less flexible and more hegemonic in this regard as it encompasses all applications of ML and explains away the differences between them as merely being changes in what a program learns. The next account proceeds in a similar fashion to Esposito by trying to capture the political dimensions which accompany the creation of any algorithm or software program.

8.5 Algorithms as Politics

One framework which attempts to move beyond the current ethical paradigm is that of political geographer Louise Amoore, who in her book *Cloud Ethics* takes the incomprehensibility and unattributability of these sorts of programs as a given and considers attempts at constructing the necessary pillars of accountability to be misguided in the first place (Amoore 2020). She focuses on ML programs not as cognitive or communicative agents, but as inherently ethico-political instruments of control which define the boundaries of what can be considered up for debate, and even of what can be considered a political question in the first place. A key part of her argument is that the dominant framework of ethics not only fails to address the already acknowledged problems within the field, especially those of discrimination and bias, but she views the current paradigm as, in some cases, actively enabling and permitting the most dangerous and harmful usages to continue relatively unimpeded.

Amoore argues against this complex of ethical oversight mechanisms on a few fronts: First, by acknowledging that any given piece of data is equally as usable by the algorithm as any other piece of data because all data is inherently meaningless to it. She expands on this by arguing that the most pressing ethico-political problem posed by the usage of machine learning algorithms is the reduction of reality and contingency to mere "feature spaces". This reduction, in her view, is the most politically charged aspect of these devices because it is the mechanism by which possibilities are foreclosed in the name of bureaucratic management. Second, she argues that the most challenging feature of the technology is impossible to grapple with under the framework of accountability: that there is no clear singular agent who could even be held accountable for a given output. In effect, the diffuse nature of data collection, distribution, and filtering, as well as the inherently distributed development, deployment, and usage processes themselves make the actual administration of these measures so intractable that she believes the paradigm itself should be abandoned, at least in its current form. To make one thing clear, Amoore does not reject that the corporations who make the programs can and should be held legally liable for the harms they cause, but rather she argues that it is an exercise in futility to attempt to build a framework of ethics and public policy with the aim of imputing any single element as being responsible for some output the model produces.

Amoore takes attributability to be the primary issue which must be addressed before constructing any means of ethical oversight, something which an ethics premised on accountability presupposes is possible. In order to hold some entity accountable for an outcome, there must be some entity we can attribute the outcome to. Amoore argues that there is no such entity which has the requisite level of attributability for the outcome, at least whenever one is looking at any particular output of a program. The fact that this parole applicant was deemed "high-risk" cannot be primarily attributed to features of the training data, a particular decision made during the design process, nor to the specific traits of the individual in question, but can only be properly attributed to the totality of actions which led the particular program to be constituted as it is. This would include not only the process of "training" the ML program, but also the prerequisite acts of data collection, refinement, and labeling which turn the organizational reality into a ready-made table of acceptable and unacceptable outcomes for the machine to optimize against. While it is possible to disaggregate these things from one another conceptually, in practice the decisions made in any one domain will require coordinated changes in the others: changing the types of data used for training data will quite often require changes in model architecture, and vice versa. The only entities which we could feasibly attribute the outcomes to are the organizations which create and implement the ML programs, but even they are only indirectly responsible for any particular outcome because the developers in question do not have direct control over how specific cases are decided by the program.

As a very concrete example of how this shift in understanding affects the ethico-political priorities, we turn to a discussion Amoore had at the Centre for Digital Inquiry at The University of Warwick. She mentions the case of Amara Majeed, an American born daughter of Sri Lankan immigrants who was misidentified as one of the perpetrators of the 2019 Sri Lanka Easter Bombings, supposedly due to an "error" in a facial recognition algorithm that the authorities had used. Amoore argues that this attempt to "shelter the output of the algorithm in an error" distorts the actual mechanisms which led to the incident being possible in the first place. Amoore argues that if we expand our notion of algorithms to understand them as being already ethico-political, then:

> You might actually conclude that she was not wrongly identified at all, coming to the attention of the authorities [through] a series of risk algorithms in which the attention to facial recognition is just one element of a much broader set of practices that are really not about 'Is this person the same person we have here?', but actually worse than that, 'What is their propensity to be posing a problem to the state?'. But actually, on those measures, you might say that this problem of recognition and misrecognition is entirely redefined. (Centre for Digital Inquiry 2020)

For Amoore, it does not necessarily matter which specific development process or operational mechanism within the algorithm caused this to happen, nor whether the device could be made "more accurate" or "transparent". What matters is the instrumental function that the algorithm plays in effecting the administrative requirements and desires of the state. It enables the controlling body to have individuals labeled in real time as risks to be assessed, surveilled, and controlled. Aiming to create more accountability and transparency in these sorts of devices amounts to nothing more than permitting and ethics-washing what she views as an incredibly fraught, reductive, and harmful political practice. Regardless of what safeguards or oversight mechanisms are put in place, this fundamentally ethico-political reality remains at the root of the entire practice.

8.6 Application of Paradigms

This sentiment about "error" expressed by Amoore is in line with one aspect of the recent book by Herman Cappelen and Josh Dever called *"Making AI Intelligible"* which aims to outline a theory of semantic content capable of explaining the outputs of ML programs (Cappelen and Dever 2021). They use the example of a hypothetical program they call "SmartCredit" which weighs in on the riskiness of a given loan applicant that a bank is considering lending money to by producing a 'Yes/No' answer which corresponds to some propositional content about the applicant. They recruit and modify various externalist theories of semantic content, especially that of Saul Kripke, and apply it to the "SmartCredit" case. Under their analysis, the program's output is anchored in some trait of the applicant's profile such as "high-risk" by virtue of the fact that it produces a similar distribution of outputs as does a procedure of hand-coding applications for this trait. To quote the slogan version of their proposal, "SmartCredit is anchored in high risk via a scoring function that scores well for matching high risk hand coding and low for not matching," (Cappelen and Dever 2021, p. 89). They additionally offer an analysis which explores how changes in the training process—such as including the loan amount as an additional input variable in order to weigh the relative financial costliness of a potential default—can produce outputs which are not only entirely different for the same input data, but are also absolutely worse at predicting whether any given loan will actually be defaulted on, an outcome which seems to undermine their thesis. To rescue their argument, they say:

> But the new patterns are well coordinated with something like *costly default*. When we think of it in this way, there hasn't really been an increase in the error rate. It's true that more often now the machine learning system says 'Yes' for a loan that goes on to be defaulted on, and 'no', to a loan for which there isn't any subsequent default. **But those are only *errors* if we think that the programs 'yes' means 'yes, this loan is safe from default' and it's 'no' means 'no, this loan isn't safe from default.'** If the machine has changed, by virtue of the new reinforcement pattern, from being a default detector to being a costly default detector, then its 'yes' now means 'yes, this loan is safe from costly default,'

> and it isn't, in fact, making more mistakes. (Cappelen and Dever 2021, p. 142; bolding my own, italics present in original)

In other words, even if one were to construct a concrete externally grounded notion of what propositional content lies within the output of a given ML program, it can only be "wrong" in the sense that the propositions we project onto the output are not the same as those that the program is actually producing. Despite immense areas of disagreement between myself and these authors on the nature of both content and representation, especially many of the ways they apply those concepts to the outputs of computer programs, they have touched upon the very same thing as Amoore by describing the outputs of "SmartCredit" in this way.

Taking the same toy case study as an example, we can see how Esposito's framework would behave in a situation like this. Loan determinations can be understood to function in much the same way as the taste profiles used to serve up music and video recommendations. To strain the analogy only slightly, the analogue of an individual song would be the individual applicant, and the analogue of the 'Yes/No' which determines whether or not a song is recommended would simply be the 'Yes/No' output used to decide whether a loan is granted. The profiles it is constructing are basically the same, just oriented around the financial institution's monetary interests rather than musical taste. It is important to emphasize that just as the actual songs that are recommended are reduced to a list of metadata—including genre, artist, mood, tempo, etc.—so too are the applicants reduced to a list of metadata, in this case consisting of payment history, open lines of credit, revolving utilization, loan amount, and potentially even the average battery percentage of an applicant's phone (King 2016) or their posting habits on social media (Hiller and Jones 2022). In this case, the applicants themselves are not the communicators of information, but have been compressed into the information being communicated. The "artificial communication" must then be occurring between the institutions which record that information about applicants and the lending institution itself. While this framing does capture the communicative arrangements and information flows that enable the program to function in the first place, it still obscures what the program itself is actually doing and, in my view, misleadingly characterizes what information most centrally determines the program's functions. It is not the metadata—the "raw material"—which "SmartCredit" is given that plays the decisive role in producing the output, but primarily the "taste profile" of the lender that does so. This is the thing that "SmartCredit" is shaped to fit and which determines its ultimate operating form.

Augmenting Esposito's account with aspects of Amoore's, it could be said that what the "SmartCredit" model is doing is creating different "risk profiles" which reflect the willingness of the lender to approve a loan for that genre of applicant. These clusters are then used to reduce a given input to a simple 'Yes/No' determination within the output such that the lender is satisfied with the answers that it gives. This allows them to carve out ostensive definitions of what they deem to be unacceptable applicants by effectively doing the exact same thing as someone pressing 'like' or 'dislike' when a song is recommended. This is precisely what the "training" process is. The metadata are simply used as a sort of justifying rationale, as dials that the program can adjust to produce the "correct" output. The metadata cannot represent the *reasons for* a contingent output appearing, but only the *means used to ensure* that an already pre-determined output consistently appears.

8.7 Conclusion

Now, to hint at a path forward I want to briefly mention two areas which I believe could prove fruitful if given a serious and varied re-examination in order to separate the wheat from the chaff: (1) the social conception of these devices in particular, computation in general, as well as the act of their creation, and; (2) the functional conception of their operations which can explain their outputs independently of any specific use case. This second point is not meant to imply that the context and implementation are irrelevant, but rather is just meant to indicate that there has so far not been a way to speak about these devices in general which does not rely on misleading metaphorical heuristics, overlapping field-specific terminologies, or both.

In regards to the social conception of computers, it is worth reconsidering whether understanding them as simple tools for "information processing" accurately reflects our relationship to them and the ways in which they are empowered to act upon the world. Looking at these things as simple tools, as we would something like a hammer, a camera, or a pencil is certainly tempting. This is how the problem is currently approached and it is how computational devices are currently understood. The problem, however, is that those traditional tools have pre-defined and delimited possible usages. One cannot use a hammer for anything that a hammer is not capable of doing, which limits us to smashing and prying. In the case of machine learning, and computation more generally, the delimitation of their possible usages has a qualitatively different character than in the case of the hammer. The limit lies more in our capability and willingness to abstract a given task into computationally tractable sets of information, definitions, languages, and operations. In other words, the possible capabilities of these devices are determined by our willingness and ability to enframe a given problem in a machine-readable way and thus the limit exists not within the device, but within our own technical and mathematical capacities as well as our collective will to restrain or enable their governance over various areas of our collective social life. While not everything can be made into a nail, everything can be made into "information" if we are so willing.

In regards to the second item, the context-independent conception of these device's functional operations: many of the issues discussed here are not necessarily unique to ML and can be understood as problems inherent to modeling more generally. While this may be correct in some sense, it is important not to move too quickly to thinking that the problem has then been solved, and all that is required of us is to make "more accurate" models. This quick jump misses the fact that they are suited to very distinct ends from those of traditional modeling and can be readily applied to a vaster and more varied array of systems and subjects. These devices do not actually "model" a system in the way we typically use the word in science, nor do they "predict" the most likely outcome in the same sense that a meteorologist predicts the most likely path a hurricane will take based on known, relevant causal indicators present in the world. Rather, they construct ostensive definitions of the bins they are told to place specific inputs within. It categorizes any and all input with a label that has been provided beforehand, with definitions that have been impressed upon it through "training" on an ordered collection of specific data-points, and it is ensured by design that there is no input which is out of its power to determine. As Amoore says, there is no space for "errors" here, it simply uses the recorded features of enframed past problems and their corresponding "correct" solutions to filter all possible outputs it could produce to only those that its designer deems to be acceptable. Its "proper" functioning does not necessitate that it has developed any cognitive capacity or even a representation of the system that the designer wants it to determine, but only that the problem it was given was defined in such a way that the designers got the results they wanted out of it and thus stopped trying to "train" it. In the worst case, they serve as palliative enablers of wishful thinking. In the best case, they play a damn good game of chess. The cognitive conception obscures this entirely by labeling any undesired output a "hallucination", a designation which implies that something about the process which produced that output is qualitatively different from the process that produced the "correct" output. This is not the case, as the same exact mechanisms which underlie both processes must necessarily be the same, or else the thing would not work at all. To describe the outcome in this way is like saying that a hammer must be broken if it is used to murder someone rather than to drive a nail.

My biggest overall concern with the current ethical paradigm is that it provides a weak foundation for challenging specific usages of the technology which are ethically fraught and likely unjustifiable, such as the case of programs used in the carceral system. If the primary way these usages are understood to be ethically problematic relies on the fact that their "behaviors" are insufficiently rigorous, transparent, or "fair", then whenever those problems feel as though they have been sufficiently solved by introducing various metrics, the contention evaporates. There is a case to be made that the modus operandi of these devices precludes their usage in certain contexts and domains entirely, with no qualification of transparency or accuracy whatsoever, but arguments of this sort are less common and remain relatively underdeveloped because these questions have been bypassed by much of the literature. Finding strong grounding in the ethical debates surrounding these devices is of paramount importance, as I'm sure anyone who has been keeping up with the news in this area understands.

To close I'd like to leave you with a quote from Sebastien Bubeck's talk that I was surprised to actually agree with, though I believe my reasons are quite different from his.

> For us, as a society, to confront this question we have to go beyond the discussion of whether this is copy/paste or statistics. We have to leave this discussion behind us. The train has left the station. If we keep getting bogged down by this version of the question, we're going to miss all the real important questions, so I think it's important to move on.

Rather than dismissing the conversation because we are so obviously in a new domain of "Artificial General Intelligence", as he suggests, I argue that we must abandon the cognitive conception entirely so that we can better understand the actual nature of what is at stake.

References

Abend, G. 2014. *The Moral Background: An Inquiry into the History of Business Ethics*. Princeton: Princeton University Press.

Amoore, L. 2020. *Cloud Ethics: Algorithms and the Attributes of Ourselves and Others*. Durham: Duke University Press.

Benjamin, R. 2019. *Race After Technology: Abolitionist Tools for the New Jim Code*. Medford: Polity.

Biddle, J. 2022. On Predicting Recidivism: Epistemic Risk, Tradeoffs, and Values in Machine Learning. *Canadian Journal of Philosophy* 52 (3): 321–341.

Bubeck, S., V. Chandrasekaran, R. Eldan, J. Gehrke, E. Horvitz, E. Kamar, P. Lee, Y. Lee, Y. Li, S. Lundberg, H. Nori, H. Palangi, M. Ribeiro, and Y. Zhang. 2023. Sparks of Artificial General Intelligence: Early Experiments with GPT-4 (arXiv:2303.12712). *arXiv*.

Cappelen, H., and J. Dever. 2021. *Making AI Intelligible: Philosophical Foundations*. Oxford: Oxford University Press.

Centre for Digital Inquiry, University of Warwick (Director). 2020, July 22. Inquiries: Cloud Ethics with Louise Amoore and Nick Seaver. https://www.youtube.com/watch?v=NhUllrTDxf0.

Esposito, E. 2017. Artificial Communication? The Production of Contingency by Algorithms. *Zeitschrift für Soziologie* 46 (4): 249–265.

Esposito, E. 2022. *Artificial Communication: How Algorithms Produce Social Intelligence*. Cambridge: MIT Press.

Greene, D., A. Hoffmann, and L. Stark. 2019. Better, Nicer, Clearer, Fairer: A Critical Assessment of the Movement for Ethical Artificial Intelligence and Machine Learning. In *Proceedings of the 52nd Hawaii International Conference on System Sciences*.

Guidotti, R. 2022. Counterfactual Explanations and How to Find Them: Literature Review and Benchmarking. *Data Mining and Knowledge Discovery*.

Hagendorff, T. 2020. The Ethics of AI Ethics: An Evaluation of Guidelines. *Minds and Machines* 30 (1): 99–120.

Hiller, J., and L. Jones. 2022. Who's Keeping Score?: Oversight of Changing Consumer Credit Infrastructure. *American Business Law Journal* 59 (1): 61–121.

King, H. (2016, August 24). This startup uses battery life to determine credit scores. CNNMoney. https://money.cnn.com/2016/08/24/technology/lenddo-smartphone-batteryloan/index.html.

Kosinski, M. 2023. Theory of Mind May Have Spontaneously Emerged in Large Language Models (arXiv:2302.02083). *arXiv*.

Matsakis, L. 2020, May 29. Walmart Employees are Out to Show its Anti-Theft AI Doesn't Work. *Wired*. https://www.wired.com/story/walmart-shoplifting-artificial-intelligence-everseen/. Accessed 22 July 2023.

Noble, S. 2018. *Algorithms of Oppression: How Search Engines Reinforce Racism*. New York: New York University Press.

O'Murchu, C., and Kinder, T. 2025, February 7. Palantir's 'revolving door' with government spurs huge growth. *Financial Times*. https://www.ft.com/content/a65c93af-abf9-427c-a2ee-a9f5b3d518b9.

O'Neil, C. 2016. *Weapons of Math Destruction: How Big Data Increases Inequality and Threatens Democracy*. 1st ed. New York: Crown.

Office of Science and Technology Policy. 2022, October. Blueprint for an AI Bill of Rights. *The White House*. https://www.whitehouse.gov/ostp/ai-bill-of-rights/. Accessed 26 October 2022.

Removing Barriers to American Leadership in Artificial Intelligence. 2025, January 31. Federal Register 2025-02172. https://www.federalregister.gov/documents/2025/01/31/2025-02172/removing-barriers-to-americanleadership-in-artificial-intelligence.

Russell, S., and P. Norvig. 2021. *Artificial Intelligence: A Modern Approach*. 4th ed. Hoboken: Pearson.

Schwitzgebel, E., and M. Garza. 2015. A Defense of the Rights of Artificial Intelligences. *Midwest Studies in Philosophy* 39:98–119.

Sebastien Bubeck (Director). 2023, April 6. Sparks of AGI: Early experiments with GPT-4. https://www.youtube.com/watch?v=qbIk7-JPB2c.

Shanahan, M. 2023. Talking About Large Language Models (arXiv:2212.03551). *arXiv*.

Wachter, S., Mittelstadt, B., and Russell, C. 2020. Why Fairness Cannot Be Automated: Bridging the Gap Between EU Non-Discrimination Law and AI (SSRN Scholarly Paper No. 3547922). https://doi.org/10.2139/ssrn.3547922.

Chapter 9
An Algorithm in Doctor's Clothing: Anchoring Trust Appropriately in AI Healthcare Deployment

Emily LaRosa

Abstract It is now a near-given that AI systems will be used in a variety of healthcare scenarios. Yet, there is also an upswell of concern that such "black box medicine" may damage both patient needs as well as trust in care providers. Trust and trustworthiness have been stressed as important factors of justifiable deployment of AI when we consider these needs. Recently, there has been a tremendous amount of attention being paid to the notion of trust regarding technology, and in particular autonomous systems, with a focus on *trustworthiness* as a metric for justifiable deployment (Shneiderman, ACM Transactions on Interactive Intelligent Systems (TiiS) 10(4): 1–31, 2020; Kaur et al., ACM Computing Surveys (CSUR) 55(2): 1–38, 2022). In particular, the recent publication of the National Institute of Standards & Technology (NIST)'s Artificial Intelligence Risk Management Framework (AI RMF), as well as Newman's Taxonomy of Trustworthiness (Newman, A Taxonomy of Trustworthiness for Artificial Intelligence. CLTC: North Charleston, 2023), underscores this concern. The AI RMF is to serve as a "living document" robust enough to be used for a variety of case instances, and intended to help AI experts assess trustworthiness of systems at different phases of an AI's life cycle. This document establishes seven key characteristics of trustworthiness, without ever fully articulating what trust is. I hold we should take a step back from trustworthiness, and revisit trust itself. We ought to reflect on the trust involved when deploying Healthcare Artificial Intelligence systems.

Keywords Trust · Trustworthiness · Artificial intelligence · Autonomous systems · Medicine · Healthcare

E. LaRosa (✉)
Department of Philosophy, Michigan State University, East Lansing, MI, USA
e-mail: larosaem@msu.edu

J. Y. Tsou et al. (eds.), *Values, Pluralism, and Pragmatism: Themes from the Work of Matthew J. Brown*, Boston Studies in the Philosophy and History of Science 347, https://doi.org/10.1007/978-3-031-92958-8_9

9.1 Introduction

It is now a near-given that AI systems will be used in a variety of healthcare scenarios. Yet, there is also an upswell of concern that such "black box medicine" may damage both patient needs as well as trust in care providers. Trust and trustworthiness have been stressed as important factors of justifiable deployment of AI when we consider these needs. Recently, there has been a tremendous amount of attention being paid to the notion of trust regarding autonomous systems, with a focus on *trustworthiness* as a metric for justifiable deployment (Kiran and Verbeek 2010; Shneiderman 2020; Kaur et al. 2022). In particular, the recent publication of the National Institute of Standards & Technology (NIST)'s Artificial Intelligence Risk Management Framework (AI RMF), as well as Newman (2023)'s Taxonomy of Trustworthiness, underscores this concern. The AI RMF is to serve as a "living document" robust enough to be used for a variety of case instances, and intended to help AI experts assess trustworthiness of systems at different phases of an AI's life cycle. This document establishes seven key characteristics of trustworthiness, without ever fully articulating what trust is. I hold we should take a step back from trustworthiness, and revisit trust itself. We ought to reflect on the trust involved when deploying Healthcare Artificial Intelligence systems.

Where trust is the foundation of reliability and *mutual* expectation between agents (per Baier 1986), an AI's trustworthiness is (simply put) the *ability* of a system to be relied upon in such a manner, with qualities signaling this *ability to be trusted*. Importantly, as trust requires mutual expectation, and an AI is not an agent and cannot fulfill this role, some other form of trust grounding that is not rooted in the AI system must come into play if we want to use such systems in medical practice.

Bjerring and Busch (2021) raise a further concern: as advanced AI systems make their way into medical practice; it becomes more important than ever to think about the ways in which black-box medicine may conflict with the core ideals of patient-centered medicine. Patient-centered medicine, developed in the 1970s and becoming standard care practice in the 1990s, approaches medicine from the priorities of the patient rather than of their illness (the evidence-based approach) (see Bensing 2000). When we use AI and have black-box medicine, we have decisions guiding care which are opaque to both patient and care provider; as a result, "if a practitioner honors her epistemic obligation, she will align her medical verdict with that of the superior AI system, but in doing so, she will end up violating central tenets of patient-centered medicine" (Bjerring and Busch 2021, p. 352). I argue in this article that in order to counter such violations, we need to anchor trust appropriately via *extensional, agentially-grounded trust*. This form of trust requires an extensional anchor: a human expert or experts that can ground the trust extended over to these systems and intercede should they fail to uphold the values of the patient being treated. Agentially-grounded trust extends from the expert to the HCAI.

Such extensional trust is not unheard of in medicine. Indeed, we practice this currently in the way in which we train new experts of medicine. We trust residents, doctors with expertise, to be responsible of interns, those with less experience though still experts. We also trust attendings to be responsible for the actions of their residents, and so on. We trust that our residents will be able to assess that their interns are performing well and are acting in line with the values of their patients. While HCAI is in training, it will need such extensional trust; yet, it will also need a human agent to anchor said trust. As will be explored in Sect. 9.1, trust requires agency; these systems, as they are not true agents, cannot be trusted *alone*. Yet, they can be encompassed by trust we have in our human care providers, and justifiably so, in so long as there is the ability of a human to intercede on behalf of the patient to uphold their values as well as the continued ability to opt out of having a system used. These concepts and their importance are explored more deeply in Sect. 9.3.

Later iterations of healthcare AI may rise to a level of agency that allows it to be trusted on its own; they may someday have internal motivations and a robust capacity for respecting others' autonomy (both needed for full agency, as will be discussed in Sect. 9.1). However, at this stage of development, these systems need extensional, agentially-grounded trust. When such extensional, agentially-grounded trust manifests, and the trustworthiness limitations of the system *alone* are understood, we stand to have trustworthy deployment of healthcare AI (HCAI) as well as improve patient-centered care in our modern world.

9.2 Trust: Its Different Forms

9.2.1 Baier's Framing on Trust

The Baier framing of trust has been influential in discussions surrounding social trust. Baier's perspective emphasizes the crucial role trust plays in fostering human relationships and enabling individuals to exercise their agency effectively. Baier argues that trust is not simply a singular rational calculation or a one-off contractual agreement, but a fundamental continued element of our social existence which cannot be captured by rationality or contracts alone. Sufficient agency in trust relationships means we have the capability of shifting and aligning our own internal motivations, recognizing these motivations and acting upon them. Trust is an essential mechanism that allows us to navigate and engage with the world, enabling us to rely on others and establish cooperative relationships. According to Baier (1986), trust is built on a foundation of vulnerability and openness, as it involves placing ourselves at risk and depending on others' reliability and goodwill. Trust and mere reliance are not the same thing, and do not have the same desirability for a healthcare network that seeks to uphold the tenets of patient-centered medicine.[1] Trust

[1] Picker's *Eight Principles of Patient Centered Care* are the modern standard approach to patient-centered medicine. These principles are as follows: respect for patient's preferences; coordination and integration of care; information and education; physical comfort; emotional support; involvement of family and friends; community transition; and access to care. To Baker, these standards

requires a trustor to have an opted-in vulnerability to the trustee's likely actions; without this opt-in vulnerability, there is no trust. Rather, what we have instead is *reliance*. In other words, when someone does not have meaningful choices available to them—such as opting into or out of an AI-driven system or selecting an alternative mode of care—then what they experience is reliance rather than trust. Opt-in vulnerability is just that: opted into. It is a choice. When there is no true meaningful choice to the trustor, the relationship fails to meet Baier's criteria for trust. When we need the service of the agent we're interacting with, but don't have another viable alternative, we are forced into relying on that agent to have the desired outcome. In instances of reliance, we cannot exercise our agency and opt-in or opt-out of engagement in a meaningful way. Reliance, in this way, undermines agency, and is not the desired dynamic in patient-centered medicine.

For Baier, trust is intimately linked to agency. Trust is necessary for individuals to exercise their agency fully and provides the necessary psychological and social conditions for agency to flourish. When we trust others, we can rely on them to fulfill their commitments and respect our autonomy, thereby empowering us to act confidently and take risks. Baier also highlights the reciprocal nature of trust. Trust is not a one-sided phenomenon but operates in a web of mutual expectations and obligations. It involves a reciprocal process where trust is extended and earned, and its presence or absence affects the actions and choices of both individuals involved. This reciprocal nature reinforces the importance of trust in facilitating cooperative and collaborative relationships. Trust creates a foundation of reliability and mutual expectations, allowing individual agents to rely on others, and take risks confidently. Crucially, trust requires agency of both the trustor and trustee, which includes respect for autonomy. An autonomous HCAI system has neither full agency, nor an ability to respect a patient's autonomy in a self-motivated manner. To Baier, there is a reciprocal nature to all trust; yet, current forms of HCAI do not extend trust to patients as the trustor, nor does its absence affect a patient's decision-making processes. As we lack this bidirectionality given the limitations of current systems, trust is therefore the wrong way to characterize the relationship a patient may develop with current forms of HCAI. Yet, trust itself is desirable when we deploy these systems.

When discussing trust and HCAI deployment, we need to establish the parameters of who and what is being trusted, and what is needed to have trust germinate and take root.

---

boil down to "providing care that is respectful of, and responsive to, individual patient preferences, needs and values, and ensuring that patient values guide all clinical decisions" (Baker 2001).

9.2.2 Agential Trust in Focus: Behavioral Versus Role-Based Trust

Though we may seek a robust form of trust such as that which Baier conceptualizes in healthcare, it is also important to understand that a reciprocal nature can also exist in thinner concepts of trust which do not run afoul of our need to have reliability and our expectations met. In our interactions with others, we often have two distinct, but non-exclusive, trust types: behavioral trust, and understanding trust (see LaRosa and Danks 2018). In behavioral trust, we ground our trust in the reliability of the performance of a system or person. We can say that I trust that my coffee maker will turn on and make my pot of coffee at 6 am because it has done so every day for the past 2 years. This type of trust differs from reliability in that I do not need to use my coffee maker; indeed, I have a French press that will do the job of making coffee just fine. Yet, as I have a choice and continue to choose my coffee maker based on its performance history, I am trusting it to continue to act as it has thus far. This type of trust does not require us to know the mechanisms or values at play for the output to be justifiably reliable and the agent (or coffee pot) to be trusted.

Understanding trust, however, is grounded in the understanding of the values, desires, or motivations of another. When I take my best friend to a new restaurant, I can justifiably trust that she will order a fish dish because she believes in supporting the fishing industry. Understanding trust does not need predictive ability but does allow us to build a theory of mind surrounding the object or person in question. This differs from Baier's trust: while Baier acknowledges the interpersonal aspects of trust, her conceptualization revolves primarily around the expectation of goodwill, dependency and vulnerability, and reciprocity. In understanding trust, rather than having vulnerability or reciprocity take the forefront, we have trust based on a knowledge of the underlying ideals and motivations of another, adding a psychological dimension distinct from Baier's trust conceptualization.

Importantly, for HCAI, we exercise a particular form of role-based trust. Elsewhere, I have argued that role-based trust is not a third type of trust, but it works as a vehicle to establish one or the other kind—the nature of the particular role-based trust will be dependent on the role which grounds that trust (see LaRosa and Danks 2018). If a role is defined by reliable behavior, we will have behavioral trust. Consider our garbage collector, who reliably picks up our trash every Wednesday: her role is defined by this behavior of collecting the trash, without us possessing any insights into values, beliefs or intention. Contrastingly, if our role is defined by function or skill, then we usually have understanding trust, as the trustor can (unless countervailing reasons are provided or evident) assume that the trustee has relevant skills and ways of thought which define this role.

This differs from Baier's trust again—it is likely that Baier would term role-based trust, reliance. However, when we are opting into a relationship that is defined by its roles, as well as trust based on vulnerability and an expectation of care and goodwill (such as the dynamics present in the medical care system) there is distinct enough overlap so as not to be incongruous with Baier's account. The opt-in vulnerability is role-dependent; role-based trust works well with Baier's account of trust if the role inhabited engenders said vulnerability from the trustor to the trustee.

Doctors are a canonical example of understanding role-based trust. Their role as physician is defined by skills and knowledge, and we can trust that the doctor will have particular beliefs (such as beliefs on human physiology), values (patient health superseding personal preference), and intentions (acting to improve health to the best of their ability).

Trust, for healthcare network providers, should be understood based on both Baier's notions of vulnerability and respect for autonomy, as well as through the scope of role-based trust. Agential trust that supports autonomy within care scenarios, and upholds the doctor-patient collaborative relationship, is what we should strive for in healthcare HCAI. If we care about patient-centered care remaining the standard, and we want to preserve the integrity of the doctor-patient relationship paradigm healthcare networks have worked so hard to create, establishing appropriate trust between the patient and care provider becomes paramount. Modern healthcare works towards promoting both patient health, and patient autonomy; the current doctor-patient relationship works on having a partnership style attitude towards pursuing care. Agentially-grounded trust is the best way to preserve this partnered relationship when an HCAI becomes involved in a patient's care.[2]

9.2.3 Agentially-Grounded Trust and Its Effects

Agentially-grounded trust, as discussed previously, manifests itself between two agents with internally driven motivations that are actionable: people who can both have desires, and do something to cause them to come to pass. In the instance of healthcare, the desires center around treatment. Importantly, our current partnership dynamic between doctor and patient in modern healthcare focuses on the holistic needs of the patient, rather than paternalistically eliminating the health concern regardless of patient desires.

Agentially-grounded trust extends from the person filling the role of care provider, and it encompasses other actors they are responsible for within the care network of the patient; this includes HCAI. This form of trust is rooted in a pre-existing trust relationship; it cannot manifest alone. The extensional nature of agentially-grounded trust also does not capture mere tools or artifacts alone, which are subject to reliability concerns; it instead encompasses actors whose autonomous actions

[2] The need for agentially-grounded trust is unique to HCAI over static systems such as fMRIs or respirators, which inhabit behavioral trust and do not require role-based trust to be justifiably used. What is distinctive between these two system types (HCAI over static) is the active, autonomous component of HCAI, which lends itself to portending agency and allowing the system to be misunderstood as capable of agency itself. Such a misunderstanding must be protected against, as it can lead to potentially harmful outcomes to the patient which can also damage the doctor-patient relationship (as will be investigated in Sect. 9.3).

have direct impact on the care of the patient. In the case of the doctor-patient relationship, agentially-grounded trust stems from the role-based trust that the doctor engenders, and grows to extend both the agency and trusted performance of the doctor over to those involved in patient care provision. This means the trust in value-alignment between a patient and her doctor is also extended. Critically, the values of patient centered care, that the holistic needs of the patient will be addressed, are trusted by our patient to be met by all in the care nexus.

In encompassing the other actors in the care nexus, our patient has both a reliable source of trust, as well as an agent to refer to should their care provision go sideways. Though this form of trust is beneficial to both practitioner and patient, and appropriately grounds trust in instances where an HCAI is deployed, it can turn into a double-edged sword if the actors in the care nexus are not acting in accordance with patient needs and desires. A critical component of agentially-grounded trust is the preservation of the initial role-based trust dynamic: our doctor will fulfill their role in providing care, and maintain value alignment with our patient. This preservation must come from all agents when extensional, agentially-grounded trust is involved. HCAI poses a distinct challenge to the preservation of the initial trust, should its choices be accepted without critical evaluation. This we explore in Sect. 9.3.

9.3 Case Explorations on Trust in Healthcare Practice

When an AI is introduced, we risk upsetting the collaborative model between patient and care practitioner. Algorithms are already present in our healthcare networks (HCNs), assisting in diagnoses, making recommendations for care, and flagging patient records for review. Yet, sometimes these systems are erroneously trusted, to the detriment of both patient and provider. When the AI is seen as a full agent and treated as such in a patient's care provision, and is deployed without reflection by our care provider, it can lead to negative consequences on the doctor-patient trust dyad, and sometimes devastate our patients themselves. To flesh out how HCAI deployment can threaten our patient-centered care ideals and damage the trust dyad we desire, we should first take a look at a real-world example.

Consider Obermeyer et al. (2019): in this piece, the authors articulate the well-publicized case of a live, scaled healthcare algorithm used by a large Healthcare Network (HCN) providing care to over 70 million in the United States (Simonite 2019). The identified algorithm is just one of many commercially available risk-prediction tools that larger HCNs rely on to target patients for "high-risk management" programs, intended to provide additional, targeted resources and greater attention from trained providers (including wider resource allocation). As this is seen as a cornerstone in population health management efforts, healthcare systems and medical professionals have turned to algorithms to help identify patients for such high-risk management programs, and for care management recommendations.

In evaluating the system, Obermeyer et al. (2019) found that Black patients in this particular HCN were offered over 50% less care when compared to white counterparts. The healthcare network and associated insurance company in question deployed a care recommendation algorithm which assessed patients' health costs against the healthcare network's financial priorities, equating financial costs to care needs. In short, rather than making a prediction in health needs when recommending care plans, the algorithm made a prediction in total medical expenditures. The algorithm, in recommending a patient's long-term care plan, heavily weighted against specific long-term chronic illnesses which disproportionately affect Black communities (such as sickle cell anemia or mismanaged type II diabetes). The study further revealed that, in cases of the same illness, the system favored white patients to be flagged for additional resources over their Black counterparts. In doing so, the algorithm presented Black patients as healthier to care providers than they actually were when making care recommendations. This led to greater risk for Black patients of emergency room visits and longer hospital stays. Doctors within the healthcare network accepted the algorithm's care recommendations because they assumed the algorithm had a value alignment of patient care, leading to their own mismanagement of patient care. This case exemplifies when a system's recommendations, taken as an absolute authority by medical experts, can lead to negative outcomes against patients and damaging trust within the HCN. The core issue here is not simply that the algorithm itself was flawed through its prioritization of things irrelevant to the health of the patient, but that the procedures in place for checking the system's accuracy against human expert knowledge led to gross damage of an already historically vulnerable population. This case underscores that large-scale deployment of systems and acceptance of recommendations *without reflection* can lead to biased care outcomes. These outcomes, when harmful to patients, damage trust in the healthcare network itself.

More intimately, we can see an overextension of agency leading to harm of the autonomy of a patient and damaging their trust in their care provider. Let us consider a thought experiment for our second example. Imagine an autonomous anastomosis procedure bot (AAP) being used on a human patient, Melody. An anastomosis is when two blood vessels or channels of the body are connected; the fictional AAP mentioned can either create such connections, or sever them, depending on the requirement of the surgery and patient needs.

Melody has been shown the success rates of the system and its ability to outperform a surgeon's manual technique. She has been cursorily informed that the AAP may make executive decisions in line with prioritizing her health. She willingly agrees to the surgery, as her medical team endorses the AAP. However, what her team has failed to catch is that Melody is suffering from an ectopic pregnancy. The system has been programmed to prioritize the health of the patient, and noting the ectopic pregnancy, severs Melody's fallopian tube. This act falls under the notion of prioritizing patient health, but damages Melody's trust—this was not an action she condoned prior to surgery, and she would have opted for pharmaceutical termination. This is also unique from what the protocols of a human surgeon are: upon finding something wrong which is not life-threatening, and knowing Melody's values, a human surgeon would have ended the procedure and waited until Melody was alert to discuss care options.

Systems which are highly specialized come at a price: they are crafted to focus on particular data points and to exclude others. In the case of Melody's system, it does not take her desire to have a child into account when making surgical decisions. This value is something Melody prioritized over all else—though her medical team may have been aware of this, the system itself certainly was not. Melody's case is not intended to be a point of controversy, nor alienate the reader; rather, the goal is to represent an illness which affects over 10% of people with uteruses (endometriosis); and illustrate a case study in which we have a life-threatening medical emergency with resolutions which are either pharmaceutical in nature or surgically performed, and is laden with both physical and social value.[3]

Directed ignorance in systems is nothing new, and not necessarily a negative: indeed, attempts to craft a focused and robust algorithm require such choices. The algorithm does not and cannot (at this stage in machine learning) have comprehensive information on every data point about Melody and come to a conclusion which meets both her physical and emotional long-term goals. False or unforeseen correlations may be drawn by a system. While doctors attempt to make the best recommendations for a patient (those which prioritize them holistically), such value-laden recommendations may seem to be second-best to an HCAI. Granted, patient preferences do not equate to best medical practices. However, if we strive to have appropriate patient trust in our medical system, we must take them into account.

We need to strive for a continuation of collaborative care between a patient and her medical team. This case may seem, at first, to be that of medical malpractice: the patient trusted her doctors, who in turn overreached in their assumption of abilities of the AAP. Yet, there is something richer at play than simple malpractice. The system's success, as it is also an expert in the care nexus, becomes important to how the patient views her care providers: the human and the HCAI both become part of the broader HCN to Melody, and one's failure colors her understanding of the other's ability. Melody's is a single-deployment use case; she will likely not need to interact with this particular AAP again. However, its performance will have an impact on her relationship with her healthcare providers. In short, Melody may choose to walk away from her HCN entirely because of the AAP's error. We are often more forgiving of humans over systems when there has been a mistake made, regardless of the gravity of the error: people discriminate against machines much more readily than their human counterparts, with a lack of empathy for errors made (see Chen et al. 2021). The failure of the system will, invariably, color our patient's view of her doctor. If we are concerned with the welfare of our HCN and continued patient trust, we must consider more than the system in our assessment of trustworthy deployment. We need to anchor the trust surrounding an HCAI in something deeper.

[3] Many thanks to both Dr. Carla Fehr and Dr. Gabriele Contessa for their invaluable comments surrounding this thought experiment.

The need to trust the AAP's performance to be in alignment with her desired care outcomes in a way that does not violate her patient autonomy is imperative for Melody's continued trust in her healthcare providers. The system cannot violate her patient-centered care provision if we want to preserve this trust dyad.

Melody's case is not as fantastical as we might first think; given the recent success of porcine intestinal laparoscopic surgeries performed by autonomous surgical bots (see Saeidi et al. 2022), we may well see such bots in our near future. And, as we see through the Black patient biased algorithm example, HCNs are already deploying algorithms to guide care decisions; though this had a negative outcome, there are still algorithms which are marketed and used by HCNs around the world, performing such tasks as diagnostic recommendations and filtering MRIs for review. As they continue to be in use, we ought to consider what patient-centered care can (and should) look like when we deploy such systems.

9.4 Patient Centered Care and Trustworthy HCAI: Extensional, Agentially-Grounded Trust

Importantly, we as patients should want these systems to be used: they are proven to be effective in a myriad of ways, and often surpass their human counterparts in specific tasks. Following Melody's AAP, a real-world example of a system surpassing their human counterparts is the Smart Tissue Autonomous Robot (STAR) (Shademan et al. 2016). This autonomous robot, deployed to conduct a series of soft-tissue surgeries on the porcine intestine, outperformed human surgeons in dexterity and outcome (ibid). STAR is unique in that, rather than being either dependent upon human control (think of the DaVinci surgical device) or partially autonomous (such as the CyberKnife deployed by Accuray), this system operates entirely autonomously via an HCAI, and has the same black-box issues as systems mentioned in Sect. 9.2.

These systems may be desirable in scenarios where we have no alternative care structure in place (such as warzones or remote communities); yet, these deployment scenarios are akin to reliance, as any alternative is better than none. As such, assessing their value in such deployment scenarios goes beyond the scope of this discussion of trust. What we are most interested in when these systems do have alternatives is, how can we deploy them and have an appropriate amount of trust when they are still learning? I propose we build off of role-based trust, and have trust in these scenarios to be seen as *extensional, agentially-grounded* trust. Where non-extensional trust is grounded directly in the actions of the trustor and only has to do with their distinctive choices and actions (again, consider the case of my trusting my best friend to order seafood), extensional, agentially-grounded trust encompasses more than just the trustor. It captures the other agents who fall under their purview, or tools that they are themselves responsible for.

Agentially-grounded trust is extended from the human expert, and it encompasses the HCAI. This form of trust is parallel to the HCAI's human counterpart learners in medicine: just as we do not allow residents to practice medicine without attendings to assist and oversee their training, we also expect residents to conduct themselves as doctors and perform medicine autonomously. To build off of this parallel, we also trust the attending will be able to assess if the resident is doing their job *well*, and performing medicine *appropriately and in line with their patient's values*. As such, while an HCAI is still in training, to have a justifiably trustworthy deployment, we need a human expert to have the *ability* to intercede on the social values of the patient, and to act as interlocutor and imbue the patient with trust which encompasses the system itself. We also need to ensure that patients have (while HCAI are still becoming the 'technology de arte') the ability to opt-out of their use in direct care provision (such as Melody's case), should it be against their values. Having extensional trust, grounded in a human care provider who is an expert and respects the need for choice and autonomy, allows patients to justifiably participate in the deployment of an HCAI as its subject.

In having extensional, agentially-grounded trust, we can establish systems which exemplify the National Institute of Standards and Technology's AI Risk Management Framework's concept of trustworthiness: the human anchor is not, importantly, a human in the loop. They do not aid or second-guess, but rather review and assess based on *value alignments of the patient.* Following Topol (2019), though it may be possible to achieve fully autonomous robotic surgical procedures, such a system may well continue to be undesirable should that HCAI not allow for robust forms of human input once a procedure commences. In extensional trust dynamics with an HCAI, though the system may be capable of performing a wide-range of processes and procedures largely unaided, humans may be present if complications unfold, and to supervise and assist should the need arise. Rather than overriding the system, the human expert and the patient use the system; in instances of novel cases, they are presented with the system's recommendation, thereby reinforcing the agency of the patient needed for trust to manifest.

The purpose of how modern healthcare functions is to care for the patient with a patient-centered focus. If we involve an HCAI in that care, we need extensional, agentially-grounded trust to ensure the patient and their needs remain at the center of that care. This means moving away from the concerns of how a particular system can and will operate in alignment with our values, and focusing on how care providers can act responsibly when these systems are in place. Coming to a consensus of what adequate agency is in a system is beyond us at this time; instead, we must work with the real-time limitations of the systems we *have*. In accepting the limitations of our already existing systems, and moving forward by anchoring the system in agentially-grounded trust, we allow ourselves to have trustworthy systems and to protect the values of patient-centered care at the same time. Through agentially-grounded trust, any errors in care are the responsibility of the care provider, not the HCAI. This allows for further trust to take place via an understanding of vulnerability and consequence on the part of the trustee, our patient.

9.5 Benefits of Agentially-Grounded Trust in the Healthcare AI Context

There are many benefits of extensional, agentially-grounded trust to our patient and our healthcare network.

1. *Extensional, agentially-grounded trust restructures the role of the AI from that of an arbiter of care, which erodes the epistemic knowledge of a practitioner due to black boxing, to that of an expert.*

   As AI is used, practitioners will have an epistemic obligation to align the outputs of the AI with their own medical verdicts, but it respects the role and relationship of the practitioner as a healthcare provider, and a lesser expert. As Bjerring and Busch articulate:

   > If a general practitioner is unsure about how to interpret, say, a MR image, it seems perfectly reasonable to say that he should adopt the opinion that an expert in MRI interpretation would form after inspecting the image. In part he should do so because the expert is more knowledgeable about the medical domain, more reliable in her diagnostic work, and better at making accurate predictions based on the image. But these platitudinous claims also seem true when we focus on the relationship between practitioner or expert and AI system. For on the assumption that the AI system outperforms the expert on all parameters on which the expert outperforms the general practitioner, the AI system will play the expert role in the relationship. (Bjerring and Busch 2021, p. 354)

Were the system seen as a fellow expert (albeit a very different one) in the Obermeyer et al. case, we may have seen a more thoughtful assessment of the recommendations by the human providers who were told to accept these recommendations. In restructuring the relationship between the provider and the HCAI to that of provider and expert due to the trust being anchored in the provider, allowing them to embrace their epistemic obligation to align their medical verdicts with those of the HCAI, we may see more critical assessments of recommendations made by systems training in vivo that benefit all.

2. *Allow trustworthy deployment of a system, in line with the NIST RMF framework*

   The most notable of recent attempts to capture the important dynamics of trustworthiness include the National Institute of Standards & Technology (NIST) 2023 publication of the Artificial Intelligence Risk Management Framework (AI RMF) as well as Newman (2023)'s Taxonomy of Trustworthiness. In particular, the RMF is to serve as a "living document" that is use-case agnostic, intended to help AI organizations and teams assess trustworthiness of systems at different phases of an AI's life cycle. This document establishes seven key characteristics of trustworthiness:

   (a) Valid and reliable
   (b) Safe
   (c) Secure and resilient
   (d) Accountable and transparent
   (e) Explainable and interpretable

(f) Privacy-enhanced
(g) Fair, with harmful biases managed

The goal of the RMF is to verify and ensure outputs of a system are as expected, and not intentionally deceptive, and to have a functional system with the achieved objective function. NIST AI RMF, having previously set the industry standard in Cybersecurity Framework internationally, may well have the same impact in AI and significantly alter industry approach to both AI deployment and its characteristics *a priori*. In leveraging extensional trust, healthcare providers stand to embody these standards when deploying HCAI. The NIST AI RMF standards also dovetail with many of the tenets of patient-centered care, important for our current care standards. Notably, the NIST AI RMF does not address the need for a standard of the ability for human intervention throughout an AI's deployment lifecycle—though this may be a desirable standard, it goes beyond the scope of this paper to address how to accomplish such a standard and situate it in the appropriate framework.

3. *Open up avenues for patient-centered care tenets to continue to manifest themselves*

   Extensional trust, as an anchored trust approach, is a contemporary assistive measure to our HCAI. As AI grows and learns, it may raise to a sufficient level of agency to be trustworthy on its own. However, at this stage of deployment, and in order to ensure in vivo deployment that is necessary for its training, we want to have appropriately grounded trust that provides us with respect for agency and autonomy. While an HCAI is still in training, to have a justifiably trustworthy deployment, we need our human expert to be able to intercede and act on behalf of the patient's values. In doing so, our care provider (or human expert) will ensure that patient-centered care remains the care standard, and that we do not see care devolve to the paternalistic care standards of the 1970s due to HCAI deployment.
4. *Ensure an opt-out mechanism that values patient choice and prioritizes ethical standards of care*

   Patients ought to have an overt mechanism to be able to decline AI-driven interactions when it conflicts with their personal values. The care provider (human expert) here plays a critical role in articulating and facilitating informed decision making, explaining AI-generated recommendations, and discussing alternatives. It is patient education which will likely be crucial in playing an underlying role in if HCAI are accepted and adopted—the development of an appropriate opt-out mechanism that respects the autonomy of patients as well as the expertise of medical professionals is an area for future research.

9.6 Conclusion

As patient-centered care remains the standard of modern medicine, care providers must consider how to best preserve this in the face of HCAI deployment. When care providers deploy HCAI, it becomes critical to reframe how we view trust between providers and their patients as these systems operate autonomously. One way will be to anchor the necessary trust for care appropriately, within the care provider. Trust surrounding HCAI deployment should be extensional, agentially-grounded in the human expert, and deployment should respect the autonomy of the patient and their needs. In anchoring HCAI via extensional, agentially-grounded trust, we protect the collaborative care dynamic and preserve the existing doctor-patient trust relationship. We change the perception and use of the system from that of an arbiter on medical decisions, to a fellow collaborative medical expert in patient care decisions. Lastly, anchoring HCAI use in agentially-grounded trust allows for truly trustworthy deployment of such systems.

In articulating the trust as extensional and agentially-grounded, we prevent mere reliance, and keep medicine from devolving to the paternal care standards healthcare providers have worked so hard to move away from. We also allow care providers to see the systems for what they truly can be: fellow experts, assisting in the care of their patients, who need that extra boost of human value alignment only they can provide.

Acknowledgements Many thanks to my multiple reviewers and commenter for their invaluable feedback, including Dr. Carla Fehr, Dr. Jamie Shaw, Dr. Jonathan Tsou, Dr. Kevin Elliott, and Dr. Samantha Noll.

References

AI, NIST. 2023. *Artificial Intelligence Risk Management Framework* (AI RMF 1.0).

Baier, A. 1986. Trust and Antitrust. *Ethics* 96 (2): 231–260.

Baker, A. 2001. Crossing the Quality Chasm: A New Health System for the 21st Century. *BMJ* 323 (7322): 1192.

Bensing, J. 2000. Bridging the Gap: The Separate Worlds of Evidence-Based Medicine and Patient-Centered Medicine. *Patient Education and Counseling* 39 (1): 17–25.

Bjerring, J.C., Busch, J. 2021. Artificial Intelligence and Patient-Centered Decision-Making. *Philos. Technol. 34*, 349–371.

Chen, N., S. Mohanty, J. Jiao, and X. Fan. 2021. To Err is Human: Tolerate Humans Instead of Machines in Service Failure. *Journal of Retailing and Consumer Services* 59:102363.

Kaur, D., S. Uslu, K. Rittichier, and A. Durresi. 2022. Trustworthy Artificial Intelligence: A Review. *ACM Computing Surveys (CSUR)* 55 (2): 1–38.

Kiran, A., and P.-P. Verbeek. 2010. Trusting Our Selves to Technology. *Knowledge, Technology & Policy* 23 (3): 409–427.

LaRosa, E., and D. Danks. 2018. Impacts on Trust of Healthcare AI. In *AIES '18: Proceedings of the 2018 AAAI/ACM Conference on AI, Ethics, and Society*, 210–215. New York: Association for Computational Machinery.

Newman, J. 2023. *A Taxonomy of Trustworthiness for Artificial Intelligence*. North Charleston: CLTC.

Obermeyer, Z., B. Powers, C. Vogeli, and S. Mullainathan. 2019. Dissecting Racial Bias in an Algorithm Used to Manage the Health of Populations. *Science* 366 (6464): 447–453.

Saeidi H., Opfermann J. D., Kam M., Wei S., Leonard S., Hsieh M. H., Kang J. U., Krieger A. (2022). Autonomous robotic laparoscopic surgery for intestinal anastomosis. *Science Robotics* 7(62). https://doi.org/10.1126/scirobotics.abj2908

Shademan, A., Decker R. S., Opfermann J. D., Leonard S., Krieger A., Kim P. C. (2016) Supervised autonomous robotic soft tissue surgery Hands-free. *Science Translational Medicine* 8(337). https://doi.org/10.1126/scitranslmed.aad9398

Shneiderman, B. 2020. Bridging the Gap Between Ethics and Practice: Guidelines for Reliable, Safe, and Trustworthy Human-Centered AI Systems. *ACM Transactions on Interactive Intelligent Systems (TiiS)* 10 (4): 1–31.

Simonite, T. 2019, October 24. A Health Care Algorithm Offered Less Care to Black Patients. *Wired Magazine*.

Topol, E. 2019. High-Performance Medicine: The Convergence of Human and Artificial Intelligence. *Nature Medicine* 25 (1): 44–56.

Chapter 10
Hermeneutical Pluralism in Psychiatry: Lessons from *Spectrum 10K*

Bennett Knox

Abstract In this chapter I propose and support a view about the relationship between psychiatric science and activist movements such as the Neurodiversity Movement, which I call "hermeneutical pluralism." Using the moral imagination framework from *Science and Moral Imagination* (Brown, Science and Moral Imagination: A New Ideal for Values in Science, University of Pittsburgh Press, Pittsburgh, 2020), I explore a contemporary case study in psychiatric science: the controversy over the *Spectrum 10K* genetic study on Autism. Through the application of this framework to a case involving significant interplay between psychiatric science and the Neurodiversity Movement, I draw out some lessons about how these knowledge systems might relate to each other in a productive manner. Ultimately, I argue that this case illustrates the attractiveness of my hermeneutical pluralism view: the Neurodiversity Movement (and other movements like it) ought to be maintained as knowledge systems that are independent of psychiatric science, even as these movements and psychiatric science ought to foster some form of engagement with one another.

Keywords Autism · *Spectrum 10K* genetic study · Moral imagination · Hermeneutic pluralism · Neurodiversity · Science and activism · Philosophy of psychiatry

10.1 Introduction

In this chapter, I propose and support a view about the relationship between psychiatric science and activist movements such as the Neurodiversity Movement, which I call "hermeneutical pluralism." Using the moral imagination framework from

B. Knox (✉)
Department of Philosophy, University of Utah, Salt Lake City, UT, USA
e-mail: bennett.knox@utah.edu

J. Y. Tsou et al. (eds.), *Values, Pluralism, and Pragmatism: Themes from the Work of Matthew J. Brown*, Boston Studies in the Philosophy and History of Science 347, https://doi.org/10.1007/978-3-031-92958-8_10

Science and Moral Imagination (Brown 2020), I explore a contemporary case study in psychiatric science: the controversy over the *Spectrum 10K* genetic study on Autism. Through the application of this framework to a case involving significant interplay between psychiatric science and the Neurodiversity Movement, I draw out some lessons about how these knowledge systems might relate to each other in a productive manner. Ultimately, I argue that this case illustrates the attractiveness of my hermeneutical pluralism view: the Neurodiversity Movement (and other movements like it) ought to be maintained as knowledge systems that are independent of psychiatric science, even as these movements and psychiatric science ought to foster some form of engagement with one another.

10.2 Hermeneutical Pluralism and the Neurodiversity Movement

In previous work (Knox 2022a), I have called for the inclusion of the community I term the "psychopathologized" in the processes of psychiatric science, specifically in the revision process of the *Diagnostic and Statistical Manual of Mental Disorders* (DSM). I argued that meaningful inclusion of this population is crucial in order for that process to achieve what has been termed "social objectivity" (Gagné-Julien 2021a, b, c).[1] Further, I insisted that this must include input from those among the psychopathologized who understand themselves using concepts quite different from those of psychiatric science—for instance, the Neurodiversity Movement, the Mad Pride Movement, and the Psychiatric Survivor Movement.[2]

However, the details of the right relationship between these movements and psychiatric science have yet to be fully worked out. Owing to commentaries on my work (Aftab 2022; Bueter 2022), I came to endorse a form of pluralism about this relationship, which I have termed "hermeneutical pluralism." This view states that we "ought to maintain… a plurality of interpretations of the phenomena that psychiatry deals with, some of which are importantly extra-scientific and should not be reduced to psychiatric understandings" (Knox 2022b, p. 277). In this chapter, I focus primarily on the relationship between two different hermeneutical frameworks as they come to bear on issues related to Autism: the framework of psychiatric science and that of the Neurodiversity Movement.

Some elucidation of the Neurodiversity Paradigm and Movement is in order. It should be noted that there is a great diversity of views within the Neurodiversity

---

[1] The "social objectivity" view draws on the work of Helen Longino (Longino 1990, 2002), assessing the objectivity of a scientific process by how well it conforms to Longino's criteria which are meant to promote transformative intersubjective criticism.

[2] Though this piece was focused on the specific case of the *DSM* revision process, I think the points about inclusion being necessary for social objectivity apply to psychiatric science in general.

Movement, and some of these disagree with one another quite strongly.[3] Nevertheless, it will be helpful for my purposes to cover in broad strokes some of the key tenets of the Neurodiversity Paradigm which are most relevant here.

Nick Walker—an Autistic academic who has been writing about Neurodiversity for many years—defines the "Neurodiversity Paradigm" in contrast to a "Pathology Paradigm," which assumes that "There is one 'right,' 'normal,' or 'healthy' way for human brains and human minds to be configured and to function… [and] if your neurological configuration and functioning… diverge substantially from the dominant standard of 'normal,' there is Something Wrong With You" (Walker 2021, p. 18). By contrast, the Neurodiversity Paradigm is based on the principles that "Neurodiversity—the diversity among minds—is a natural, healthy, and valuable form of human diversity… there is no 'normal' or 'right' style of human mind… [and] the social dynamics that manifest in regard to neurodiversity are similar to the social dynamics that manifest in regard to other forms of human diversity" (p. 19).

Though the Neurodiversity Movement first emerged from discourse surrounding Autism, many would now group various other forms of neuropsychological difference under the banner of "neurodivergences": forms of neurocognitive functioning that "diverge significantly from the dominant societal standards of 'normal'" (p. 38). Many of these differences in neurocognitive functioning are understood under the Neurodiversity Paradigm as natural forms of diversity and not inherently pathological. There is some debate about the boundaries of this concept, however, and Walker herself does not reject pathologizing all forms of neurodivergence (p. 39). But there is broad consensus within those operating under the Neurodiversity Paradigm that Autism specifically is not to be understood as inherently pathological. Thus, from a Neurodiversity Paradigm perspective, it makes no sense to speak of "cure" in the context of Autism. Since Autism is seen as an essential identity of an individual, to "cure" them of it would just be to destroy that person and replace them with someone different. Similarly, Neurodiversity advocates strongly oppose efforts at "prevention" of Autism—viewing these as harmful forms of eugenics.[4]

Although the view that no neurodivergences can rightly count as disorders is commonly attributed to the Neurodiversity Paradigm by its critics, the view does not necessarily imply eliminationism about the concept of neurodisorder. Autistic philosopher and Neurodiversity theorist Robert Chapman points to the case of

[3] For example, Autistic people working under a broad Neurodiversity framework may disagree about the concept of "functioning labels"—whether they ought to be used, replaced, or done away with all together (Keates et al. 2024).

[4] It is not my intention in this chapter to adjudicate claims for or against eugenics of various kinds (though I do tend to fall fairly strongly into the "against" camp). Similarly, I am not making any claims here about either what public policies should or shouldn't be enacted in order to prevent eugenics, nor about the moral status of reproductive choices made by individuals. Rather, I am exploring what responsibilities *scientists* might have to consider potential uses of their work for eugenic purposes, in cases (like that of *Spectrum 10K*) where scientists explicitly state that eugenic efforts aimed at "curing" or "preventing" Autism are a harmful outcome to be avoided.

anencephaly[5] as one which obviously rightly counts as a disorder, and it asserts that if any Neurodiversity proponents deny that it is, "then this would be a good reason for not taking their specific understanding of the neurodiversity paradigm seriously" (Chapman 2019, p. 377). However, philosopher and bioethicist Ryan Nelson points out that if the Neurodiversity view does not wish to dispense with the concept of neurodisorder entirely, "then the view needs to come up with a way of drawing this distinction that is not subject to its own criticisms" (Nelson 2021, p. 336). Chapman has provided a social-ecological conception of mental function to begin answering this type of concern (Chapman 2021). Though Chapman admits that this concept cannot be used to clearly divide disorder from mere difference in all cases, they assert that it represents an improvement on other accounts of disorder and dysfunction who likewise cannot uncontroversially make this division (Chapman 2021, p. 1368). Rather than attempting to dispense with the concept of neurodisorder entirely, the Neurodiversity Paradigm rejects the Pathology Paradigm's assumption "that human value should be judged against some kind of species-standard norm, rather than accept that there are different ways of being human, each with their own norms, and potential values and ideals" (Chapman 2019, p. 382).

In the specific case of Autism, Neurodiversity proponents insist that Autism itself is not a disorder, even if they count some conditions that are more prevalent in Autistic people (e.g., epilepsy) as genuine disorders. However, many Neurodiversity proponents do conceive of Autism as a disability. This distinction is meant to highlight the fact that Autistic people do face certain hardships that others do not (and sometimes require access to medical resources to help address these hardships). However, Neurodiversity advocates would also highlight that many (if not all) of the difficulties faced by Autistic people are a result of hostile and unaccommodating social circumstances. There is some disagreement about the exact status of Autism in relation to concepts of disability,[6] but Neurodiversity proponents are united in rejecting the idea that Autism is a disorder to be understood in terms of pathology.

My contention is that rather than attempting to merge the Pathology and Neurodiversity Paradigms, or have one triumph over the other, we ought to aim for (versions of) these paradigms to exist alongside one another as distinct yet interacting hermeneutical frameworks. Perhaps the Pathology Paradigm can be changed in response to arguments from the Neurodiversity Paradigm—this might look like psychiatric/medical science abandoning the idea that Autism is a disorder to be cured

[5] Anencephaly is a disorder where one is born without a prefrontal cortex and/or other parts of the brain. Though it seems obvious that anencephaly is a disorder, it is less clear that it counts as a neurodivergence. This is because in this case there is arguably no mind at all, so debates about whether the mind is disordered or not seem misplaced. Likewise, attempts at "curing" anencephaly would seem to amount to attempts to bring a new mind into existence, not cure a previously existing disordered mind. None of this undermines the point that many Neurodiversity theorists admit that there are some genuine neurodisorders, however.

[6] For instance, Walker makes a distinction between the idea that "autism is a disability" (which she denies) and that "autistic people are disabled" (which she affirms) (Walker 2021, p. 68). This is meant to track the idea that Autistic people are disabled due to social failures of accommodation, rather than inherent pathology.

or prevented. Nevertheless, a medicalizing framework might be maintained since Autistic people can benefit from access to certain kinds of medical resources—perhaps this can be understood in terms of disability. The medical and Neurodiversity approaches would still be different interpretations of Autism, since one interprets it as a medical phenomenon while the other interprets it as a social phenomenon. And these different interpretations lead to different methods (or at least different priorities between methods) for improving Autistic people's well-being.

Further, the productive interactions between these frameworks require that they exist at least somewhat independently of one another. The Neurodiversity Movement can be productively informed by the best available science, while psychiatric science can be improved by science-oriented activism coming from the Neurodiversity Movement (Arnaud 2023; Arnaud and Gagné-Julien 2023). But medical science should maintain some independence from the Neurodiversity Movement, since there are considerations from other groups in society that it must be responsive to. And the Neurodiversity Movement should maintain some independence from medical science, or it risks being co-opted and robbed of its radical potential.

My vision is one of multiple, partially overlapping yet partly independent knowledge systems. Psychiatric science must engage with and substantively respond to criticisms from the Neurodiversity Movement, and Neurodiversity understandings must be informed by the best available empirical evidence. Yet there is value in maintaining a Neurodiversity knowledge system that is not fully assimilated into psychiatric science. My analysis of *Spectrum 10K* using the moral imagination framework will highlight some of the reasons why I endorse this view.

10.3 The *Spectrum 10K* Controversy

Launched in August of 2021, *Spectrum 10K* promises to be the largest genetic study on Autism in the history of the United Kingdom. The study intends to collect around 10,000 DNA samples from Autistic individuals, along with survey data and medical record data from these same participants. These will be used to conduct a Genome Wide Association Study (GWAS),[7] with the goal of increasing understanding of the genetic basis of Autism and associated phenomena. In their own words, the ultimate aim of *Spectrum 10K* is "to better understand how genetic and environmental factors affect the wellbeing of autistic individuals, including their physical and mental health" (Spectrum 10K 2021). It is also key to note that *the Spectrum 10K* team explicitly endorses a Neurodiversity view with regard to Autism, and states that they have no intent to contribute to attempts to "cure" or "prevent" Autism.[8]

[7] GWAS is a common type of genetic study, which aims to correlate single-nucleotide polymorphism (SNPs)—sites where the nucleotides differ between individuals—with a trait of interest.

[8] Many activists question the degree to which *Spectrum 10K* really is committed to these things, rather than simply paying lip-service to them (e.g., Gray-Hammond 2023). Regardless, I will take

Shortly after *Spectrum 10K* was announced, there was a significant outcry against the study, largely spearheaded by Autistic individuals—both within the academy and without—who understand Autism using a Neurodiversity framework. This led to a "Boycott *Spectrum 10K*" movement, and calls for the study to be shut down and/or subjected to greater ethical scrutiny. A group of Autistic academics associated with "Boycott *Spectrum 10K*" also released a statement which detailed the objections to the study (Aucademy 2021). These included: problems with the informed consent process for the study, especially as it pertains to minors and individuals with intellectual disabilities; concerns about the investigators leading the study[9]; issues related to the data privacy and data security plans; and, perhaps most significantly, the potential for data from the study to be used for the purpose of preventing Autistic individuals from being born (via embryo selection or selective abortion).[10]

Following this outcry, *Spectrum 10K* was paused in order to conduct a consultation with the Autistic community and their allies, with the goal of improving the study to address some of the concerns raised. As of this writing, the consultation process has concluded, and the *Spectrum 10K* team (along with the co-leads of the consultation: the independent contractor Hopkins Van Mil and Autistic co-lead Leneh Buckle) is in the process of compiling a report on the consultation, and assembling a co-production team which will translate feedback gained from the consultation into recommendations for changing the study (Spectrum 10K 2023).

Hopkins Van Mil and Leneh Buckle have released a report on the process of co-designing the consultation itself, which gives some insight into how the process was conducted (2022). This process—which involved eliciting feedback from a broad range of Autistic and non-Autistic individuals about how the consultation should be conducted—led to the articulation of four broad principles for the consultation:

1. Build trust in the process
2. Make real and positive change to the *Spectrum 10K* study
3. Show that the consultation involves, and has been designed by, autistic people

---

them at their word that they do view eugenics as a harm to be avoided, in order to explore what the relationship between these two groups should look like if they do agree on this.

[9] The principal investigator for *Spectrum 10K* is Simon Baron-Cohen, who is quite a polarizing figure in the Autism community. His history of research on Autism has been criticized by activists (e.g., Boon 2022); and other scholars have questioned his most influential theories, including: that Autistic people have deficits in theory of mind (Gernsbacher and Yergeau 2019), that Autistic people have less empathy than non-Autists (Milton 2012; Milton et al. 2022), and that Autistic people have an "extreme male brain" (Ridley 2019). Nevertheless, he and his theories are still quite influential in the Autism research community.

[10] One thing that GWAS studies are particularly good at is generating Polygenic Scores (PGSs)—often referred to as "Polygenic Risk Scores (PRSs). Essentially, data from the GWAS can be used to produce a model which can analyze a new genome not present in the original study, and produce a score which is supposed to predict how likely the person with that genome is to display the trait of interest. The concern is that *Spectrum 10K* could be used to produce a test that could yield a PGS for Autism, which could be used to screen embryos with a high score.

4. Create a space for constructive discussion where people can hear, learn from and respond to a range of views and experiences (Hopkins Van Mil and Buckle 2022, p. 10)

However admirable this attempt to garner input may be, *Spectrum 10K* has also been clear that there are certain aspects of the study which are not open to being changed—since the study has received grant funding which obligates it to do certain things.[11] The study will attempt to collect around 10,000 DNA samples, along with questionnaires, and use these for a GWAS; it will enroll only Autistic people with a formal medical diagnosis; and it will publish its results, and share data with other researchers—though which other researchers have access to the data will be governed by a "Data Access Committee" (Hopkins Van Mil 2023a). The *Spectrum 10K* team has not said specifically who will make up this Data Access Committee, though they have committed to a majority of its members being Autistic people.

Although the final report has yet to be released,[12] given the nature of the consultation and the specified limitations, we can make some reasonable speculations about what the results of the consultation will be. It is likely that the consultation process will result in some substantive changes to *Spectrum 10K*. My impression is that areas which are most likely to see some significant change include the informed consent process, the data sharing and data privacy plans, and (potentially) the articulation of the specific aims of the study. Given that these were some of the central concerns raised by the Boycott *Spectrum 10K* movement, to the extent that these changes are made, they are encouraging.

However, given what *Spectrum 10K* has said about the limitations of the consultation, there is one central concern of objectors to the study which will go largely unaddressed: the potential for the results of the study to be used for eugenic purposes. Heini Natri, an Autistic geneticist who is currently on the *Spectrum 10K* co-production team (and who has been vocally critical of aspects of the study) puts the point quite starkly:

> [The *Spectrum 10K* investigators] claim to oppose eugenics but fail to acknowledge that once the autism-associated GWAS variants are published, anyone can use the information for prenatal testing and embryo screening. The project also promises to safeguard data and to only release data to collaborators who receive ethical approval but does not acknowledge that this does not prevent anyone from using the published GWAS results for screening. It is likely that eventually autism risk scores will be used for embryo selection, no matter how inaccurate the underlying models may be. (Natri 2021)

---

[11] Of course, it is not clear that the *Spectrum 10K* team actually could not change any of these things, this is just what they have *said* they cannot change.

[12] On January 30th, 2025—significantly after this chapter was written and accepted for publication—*Spectrum 10K* announced the cancellation of the study. Their reasoning was that "there are now other very large health and genetic databases in the UK and internationally that have become available as a resource for autism health research. This means that we no longer need to collect new DNA samples as we can use existing health and genetic databases to better understand the health challenges faced by autistic people" (Autism Research Centre 2025). They have also released the final report from the consultation (Hopkins Van Mil and Buckle 2023).

There are two related but distinct points being made here by Natri. First, the Data Access Committee proposed by *Spectrum 10K* is manifestly insufficient to prevent the results of the study being used for eugenic purposes. This is because one would not need access to the full data of the study in order to use it to develop a potential genetic test for Autism. All that is needed is the results of the GWAS, which would almost certainly be published in any paper produced by *Spectrum 10K*.

Second, Natri is answering a response that *Spectrum 10K* investigators themselves have given to concerns about possible prenatal genetic testing for Autism. In an interview with *Nature* about the pausing of the study, Simon Baron-Cohen rebuffed concerns about possible development of a genetic test to be used for eugenic purposes by pointing out: "[y]ou could never diagnose autism prenatally, and that's because, even if we knew the biology, diagnosis rests on behaviour. That's only possible to observe postnatally" (Sanderson 2021, pp. 17–18). Baron-Cohen is correct here, but Natri points out that this is largely a *non sequitur*. It is true that a fully accurate genetic test for Autism is likely a conceptual impossibility, since Autism is certainly not 100% genetic. But this does not change the fact that those who wish to prevent births of Autistic people could (and likely would) use an inaccurate genetic model to screen out embryos with a high polygenic score for Autism. And this would still amount to eugenics of precisely the sort that Neurodiversity advocates most fear, albeit with limits on its effectiveness.

10.4 The Moral Imagination Framework

Matt Brown's moral imagination framework is an approach to science which aims to embrace the role that values play in the scientific process, in a way that can improve scientific inquiry. Given that there are many different and competing values coming to bear on the *Spectrum 10K* controversy, I believe that this framework can help us to understand this case and provide insight into how psychiatric research might proceed in a productive and ethical manner. In this section I will present the aspects of Brown's account that are most relevant to my case, and then apply the moral imagination framework to the case of *Spectrum 10K*.

According to Brown's account, scientists face what he calls "contingencies" throughout the scientific process. These contingencies are unforced choices, where "there are multiple options, and no factor decisively settles the matter from the perspective of the scientific inquirer at the moment the choice is made" (Brown 2020, pp. 57–58). Brown argues that these contingencies are ubiquitous throughout the scientific process, such as when choosing which questions to pursue, which hypotheses to test, which methods to utilize, how to gather and interpret data, and how to weigh inductive risk when deciding whether or not to endorse a hypothesis. Further, he argues that often when scientists are faced with these contingencies, they can and should make value judgments to settle the contingencies. He calls this the "contingency argument," which essentially says:

> The responsible thing to do is to carefully consider our decision and make our best judgment about the best course of action—this is a value judgment. It is here that the need for value judgments in the scientific process arises. Value judgments play a crucial role in settling contingencies in science, where the upshot of how the contingencies are settled is significant. (Brown 2020, p. 60)

Rather than shy away from making value judgments in the scientific process (which in practice will just mean making implicit value judgments anyway), scientists ought to aim to make better value judgments when faced with contingencies.

How can scientists improve the value judgments that they employ in resolving contingencies? Brown proposes that they do this by aiming for the "ideal of moral imagination," which he defines as follows:

> Scientists should recognize contingencies in their work as unforced choices, discover morally and epistemically salient aspects of the situation they are deciding, empathetically recognize and understand the legitimate stakeholders and their interests, imaginatively construct and explore possible options, and exercise fair and warranted value judgment in order to guide those decisions. (p. 186)

The idea is that by going through this process, scientists will expand their moral imaginations and be able to make better value judgments when faced with contingencies in their work. Along with giving us an ideal for value-laden science, the moral imagination framework also gives us a new way to conceptualize some ways in which science can go wrong. That is, it allows us to think of some bad outcomes in science as the result of failures of moral imagination. These occur when scientists fail to exercise their moral imaginations to the degree that they should when dealing with contingencies. For instance, they may fail to consider options that are available to them, or they may fail to respond to the values of those who have a stake in the research. In the following section, I will examine how each of the considerations in the moral imagination framework (goal/task, options/alternatives, stakeholders, and values) apply to the *Spectrum 10K* case, and then argue that Brown's account helps us make sense of some of the failures of *Spectrum 10K* in a way that a standard research ethics framework cannot.

10.5 Applying the Framework to *Spectrum 10K*

Both the investigators and the activist critics of *Spectrum 10K* seem to agree that the relevant ultimate goal in this case is to conduct genetics research on Autism in an ethical and productive manner. In their statement, Aucademy puts this front and center: "We fundamentally recognise the need for good, robust genetic and biological research in order to aid greater understanding of the impact of conditions that have genetic roots and can have life-limiting effects or impact on the well-being of Autistic people" (Aucademy 2021, p. 2). Though perhaps some objectors to *Spectrum 10K* may question whether GWAS has much to offer (or whether its benefits could outweigh potential dangers), virtually no one in this debate is arguing for a blanket ban on genetics research on Autism. Similarly on the side of the investigators, the willingness of the team to engage in this extensive consultation process indicates that they do want their research to be attentive to the ethical concerns of the Autistic community (at least to some degree). There may be great disagreement about what would constitute ethical genetics research on Autism, but there is some consensus that this is the primary goal.

As far as the options/alternatives available to pursue this goal, evidently the *Spectrum 10K* team has chosen GWAS as their primary methodology. However, there are other methods of conducting relevant genetics research on Autism that are available to researchers. For instance, the *Spectrum 10K* team themselves express a desire to "…identify genetic variants that are more or less frequently seen in autistic people and investigate how these are linked to co-occurring health conditions" (Hopkins Van Mil 2023b, p. 6). They cite epilepsy and gastrointestinal disorders as examples of co-occurring conditions. However, it is unclear how a broad GWAS on Autism itself would contribute to this kind of research. Instead, one might conduct molecular genetics research in an attempt to uncover the mechanisms of genes already implicated in these conditions in Autistic people (*Spectrum 10K* even cites the example of the CHD8 gene which is implicated in gut issues in Autistic people). Or one might conduct a candidate-gene study or a hypothesis-driven GWAS on these co-occurring conditions, in an attempt to uncover and increase understanding of genetic factors of these conditions (which would not necessarily require conducting a GWAS on Autism itself). The broader point here is that there are methodological options available to genetics researchers other than the approach that *Spectrum 10K* has chosen to take—there is genuine contingency here.

Identifying the relevant stakeholders (and identifying and weighing their values and interests against one another) is evidently of central importance in the case of *Spectrum 10K*. Quite obviously, Autistic people themselves are primary stakeholders in this research. Further, though *Spectrum 10K* has said they will only enroll those with a formal diagnosis in the study itself, they have stated that the consultation "…should operate with an inclusive definition of autism—including those who are awaiting diagnosis or who are self-diagnosed, for instance—rather than for it to distinguish between those with and those without a formal diagnosis" (Hopkins Van Mil and Buckle 2022, p. 16). This indicates that the *Spectrum 10K* team regards Autistic people without a formal diagnosis as genuine stakeholders in their research. Other stakeholders include families, friends, and caregivers of Autistic people; Autism researchers themselves; and those agencies and individuals who fund the research.[13] There is also an extended sense of stakeholder in which all members of society would count as stakeholders in this research, since the benefits or harms that it produces will have impacts that ripple through the whole of society—albeit not always directly.

---

[13] We might also count individuals who are *potential* friends, family, and caregivers of Autistic people as stakeholders—though this category may be so expansive as to extend to virtually all members of society.

Having identified the stakeholders, we must now consider the various values[14] at play in *Spectrum 10K* and genetics research on Autism more broadly. One primary value held by virtually all stakeholders is that of greater scientific understanding of Autism and genetics. Further, all parties seem to place high value on how this understanding may be used to improve the well-being of Autistic people. Since the *Spectrum 10K* team themselves state explicitly that they endorse a Neurodiversity framework—and do not wish to "cure" or "prevent" Autism—the relevant stakeholders all place central value on avoiding the use of this research for eugenic purposes. This broad agreement about central values would seem to make the conflict between *Spectrum 10K* and its critics rather mysterious. However, Brown points out that even when different parties can agree on what the relevant values are, they may still interpret these values in vastly different ways, and they may weigh these values against one another in different ways. One way of reading this conflict is that different parties have different understandings of what constitutes valuable scientific understanding of Autism, and of what constitutes adequate avoidance of eugenics. Or they may also disagree about how to weigh these various values against one another, and thus come to a disagreement about how research should proceed.[15]

There are some aspects of the controversy that can be well accounted for within a standard research ethics framework. For instance, issues related to informed consent, data sharing and security, and the risks and benefits individuals can expect from participation in the study are all standard considerations taken into account when ethically evaluating research. We might think that in this case that ethical evaluation failed to be sufficiently stringent, but that is a problem we can explain without appealing to a new conceptual framework. So what does applying the moral imagination framework to *Spectrum 10K* get us? Well, some of the central problems arising in this controversy cannot be productively captured by a standard research ethics framework, and are better understood as failures of moral imagination.

One of the major ways that scientists can fall short in their use of moral imagination on Brown's account is to fail to see contingencies they face as genuine unforced choices. I suggest that this was part of what went wrong in the *Spectrum 10K* case. Researchers failed to seriously consider all of the possible methodologies for conducting genetics research on Autism, instead zeroing in on a methodology that has garnered a great amount of attention and funding in the past several years: GWAS. Since the researchers (by their own admission) did not incorporate stakeholders' viewpoints (such as those of Autistic people) from the start of their

[14] Brown presents quite a rich and expansive notion of values and value judgments, the details of which I cannot do justice to here. For my present purposes, the values I am concerned with are things that are held to be of primary value by stakeholders in the research, and which are relevant to how that research is conducted.

[15] In perhaps a more cynical vein, there is also another implicit value at play in the *Spectrum 10K* project, which is not shared by all stakeholders. That is the value of professional incentives that are tied to conducting this research. GWAS studies have attracted much attention and funding in the past several years, and there is a continuing push for larger and more diverse sets of participants (Mills and Rahal 2019). Given this attention, it serves the professional interests of Autism researchers to pursue this particular methodology, which could also be a source of conflict.

consideration of which research methods to pursue, they could not respond to stakeholders' values in the way that the moral imagination framework calls for. As a result, they failed to fully consider the broader harms that could result from a GWAS of the kind they were proposing—including the potential for (perhaps inadvertently) creating a tool that could be used for eugenic purposes. These are issues that would not come up in a standard ethical review of a proposed study.[16]

Various explanations for what in particular went wrong in this case all lead to the conclusion that this was a failure of moral imagination. One might imagine that the scientists of *Spectrum 10K* simply did not understand the ways in which their research might be used for eugenic purposes, but Baron-Cohen himself has been aware of this potential issue for over a decade (Baron-Cohen 2009). And if the team had not understood these dangers, that would be a glaring indication that they did not properly consider stakeholder perspectives and values—since this has been a concern for many Autistic advocates for many years. Baron-Cohen's abovementioned statements to *Nature* regarding the impossibility of an accurate prenatal test may seem to indicate that perhaps he simply estimated the likelihood of *Spectrum 10K* data being used for eugenic purposes as very low. But given Natri's clear and convincing response to that line of argument, such an underestimation would also represent a failure to engage fully with stakeholder concerns. Assuming the *Spectrum 10K* team was aware of the potential use of their data for eugenics, as well as the likelihood that if available it would be put to those uses, we might think that the team simply judged that this risk was outweighed by the knowledge that could be gained through conducting their GWAS. But the failure of *Spectrum 10K* to articulate specific ideas for how a GWAS on Autism could be used to generate knowledge relevant to improving Autistic well-being makes that judgment difficult to defend. And given the outpouring of calls from the Autistic community to boycott or cancel the study because of these concerns, such a judgment would also seem to fly directly in the face of the values of key stakeholders in the research. Each of these potential explanations suggests that the *Spectrum 10K* researchers at some point did not "empathetically recognize and understand the legitimate stakeholders and their interests," which is a key requirement of the ideal of moral imagination (Brown 2020, p. 60).

10.6 Conclusions and Lessons

My analysis of the *Spectrum 10K* controversy suggests certain lessons that researchers might take from this incident, if they truly wish to conduct ethical genetics research on Autism (and ethical psychiatric research more broadly). It is worth noting firstly that if the consultation process that *Spectrum 10K* has conducted does

[16] A possible exception is that in the UK "Patient and Public Involvement (PPI)" has been somewhat incorporated into many research contexts. However, in the *Spectrum 10K* case, there was no PPI at the stage of choosing research methodologies.

lead to substantive changes in the study, then this consultation process may be able to serve as a good model for gathering stakeholder input in psychiatric research in general.[17] The *Spectrum 10K* team does seem to have attempted to make the consultation process at least partly co-designed and governed by independent entities (i.e., Hopkins Van Mil and Leneh Buckle). They seem to have genuinely made an effort to include as many different perspectives of Autistic people as possible—for instance, by allowing input via a diversity of different modalities, so as to allow Autistic people with differing communication styles to participate. And they seem to genuinely be open to the study design changing (at least in some ways) as a result of the consultation. In line with Brown's framework, consultations such as this could be seen as a method for scientists to expand the range of their moral imaginations, consider new options and alternatives, and respond to the values of stakeholders in their research.

However, the impact of the consultation was also critically undermined by the lack of stakeholder inclusion earlier on in the process, when methodologies themselves were being considered. If research is truly to respond to the priorities of Autistic people themselves (Roche et al. 2021), then their perspectives must not be limited to providing feedback after a study has already been funded and approved. Instead, Autistic people must be incorporated into the decisions about what kinds of research to pursue in the first place—for instance by being given a role in choosing which studies on Autism are funded. There is also a case to be made for making active efforts to promote the inclusion of Autistic and Neurodiversity-affirming scientists in genetics and related fields. For instance, Natri's vocal critiques of *Spectrum 10K* as well as her work on the co-production team have certainly brought considerations to the surface that researchers might not have encountered otherwise. And there seems to be a great deal of enthusiasm among Autistic people to be involved in guiding and conducting research in the ways I have recommended (e.g., Pukki et al. 2022).

A recent publication from "a workgroup of autistic and non-autistic geneticists, bioethicists, autism researchers, clinicians, and advocates" (of which Natri is a member), has presented a framework for conducting ethical genetics and genomics research on Autism (Natri et al. 2023, p. 2). The workgroup calls for "transparent and inclusive practices… to be applied at every stage of research, from study design and prioritization to clinical translation" (p. 3). In concrete terms this calls for increased co-production and community-based participatory research practices by researchers; but also institutional interventions such as targeted funding and

[17] I do not mean to suggest that the consultation can only serve as a good model if it achieves a certain substantive outcome. Brown's account does not demand any specific substantive outcome in order for the ideal of moral imagination to be met. Rather, I mean to say that if *Spectrum 10K* does make some substantive changes, then this would be a sign that it has conformed to the requirements that scientists genuinely respond to the concerns and values of stakeholders, as opposed to merely paying lip service to them. Changes in line with stakeholder values would therefore be a sign that procedural norms have been met, even if the outcome ends up not being exactly what stakeholders would like. Thank you to an anonymous reviewer for pointing out this distinction.

publishing, as well as federal, local, and institutional oversight of research practices. At both individual and institutional levels, the inclusion of Autistic voices in shaping research priorities and practices is essential. They call for greater attention to the potential for indirect group harms when subjecting studies to ethical scrutiny, and demand that genetics and genomics researchers be forthcoming and transparent about the possible harmful (including eugenic) uses that their research may be put to, even if they do not intend these harms themselves (p. 4). And—in contrast to the vagueness of the aims articulated by *Spectrum 10K*—they "recommend that researchers ground their work on solid hypotheses for why and how studying differences at the genetic and molecular level could promote autistics' well-being. They should clearly communicate their hypotheses to the autistic community and other AIPs and demonstrate… that their proposed methods to test those hypotheses are appropriate to their research objective" (p. 4). I wholeheartedly concur with all of these recommendations.

But I also think there is another important point to be made here about what is needed for ethical genetics research on Autism to proceed. Namely, I believe my analysis of this case using Brown's framework speaks in favor of the hermeneutical pluralism view that I began this chapter with. That is, I believe that the *Spectrum 10K* case shows why it is necessary to maintain a Neurodiversity knowledge-system that is (at least partly) independent of scientific (and, more generally, academic) institutions. For one, the *Spectrum 10K* consultation itself simply would not have been conducted if not for the existence of a robust Neurodiversity activist culture that was able to both publicly pressure researchers and articulate clear criticisms in a manner that could be legible to scientists. This required both insider-outsiders (such as Natri) who could understand and respond to the details of a scientific research study design, as well as activists who were well-versed in the Neurodiversity paradigm but were not beholden to scientific and academic institutions (and thus could voice their unvarnished objections without worrying about career consequences).

Further, recall that the *Spectrum 10K* team had themselves endorsed the Neurodiversity framework, and explicitly denied that they would allow their research to be used for eugenic purposes. Nevertheless, the research team was not forthcoming about the possibilities that their work could be put to use for eugenic purposes even given their intentions, and this point was not recognized until Neurodiversity advocates began to scrutinize the study. This supports a key aspect of my view: even if psychiatric science were to explicitly adopt a broadly Neurodiversity-affirming framework (which is far from the case at the moment), there would still be an important role for a Neurodiversity knowledge-system that is not beholden to psychiatric science and medicine more broadly. Such a knowledge-system serves as a critical check on a system that otherwise might adopt the Neurodiversity framework on paper but fail to actually change its practices in the ways that a thoroughly Neurodiversity-affirming approach demands.

At the same time, I would also still hold that psychiatric science and medicine ought to be themselves somewhat independent of activist movements like the Neurodiversity movement. This is not to say that they should be unresponsive to concerns that come from these communities, but fully abandoning a scientific and medicalized perspective on various neurodivergences such as Autism would be too quick as well. I would also argue that it is possible to medicalize certain conditions (in the sense that we treat them as things which can merit medical oversight, intervention, and support), without necessarily pathologizing them (which would imply that there is something "going wrong" in these individuals).[18] It is also worth noting that some Autistic people do understand themselves as having a pathology rather than a Neurodivergence. Though I somewhat bristle at the idea, I would not want to deny those people the autonomy to understand themselves in the way they believe best suits them—which requires maintaining a role for scientific and medical institutions.

The vision I have for the future of genetics research on Autism is one where there is a plurality of independent but connected knowledge-systems, interpretive frameworks, and research communities all striving to improve well-being for Autistic people in their own distinctive ways. To move towards this vision, scientists must substantively engage with the Autistic community and the Neurodiversity Movement, so that research can be made to align with their values and interests. I suggest that building on the successes and learning from the failures of *Spectrum 10K* and its consultation may be a step towards making this vision a reality. If we can build a healthy ecosystem of mutually-informing frameworks and communities throughout psychiatric inquiry, this will help all inquirers involved expand their moral imaginations, and lead to more ethical and effective research in psychiatry.

References

Aftab, A. 2022. Can the Psychopathologized Speak? Notes on Social Objectivity and Psychiatric Science. *Philosophy, Psychiatry, & Psychology* 29 (4): 267–270.

Arnaud, S. 2023. First-Person Perspectives and Scientific Inquiry of Autism: Towards an Integrative Approach. *Synthese* 202 (5): 147.

Arnaud, S., and A.-M. Gagné-Julien. 2023. The New Self-Advocacy Activism in Psychiatry: Toward a Scientific Turn. *Philosophical Psychology* 37:1–24.

Aucademy. 2021. Joint Statement from Autistic Community Members, Academics, Representatives of Advocacy Organisations, and Allies About Grave Concerns Regarding the Spectrum 10K Study Announced Tuesday 24 August 2021. https://form.jotform.com/aucademy/boycott-s10k.

Autism Research Centre. 2025. *Spectrum 10K*. Autism Research Centre, University of Cambridge. https://www.autismresearchcentre.com/projects/spectrum-10k/. Accessed 30 January 2025.

Baron-Cohen, S. 2009. Autism Test 'Could Hit Maths Skills.' *BBC News*. http://news.bbc.co.uk/2/hi/health/7736196.stm. Accessed 1 July 2009.

Boon, S. 2022. A Brief Overview of Simon Baron-Cohen's Autism Research. *Autistically Sarah* (Blog). https://autisticallysarah.com/2022/02/13/a-brief-overview-of-simon-baron-cohens-autism-research/. Accessed 13 February 2022.

---

[18] I have tried to make this point in previous work by drawing a distinction between "psychiatric conditions" and "mental disorders," though I am not committed to this particular terminology (Knox 2023).

Brown, M. 2020. *Science and Moral Imagination: A New Ideal for Values in Science*. Pittsburgh: University of Pittsburgh Press.

Bueter, A. 2022. On the Limits of Diversity. *Philosophy, Psychiatry, & Psychology* 29 (4): 271–273.

Chapman, R. 2019. Neurodiversity Theory and its Discontents: Autism, Schizophrenia, and the Social Model of Disability. In *The Bloomsbury Companion to Philosophy of Psychiatry*, ed. Ş. Tekin and R. Bluhm, 371–389. London: Bloomsbury Academic.

Chapman, R. 2021. Neurodiversity and the Social Ecology of Mental Functions. *Perspectives on Psychological Science* 16 (6): 1360–1372.

Gagné-Julien, A.-M. 2021a. Towards a Socially Constructed and Objective Concept of Mental Disorder. *Synthese* 198:9401–9426.

Gagné-Julien, A-M. 2021b. Defending Social Objectivity for 'Mental Disorder.' *Philosophy, Psychiatry, & Psychology* 28 (4): 381–384.

Gagné-Julien, A.-M. 2021c. Dysfunction and the Definition of Mental Disorder in the DSM. *Philosophy, Psychiatry, & Psychology* 28 (4): 353–370.

Gernsbacher, M., and M. Yergeau. 2019. Empirical Failures of the Claim that Autistic People Lack a Theory of Mind. *Archives of Scientific Psychology* 7 (1): 102–118.

Gray-Hammond, D. 2023. Spectrum 10K Researchers Have Given up on Hiding Their Eugenics Agenda. *Emergent Divergence* (Blog). https://emergentdivergence.com/2023/01/22/spectrum-10k-researchers-have-given-up-on-hiding-their-eugenics-agenda-2/. Accessed 22 January 2023.

Hopkins Van Mil. 2023a. Limits of the Spectrum 10K Consultation. https://www.hopkinsvanmil.co.uk/s/Limits-of-the-consultation-FINAL.pdf.

Hopkins Van Mil. 2023b. Spectrum 10K Themes and Responses to Questions. https://static1.squarespace.com/static/56f16de77da24f3e5612733b/t/6405b0df447b784f17f22c22/1678094563078/Formatted+full+Q%26A + document+HVM + FINAL.pdf.

Hopkins Van Mil, and K. Buckle. 2022. Spectrum 10K Consultation Phase 2: The Co-Design. https://static1.squarespace.com/static/56f16de77da24f3e5612733b/t/639cab3adb946720f4703368/1671211860800/Spectrum+10K+Phase+2+Report+Final+.pdf.

Hopkins Van Mil, and K. Buckle. 2023. Spectrum 10K Consultation Phase 3: The Findings. https://www.autismresearchcentre.com/content/uploads/2025/01/Spectrum10K_consultation_report.pdf.

Keates, N., F. Martin, and K. Waldock. 2024, March. Autistic People's Perspectives on Functioning Labels and Associated Reasons, and Community Connectedness. *Journal of Autism and Developmental Disorders*.

Knox, B. 2022a. Exclusion of the Psychopathologized and Hermeneutical Ignorance Threaten Objectivity. *Philosophy, Psychiatry, & Psychology* 29 (4): 253–266.

Knox, B. 2022b. Standards and Assumptions, the Limits of Inclusion, and Pluralism in Psychiatry. *Philosophy, Psychiatry, & Psychology* 29 (4): 275–277.

Knox, B. 2023. The Institutional Definition of Psychiatric Condition and the Role of Well-Being in Psychiatry. *Philosophy of Science* 90 (5): 1194–1203.

Longino, H. 1990. *Science as Social Knowledge: Values and Objectivity in Scientific Inquiry*. Princeton: Princeton University Press.

Longino, H. 2002. *The Fate of Knowledge*. Princeton: Princeton University Press.

Mills, M., and C. Rahal. 2019. A Scientometric Review of Genome-Wide Association Studies. *Communications Biology* 2 (1): 1–11.

Milton, D. 2012. On the Ontological Status of Autism: The 'Double Empathy Problem'. *Disability & Society* 27 (6): 883–887.

Milton, D., E. Gurbuz, and B. López. 2022. The 'Double Empathy Problem': Ten Years On. *Autism* 26 (8): 1901–1903.

Natri, H. 2021. Spectrum 10K and the Questionable Past, Present, and Future of Genetic Autism Research. *ResearchGate*.

Natri, H., C. Chapman, S. Heraty, P. Dwyer, N. Walker, S. Kapp, H. Dron, J. Martinez-Agosto, L. Mikkola, and M. Doherty. 2023. Ethical Challenges in Autism Genomics: Recommendations for Researchers. *European Journal of Medical Genetics* 66 (9): 104810.

Nelson, R. 2021. A Critique of the Neurodiversity View. *Journal of Applied Philosophy* 38 (2): 335–347.

Pukki, H., J. Bettin, A. Outlaw, J. Hennessy, K. Brook, M. Dekker, M. Doherty, et al. 2022. Autistic Perspectives on the Future of Clinical Autism Research. *Autism in Adulthood* 4 (2): 93–101.

Ridley, R. 2019. Some Difficulties Behind the Concept of the 'Extreme Male Brain' in Autism Research. A Theoretical Review. *Research in Autism Spectrum Disorders* 57 (2019): 19–27.

Roche, L., D. Adams, and M. Clark. 2021. Research Priorities of the Autism Community: A Systematic Review of Key Stakeholder Perspectives. *Autism* 25 (2): 336–348.

Sanderson, K. 2021. High-Profile Autism Genetics Project Paused amid Backlash. *Nature* 598 (7879): 17–18.

Spectrum 10K. 2021. Spectrum 10K—The Largest Study of Autism in the UK. https://spectrum10k.org/.

Spectrum 10K. 2023. Phase 3 Consultation Update—2 June 2023. https://spectrum10k.org/phase-3-consultation-update/.

Walker, N. 2021. *Neuroqueer Heresies: Notes on the Neurodiversity Paradigm, Autistic Empowerment, and Postnormal Possibilities*, Part IV: Moral Imagination. Fort Worth: Autonomous Press.

Part IV
Moral Imagination

Chapter 11
Matt Brown at the Funeral of the Value-Free Ideal

Janet A. Kourany

Abstract After more than four centuries of distinguished service to science, the ideal of value-free science finally passed on. Its funeral, though, far from the dignified affair proper for such occasions, was marred by controversy. On the one hand, there were the expected outpourings of grief and loss—from the public ("science is just not trustworthy anymore") and the underrepresented in the scientific community ("more worried about sexist and racist science than ever before"), and the science studies community ("now confronted with an unbelievably complex problem of demarcating legitimate from illegitimate science"). But, on the other hand, there were the sharp critiques of those outpourings along with a spirited defense of Matt Brown's suggested replacement ideal of science, the ideal of moral imagination—all offered by Matt Brown himself. It was an unseemly kind of behavior for a funeral, but it worked! The ideal of moral imagination ended up gaining a hearing and ultimately winning approval, though not without the help of yet another old ideal of science still with us.

Keywords Matthew Brown · The ideal of moral imagination · The ideal of value-free science · Values in science

11.1 Introduction

After more than four centuries of distinguished service to science, the ideal of value-free science has finally passed away. At least, that is what we have been told by the science studies community as well as many others knowledgeable about the situation. As a result, we need to seek out a replacement ideal. Fortunately, Matt Brown has put forward an attractive candidate in his *Science and Moral Imagination:*

J. A. Kourany (✉)
Department of Philosophy, University of Notre Dame, Notre Dame, IN, USA
e-mail: jkourany@nd.edu

J. Y. Tsou et al. (eds.), *Values, Pluralism, and Pragmatism: Themes from the Work of Matthew J. Brown*, Boston Studies in the Philosophy and History of Science 347, https://doi.org/10.1007/978-3-031-92958-8_11

A New Ideal for Values in Science (2020) and has detailed the many strengths the new ideal has to offer. So, our task now is to determine whether this new ideal can, so to speak, fill the shoes of the old one. The problem is that Matt has not spent much time explaining just what that shoe-filling would involve. In order to complete our task, then, it might be helpful—and in keeping with the celebration of imagination the occasion calls for—to recall some of the debate that broke out during the funeral of the old ideal. That might give us a clearer sense of the loss we have sustained with the passing of the old ideal and the gain we might expect with the adoption of Matt's new one.

11.2 The Eulogies Presented at the Funeral of the Value-Free Ideal

Start, then, with the grieving words of the funeral's three eulogists.

11.2.1 The Eulogy of the Public

The public, the eulogist that spoke first, bewailed the fact that science, at least as far as the public is concerned, is far less trustworthy now than it was in the past (as documentation see, for example, Pew Research Center 2023). And the reason? "Certainly, the misinformation and disinformation currently surrounding science is at least partly to blame for the current lack of trust," said the public. "But so is the *legitimate* information currently surrounding science. Think, for example, of all the books now widely available that expose the problems of industry science—such as nutrition scientist Marion Nestle's *Unsavory Truth: How Food Companies Skew the Science of What We Eat* (2018), or McKay Jenkins' *What's Gotten into Us: Staying Healthy in a Toxic World* (2011) about the shortcomings in the chemical industry's science. Or think of all the reports exposing the problems of government science (such as the Union of Concerned Scientists' 2017/2024 14-page compilation of "Attacks on Science"), or the science funded by private foundations (see, for example, Tucker (2007) for an exposé of the racist science funded by the Pioneer Fund). All this legitimate information about science—about science's current failings—has certainly contributed to the public's decreased trust in science. But notice that all these reported failings of science have been caused by values—by, more particularly, the commercial and political and other values that are free to contaminate science now that the ideal of value-free science is no longer on the scene. Even academic science is now being corrupted by values—by the values of cutting corners and hyping results, of bypassing replications and shortchanging peer reviews, in order to achieve ever more publications and citations and grants and other sorts of career perks rather than the depth and rigor sought in the past. And the consequences? Arizona State University's Consortium for Science, Policy, and Outcomes co-director Daniel Sarewitz lays them out well:

> Scientists are more productive than ever, pouring out millions of articles in thousands of journals covering an ever-expanding array of fields and phenomena. But much of this supposed knowledge is turning out to be contestable, unreliable, unusable, or flat-out wrong. From metastatic cancer to climate change to growth economics to dietary standards, science that is supposed to yield clarity and solutions is in many instances leading instead to contradiction, controversy, and confusion. Along the way it is also undermining the four-hundred-year-old idea that wise human action can be built on a foundation of independently verifiable truths. (Sarewitz 2016, pp. 5–6)

In short, the demise of the ideal of value-free science has left us no longer able to rely on science in any substantial way. And that is a great loss."

Such was the public's eulogy at the value-free ideal's funeral.

11.2.2 The Eulogy of the Underrepresented in the Scientific Community

The second eulogy immediately followed, offered by the underrepresented in the scientific community. Here women scientists took the lead, pointing out how very helpful to them the ideal of value-free science had always been when they tried to respond to sexism and androcentrism and racism and the other biases against them in the various areas of research in which they worked. "Any theory or method, any concept or assumption, shown to be contaminated in any way with such values was immediately rendered unacceptable by the ideal of value-free science. This ideal made it possible for us to be a part of science, as welcome in science and as valuable a contributor as anyone else, not diminished by the messages science puts forward. What assurance do we have that another ideal can ever provide the same wonderful protection from prejudice?" At this point, overcome with emotion, the women scientists abruptly ended their part of the underrepresented scientists' eulogy, and the others of the group, in a moving display of sensitivity and solidarity, set about to comfort the women rather than add any further remarks of their own.

11.2.3 The Eulogy of the Science Studies Community

So, attention turned to the contribution of the science studies community, the third eulogist at the value-free ideal's funeral. Their remarks pursued a quite different tack:

> "It has been a challenge—it has always been a challenge, even when the ideal of value-free science was still with us—to distinguish in a principled way between good science and bad science and between genuine science and pseudo-science. Indeed, this was the infamous 'demarcation problem' that for centuries resisted solution by even the most gifted thinkers

> (see for some of this history, Laudan 1983). The passing of the value-free ideal, however, has unleashed a 'new demarcation problem' that many now consider an even more challenging problem: how to distinguish between legitimate and illegitimate value influences in science (see Holman and Wilholt 2022). After all, if values may now legitimately influence science, precisely which values—or whose values—should be the ones to do this influencing? Should it be the values of the scientists doing the research (it is their research, after all)? But why should the personal values of these individuals be allowed to shape the science that shapes society? Haven't we heard enough about the sexism and racism and other limitations of these values to make us leery of such a suggestion? Then should the values that legitimately shape science be the values of the citizens of the democratic societies that support the science? But these citizens all too often fail to agree on value issues, and when they do agree (for example, on the acceptability of doing very little at present to avert global warming) their agreement does not always seem to constitute a defensible result. Then should the values that legitimately shape science be the values of the governments of those democratic societies—but those governments come and go, and the values they privilege come and go as well—or the values of some other groups—but which other groups?—or the values underwritten by some ethical or political theory—but which of the many existing theories? The death of the value-free ideal has left us in a horrible predicament."

Thus ended the eulogy of the science studies community.

11.3 Matt Brown's Response to the Eulogies

It was at this point of the value-free ideal's funeral that Matt, no longer able to contain his dismay over what was being said, inserted himself into the proceedings. Perhaps he was trying to offer words of consolation and encouragement to the mourners after the eulogies made the present scene appear so bleak, or perhaps he was just trying to prepare the way for his own replacement ideal of science (albeit not at the best time). Whatever the reason, Matt's message was clear: the death of the value-free ideal was nowhere near as great a loss as the eulogists had suggested. Directing his remarks, especially, first to the public and the underrepresented in the scientific community, Matt granted that the effect of the value-free ideal was to shield science, and thereby society, from the influence of sexist and racist and other socially harmful values. But he emphasized that the effect was also, and equally, to shield science and society from the influence of anti-sexist and anti-racist—that is, gender- and race-egalitarian—values as well as other socially *helpful* values. "This certainly puts the ideal of value-free science in a different light, does it not?" Matt enthused.

"Note that this fact has not been noticed because the ideal of value-free science presupposed that all values are biases and therefore needed to be excluded from science. But this makes unintelligible your applause for the role of the ideal in supporting the just treatment of the underrepresented in science as well as the ideal's role in supporting other socially beneficial scientific results. What was never acknowledged but needed to be was that some values are justified and can even be necessary and helpful epistemically as well as socially to science as well as society."

All that was directed to the first two eulogists, the public and the underrepresented in the scientific community. To the third eulogist, on the other hand—the science studies community—Matt said that no struggle over values need be unleashed by the death of the value-free ideal. "After all, science is a very practical affair constituted by responses to particular problem situations. The different sources of values mentioned—the individual scientists doing research, the citizens of the democratic societies that support and are affected by the research, the governments of those societies, the different theories, and so on—any of those sources of values could be relevant (for example, in some cases as stakeholders) to different problem situations. The need is to consider which values should shape research on a case-by-case basis."

Needless to say, none of this went over particularly well. All three eulogists, in fact, were taken aback and nearly speechless. But they managed to say that the ideal of value-free science, even with the possible shortcomings mentioned, still encouraged scientists to tirelessly watch for potentially damaging influences on their research and try to screen them out. And this remained a great service to science and all that could be expected. "Not so," Matt shot back with perhaps a little too much enthusiasm for the occasion. "Actually, the ideal of moral imagination would have achieved far more and rendered a much greater service to science had it been the accepted ideal. For one thing, it would have encouraged scientists to appreciate the full complexity and possibilities of the research scene—how it is filled with contingencies, points where alternative actions are possible, where more than one of those alternative actions would be a reasonable choice, and where competing value judgments inform the different choices. As a result, it would have encouraged scientists not only to watch for potentially damaging influences on their research and try to screen them out, but also to watch for potentially beneficial influences on their research, social as well as epistemic, and try to screen them in. All this takes imagination, moral imagination, as well as traditionally epistemic modes of analysis."

11.4 Conclusion

The eulogists were silent now. They seemed quite satisfied with what Matt had said. But should they have been satisfied? Consider, for example, the concerns expressed by the first eulogist regarding the public's decreasing trust in science and its connection with the corrupting values now shaping that science. The eulogist's original stand, remember, was that the ideal of value-free science could have effectively dealt with the situation had it not passed away. And our question is, can Matt's proposed replacement of that ideal, his ideal of moral imagination, do at least an equally adequate job. Make the question more concrete. Consider, in particular, the processed food industry science described in Pulitzer Prize winning New York Times investigative reporter Michael Moss's *Salt Sugar Fat: How the Food Giants Hooked Us* (2013) as well as his more recent *Hooked: Food, Free Will, and How the Food Giants Exploit Our Addictions* (2021). This science, according to Moss, has massively contributed to the growing epidemic of obesity, diabetes, heart disease, and other health problems in the United States and many other parts of the world. How has it done this? By encouraging people to consume foods filled with sugar, salt, and fat—indeed, by rendering it nearly impossible that people fail to consume these foods. Processed food industry scientists have produced this unfortunate result by their impressive research accomplishments—for example, by specially engineering sugar molecules to produce the optimal "bliss point" at just the right time of ingestion; by specially engineering fat molecules to produce the optimal "mouth feel" in conjunction with the sugar; by manipulating consumers' taste sensations so that they expect and demand high levels of salt in their diets; by exploiting consumers' evolutionary vulnerabilities to addiction and the emotions associated with food; by marketing food products in the empirically-based most attractive and compelling ways— located in just the right aisles in food stores with just the right words, just the right colors, for just the right prices, and so on. Some of the scientists who did this research eventually regretted it and tried to change the research agendas of their companies that they themselves had helped to create. But their change of heart came too late: the damage had already been done. Meanwhile, the harmful research still continues without them. In short, the research of processed food industry scientists has put profits far above the health and wellbeing of the public when it should have done just the opposite—to cite the conclusion put forward by Moss as well as legions of others familiar with that industry's science. And, of course, all this has contributed to the public's decreasing trust in science. Would this unwavering condemnation of the processed food industry science be the conclusion put forward by Matt's ideal of moral imagination as well? Consider exactly what this ideal requires of scientists:

> Scientists should recognize contingencies in their work as unforced choices, discover morally and epistemically salient aspects of the situation they are deciding, empathetically recognize and understand the genuine stakeholders and their interests, imaginatively construct and explore possible options, and exercise fair and warranted value judgment in order to guide those decisions. (Brown 2020, p. 186)

Did the processed food industry scientists fulfill these requirements? They could certainly say they did. After all, they could say they had empathetically recognized and understood the relevant stakeholders and their interests: the parents, for example, whose time constraints require that their children eat their breakfast cereals and sandwiches and snacks and such with no coaxing or delays, and whose financial constraints require that those foods be far less expensive than the whole grain, organic, minimally processed alternatives; the processed food industry shareholders who are interested in excellent returns on their investments, and the processed food industry executives who are interested in being well-treated by their companies both financially and in other ways; and, of course, the scientists themselves, who are interested in expanding their career prospects and reputations. The processed food industry scientists could also say they had recognized the contingencies in their research situations—for example, the different possibilities for adding salt, sugar, and fat to their products. And finally, the processed food industry scientists could certainly say they had recognized the morally and epistemically salient aspects of the situations they confronted—for example, how to make their products full of the necessary vitamins and minerals when their production processes robbed these from the foods (thus motivating research on synthetic vitamins and minerals to be added back to the foods).

These are the aspects of the processed food industry scientists' research that Matt's ideal of moral imagination places at the center of its moral analysis. What is rendered invisible, therefore, is what was emphasized above, that this research put profits far above the health and wellbeing of the public when it should have done the opposite. This is the kind of larger scale normative assessment that is eschewed by Matt's Deweyan framework, the kind of moral assessment that is related in Matt's mind to principles and moral abstractions and moral absolutes and the like. Is Matt right to avoid all this? What might the processed food industry scientists say to support a positive answer here? I think they might grant that the long-term health trajectory of their products' users was not stellar. But they could quickly add that that was not their fault. After all, consumers were not forced to use what the scientists had produced; consumers had made their own choices. What's more, a healthy diet with even long-term excellent consequences does not require that each individual food be excellent, only that the total array of foods be so. In addition, of course, long term health is not a matter only of diet. It also has to do with exercise and sleep and stress and other aspects of lifestyle as well as genetics. So, the conclusion to be drawn? When all the other details mentioned above are taken into account along with these details, the total integrated picture that results can easily get the processed food industry scientists off the hook. Indeed, it can give the scientists an emotional response of satisfaction—what for Matt's Deweyan framework is indicative of a warranted value conclusion. Note, however, that all this is precisely the kind of response given by the tobacco industry scientists to the charge that their research also put profits far above the health and wellbeing of the public when it should have done the opposite. And that should give Matt pause.

Does this mean that the ideal of moral imagination will not do as a successor to the ideal of value-free science? Not at all. But it does mean that much additional development will be needed to make the ideal useful for the really challenging scientific cases that right now cry out for evaluative analysis. And it also means that a significant normative (ethical/political) component must be added to and integrated with what is already there, a normative component that will function in the same kind of way that methodological and other epistemological directives now function in scientific research. Of course, Matt has resisted such "absolutes," preferring, instead, to leave his account of research and the moral imagination appropriate to it very open-ended and empirically based and free to evolve and change over time in all sorts of ways. No matter. There is another way to provide the necessary normative supplement to his account. If we look back to the dawn of modern science, we find that there were actually two general content-constraining ideals of science functioning, not one: the ideal of value-free science and what I have called the ideal of (Humanist)-value-full science (the view that science would—and should—rescue humanity from its plight of "immeasurable helplessness and poverty" caused by the Fall from Eden) (see Kourany 2021, 2024 for further details). Both ideals were introduced and supported by Francis Bacon and accepted thereafter by legions of scientists through the centuries even though they contradicted each other. What Matt proposes now is to replace the ideal of value-free science with his ideal of moral imagination. This would give us two general content-constraining ideals of science far more adequate and consistent than what we had before, with the value-full ideal supplying the needed normative supplement to Matt's ideal. All in all, a very positive turn of events.

References

Brown, M. 2020. *Science and Moral Imagination: A New Ideal for Values in Science*. Pittsburgh: University of Pittsburgh Press.

Holman, B., and T. Wilholt. 2022. The New Demarcation Problem, *Studies in History and Philosophy of Science* 91, special issue on *The New Demarcation Problem*, ed. B. Holman and T. Wilholt, 211–220.

Jenkins, M. 2011. *What's Gotten into Us: Staying Healthy in a Toxic World*. New York: Random House.

Kourany, J. 2021. Bacon's Promise. In *Science, Freedom, Democracy*, Routledge Studies in the Philosophy of Science, ed. P. Hartl and Á. Tuboly, 88–109. New York and London: Routledge.

Kourany, J. 2024. The Two Ideals Shaping the Content of Modern Science, Synthese 203:167, special issue on *The Legacy of the Value-Free Ideal of Science*. https://doi.org/10.1007/s11229-024-04554-7.

Laudan, L. 1983. The Demise of the Demarcation Problem. In *Physics, Philosophy and Psychoanalysis: Essays in Honor of Adolf Grunbaum*, Boston Studies in the Philosophy of Science, ed. R. Cohen and L. Laudan, vol. 76, 111–127. Dordrecht: D. Reidel.

Moss, M. 2013. *Salt Sugar Fat: How the Food Giants Hooked Us*. New York: Random House.

Moss, M. 2021. *Hooked: Food, Free Will, and How the Food Giants Exploit Our Addictions*. New York: Random House.

Nestle, M. 2018. *Unsavory Truth: How Food Companies Skew the Science of What We Eat*. New York: Basic Books.

Pew Research Center. 2023. Americans' Trust in Scientists, Positive Views of Science Continue to Decline. Center report (November) https://www.pewresearch.org/science/wp-content/uploads/sites/16/2023/11/PS_2023.11.14_trust-in-scientists_REPORT.pdf.

Sarewitz, D. 2016. Saving Science, *The New Atlantis* 49 (Spring/Summer): 4–40.

Tucker, W. 2007. *The Funding of Scientific Racism: Wickliffe Draper and the Pioneer Fund*. Champaign: University of Illinois Press.

Union of Concerned Scientists. 2017/2024. Attacks on Science. https://www.ucsusa.org/resources/attacks-on-science.

Chapter 12
Pragmatism, Moral Imagination, and Existential Choices

P. D. Magnus

Abstract This paper addresses the ideal of moral imagination, Matthew Brown's pragmatist account of how to reason about values in science. Pragmatism points our attention to practical consequences, but practical for whom? Brown follows John Dewey in thinking that this should be actually existing stakeholders. Extending Hilary Putnam's juxtaposition of Dewey and William James, I suggest that personal existential choices may pose a problem for such an account. In some cases, it looks as if individuals should be allowed their own determination without reference to all the stakeholders. I explore some ways in which this tension might be articulated and answered within the framework of moral imagination.

Keywords John Dewey · William James · Hilary Putnam · Matthew Brown · Values in science · Moral imagination · Existential choice

12.1 Introduction

This paper addresses the ideal of moral imagination, Matthew Brown's pragmatist account of how to reason about values in science. Extending Hilary Putnam's juxtaposition of Dewey and James, I suggest that personal existential choices are a lacuna in Brown's discussion. I explore some ways in which the gap might be articulated and filled.

Matthew Brown argues for the view that inquiry is always a response to concrete problem situations and is embedded in practical life. This means that scientists face moral responsibilities, potentially at every turn. He proposes the ideal of moral imagination as a way for individuals and groups to navigate value-laden decision making. The ideal requires that scientists creatively and reflectively consider how

P. D. Magnus (✉)
Department of Philosophy, University of Albany, Albany, NY, USA
e-mail: pmagnus@albany.edu

J. Y. Tsou et al. (eds.), *Values, Pluralism, and Pragmatism: Themes from the Work of Matthew J. Brown*, Boston Studies in the Philosophy and History of Science 347, https://doi.org/10.1007/978-3-031-92958-8_12

they understand the scientific task, what options they have for accomplishing it, and what standards are relevant. Moreover, they must identify and weigh the interests of relevant stakeholders—other members of the community who are implicated or involved in the task.[1] The method is sufficiently concrete that the book ends with a worksheet to help guide scientists through value deliberation; one quadrant of the worksheet is *Stakeholders*.

The result is a view that would make science more democratic while also making democracy more scientific. As Brown summarizes, the result is not just that "science is deeply value-laden" but also "that values are… the product of inquiry that is both informed by evidence and scientific knowledge." Science will involve debates about values, and debates about values should be "similar in structure to scientific inquiry itself" (Brown 2020, p. 231).

This is avowedly pragmatist. Brown notes that the core of his account of inquiry is given by Charles Sanders Peirce and John Dewey (Brown 2020, p. 41). As he would acknowledge, however, the greater debt is to Dewey. Brown's moral psychology and metaethics are avowedly Deweyan (e.g., p. 144 fn. 47). In this paper, I consider a problem for Deweyan pragmatism that is raised by Hilary Putnam (1990) and extend it to raise questions about Brown's moral imagination framework.

Before I get to Putnam, let's consider a general point about pragmatism.

12.2 A Free Parameter in Pragmatism

Pragmatists deny the dichotomy between belief and action. This provides a lens for examining philosophical debates, because any genuine disagreement about beliefs ought to be reflected somewhere in possible practical differences. It also means rejecting any sharp division between facts and values.[2] From that starting point, one can ask: Practical for whom? Pragmatists give different answers to that question.

Peirce argues that we should not be concerned merely with ourselves and with actual communities of which we are members, but instead with everyone who is and will ever be. Inquiry "requires a conceived identification of one's interests with those of an unlimited community." Here *unlimited* extends beyond the predictable future—rather, it is "an indefinite community" realized only in "the unlimited continuance of intellectual activity" (Peirce [1878] 1992, p. 150).

Contrariwise, William James argues that individuals should be allowed space in matters of belief to follow their passions and predilections. So each individual can look just to themselves, and inquiry requires that we "respect one another's mental freedom" (James [1896] 1912, p. 30).

[1] My description here echoes Brown's summary of the ideal (Brown 2020, p. 186) but necessarily leaves out details.

[2] It might be argued that Peirce did not reject the fact/value distinction. His account provides the resources for rejecting it in any case, as Cheryl Misak (2004) shows.

Dewey charts a middle course between these two extremes of the *indefinite anybody* and *every one for themselves*. The task of thought, Dewey writes, is to make the world "more reasonable, that is to say, more adequate to the ends which we propose for ourselves" (Dewey [1925] 1998, p. 11). The *we* here is not each of us separately. Instead, the individual features as "the vehicle of experimental creation"—not an "individual per se, an individual fixed in isolation and set up for himself, but an individual who evolves and develops in a natural and human environment, an individual who can be educated" (p. 12). The human environment is a community of other individuals, so the *we* who proposes ends is not each of us separately for ourselves but instead all of us as a community. That is, Dewey's individual is not entitled to just pursue their own passions but instead must take into account their actual contemporaries.

Brown and the ideal of moral imagination are with Dewey on this. For Brown, "science is the art of systematized problem solving" (Brown 2020, p. 43). And "value judgments in science, rather than reflecting mere personal or political preferences" should be "grounded in empirical evidence and fair consideration of the legitimate stakeholders" (p. 203).

To sum up: Pragmatism points our attention to *practical consequences*, but leaves open precisely—practical for whom? Peirce thinks that the *whom* should be the hoped-for indefinite community. James thinks that the *whom* ought to be the individual inquirer. Dewey and Brown think that the *whom* ought to be actually existing stakeholders.

This was all rather quick, and there are many subtleties which I have left out. I just hope to have highlighted a difference within pragmatism so as to better understand Brown's account.

12.3 Existential Choices

Hilary Putnam argues that Dewey has the correct social philosophy but that it breaks down when applied to "individual existential choices" (Putnam 1990, p. 1688). In such cases, Putnam thinks, James' outlook triumphs over Dewey's. To put it in terms of pragmatism's open parameter: Even if it is typically appropriate to consider things in terms of the community, sometimes an individual is entitled to take a rather more personal approach.

The ideal of moral imagination, reflecting as it does a very Deweyan social philosophy, is subject to similar worries.

Putnam begins with an example he adapts from Jean-Paul Sartre, but let's make the case more relevant to science: Imagine a scientist who works as a professor in a decent department at a middle-sized university in a mid-sized city. She has tenure and lives a comfortable life. Another university attempts to recruit her. The scientist generally thinks of herself as a city person, and the second university is in a larger, livelier city. The job offer promises an increase in pay. The new department is bigger. There are more faculty working in her area, so there would be great prospects for collaboration. Let's suppose also that there are no red flags, no clear warning signs that the new job would be a poor fit or have unexpected complications.

It is plausible to say that the offered job is objectively better than the job that the scientist has now. It fits with her prior career and lifestyle goals. The reasons all seem to align in favor of taking the job, so the scientist feels obligated to accept. After she does, she is overcome by regret. She does not feel the force of any of the reasons in favor of moving. She remembers that moving always comes with costs and that she likes the life she has. It is not that she cannot enumerate the reasons in favor of taking the job and recognize their objective weight, she simply finds that she would prefer to stay at her current job and continue where she is. So she quickly reverses her decision, before things are finalized.

One might think that the scientist has been distracted by the inertia of the life she has now and by regret at leaving rather than imaginatively exploring the promise of a new life. This can even be described as a lack of moral imagination. By considering only herself, she has failed to consider other relevant stakeholders and their perspectives.[3] Yet she has thought about what the new job has to offer. It is not that she does not recognize all the reasons in favor of moving, but rather that she does not feel those reasons as motives.

Surely this is the kind of choice that the scientist is allowed to make for herself. (At least, many people have that intuition). The choice is ultimately a decision about what sort of person she will become—or perhaps about what kind of person she already is. As Putnam writes, wanting to "become who you are" "is not the same thing as wanting to follow the 'optimal policy'" (Putnam 1990, p. 1689).[4]

Nevertheless, the choice is partly about science. The projects that the scientist will work on, the questions that she will ask, the discoveries that she will make—these will all be affected by whether she stays at her current job or takes the new one. So, the ideal of moral imagination applies, and it requires that the scientist consider stakeholders. Certainly the scientist should consider the impact on her own family and friends, but those are impacts from the move itself rather than impacts of the difference the move would make on her scientific work. The ideal demands that she considers people who will be impacted by the science that she goes on to do. Is the scientist obligated to consider the value of doing higher impact work? Must she weigh that benefit against her personal druthers?

The case is thus a challenge for the ideal of moral imagination. Although the ideal requires that the scientist consider stakeholders in the science that she will do, it seems intuitive that the scientist in this case does not need to consider stakeholders in deciding whether to take the new job or not.

[3]A failure that Brown describes as a failure of moral imagination (Brown 2020, p. 191).

[4]The language of *becoming who you are* is taken from Sartre. Brown discusses Sartre's view, but construes it as the view that "values are commitments made without a rational basis, inescapable, made freely, and held to come what may" (p. 181). Regardless of whether this is a fair reading of Sartre, it is not what Putnam and I have in mind. Rather, the case here is meant to be one in which the scientist is attempting to act in response to experience and reasons. The challenge is that the force of reasons depends on who she is and that the action will change who she becomes.

We can imagine variations of the scenario which do not pose such a challenge. For example, if the old job was developing ways to hasten genocide and the new job would be developing life-saving vaccines, then she should take the offer. Such an overwhelming moral difference would make her dithering about her lifestyle seem petty. So as not to let the ideal off easily, let's suppose that the difference is not so stark. The scientific work she does will be useful and positive at either job, but the science she would do at the new job promises to be more productive and impactful. Even though this would make a difference to stakeholders, the intuition is that she does not need to weigh that marginal impact in her deliberation.

In "The Will to Believe," James introduces these kinds of cases so as to discuss religious belief. Putnam suggests that James' reasoning on this point is "precise and impeccable" (Putnam 1990, p. 1690), but let's skip over that. My focus here is on scientific beliefs and decisions. James appeals to two distinct considerations. In recent discussions, these are given the ugly labels *doxastic efficacy* and *inductive risk*.[5] Let's take them in turn.

12.4 Doxastic Efficacy

The first Jamesian consideration is that (sometimes) an agent's belief in a state of affairs is required for it to come about. Brown briefly considers cases of this kind in the context of wishful thinking. In such cases, he suggests, wishes "allow us to make the 'leap of faith' necessary to... provide the evidence that the leap was justified" (Brown 2020, pp. 99–100). The well-worn example is an alpine climber who stands on a mountain ledge and must jump decisively in order to survive. In part, their survival depends on the fact that they believe that they *can* or even that they *will* survive—a belief that outstrips their evidence but which they are deeply invested in making true.

One might object that such do-or-die moments rarely arise in the conduct of normal science or daily life. Consider, though, this mundane example which was called to my attention by Nick Boles: A clerk working in a clinic is responsible for scheduling patient visits. They make appointments at times when doctors will be working in the clinic, but there are more patients than available hours. They are told to schedule patients anyway, because a fill-in doctor will be hired for days when there are appointments but no doctor to cover them. The clerk must act on the belief that there will be a doctor on duty, even though the current schedule indicates just the opposite. If they believe that there will be a doctor, then they schedule patients and there will be.

Note that doxastic efficacy obtains not merely for some specific question (like whether the climber will survive or whether there will be a doctor on duty) but also for a specific enquirer (the climber or the clerk). If I were to watch from a distance

[5] I elaborate this reading of James elsewhere (Magnus 2022).

as these situations unfold, all the same evidence that those enquirers have would not allow me to answer the questions. I could say conditionally that if the climber jumps with gusto, then they will survive. I might even discern based on other evidence whether they will jump or not. But the climber themself faces their own belief and action not as something to be predicted but as something to be done.

Return then to the example of the scientist deciding whether to take a job in another city. The beliefs in question are whether the considerations in favor of moving are sufficient to justify taking the job. The reasons look weighty from the outside, but the scientist herself does not feel the force of them. Insofar as the motivating force of the reasons depends on what kind of person she is and will become, then she is in a position of doxastic efficacy to make herself that kind of person. So, James or Putnam might say that it is how the reasons look from the inside that matters.

One could try to reconcile this with the ideal of moral imagination by arguing that she is really the only relevant stakeholder here, because it is a vital and monumental choice for her. I suspect Brown would eschew such a weasel move, however. He writes that "everyone who is affected by the decision" is a legitimate stakeholder. Although some stakes are more vital and momentous than others, the others cannot thereby be written off. The science that she does will have some consequences, so it implicates stakeholders. Brown writes, "[w]e cannot exclude anyone by clever acts of definition of the term stakeholder" (Brown 2020, p. 193). So Brown's ideal of moral imagination holds the same force for agents in situations of doxastic efficacy as it does for agents in other kinds of cases.

12.5 Inductive Risk

The second Jamesian consideration is that, in the face of uncertainty, there is a tension between our epistemic duties to believe truth and to avoid error. For James, as I put the point elsewhere, "The appropriate balance between these duties is a matter of value commitments rather than a matter of transcendent rationality" (Magnus 2013, p. 844).

In recent discussions, this appears as the conclusion of the Argument from Inductive Risk (AIR).[6] I have elsewhere called it the James-Rudner-Douglas or JRD thesis, since versions of it are championed by James, Richard Rudner (1953), and Heather Douglas (2000, 2009). Brown addresses all that at some length. The point I want to make here is that James uses the thesis to draw an importantly different lesson than Rudner and Douglas do. Rudner suggests that the thesis shows the need for a "science of ethics" (Rudner 1953, p. 6), while Douglas takes it to show that we need better deliberative procedures. Brown's moral imagination framework can be

[6] At least on some formulations of the AIR.

seen as an attempt to satisfy these needs.[7] James thinks instead that there is a component of individual commitment in value judgements, and so takes the thesis to show that we should not expect or demand convergence.

Arguably, disagreement about values serves as a resource for a healthy disagreement about facts in cases where scientific progress is best served by a diversity of approaches and commitments.[8] Putnam suggests that this is illustrated in the acceptance of Einstein's theory of relativity. Max Planck "was an early convert" to the theory, at a time when it looked like there was no evidence that could favor it over its rivals. Even though evidence subsequently did establish it, Planck's personal enthusiasm was crucial in the early reception of the theory (Putnam 1990, p. 1691).

Another example is provided by the discovery that peptic ulcer disease is caused by bacteria *Helicobacter pylori* rather than by a mere overabundance of acid. In the 1980s, Barry Marshall was sufficiently convinced of this hypothesis that he was willing to swallow a vial of *H. pylori* in order to show that it was a pathogen. He assessed the prior evidence differently than other members of the medical establishment did, in part because the course of his life would depend on the difference.[9] Marshall's research funding was running out. Marshall later wrote, "a successful infection with *Helicobacter* would point towards a career in clinical research" (Marshall 2006, p. 269). Absent such a discovery, he would have to move into private practice. As Alex Klein writes, this example illustrates a more general point:

> Particularly for an early-career researcher, choices about what experimental program to pursue (and thus about what hypotheses one should believe) are inevitably tied up with one's desires and fears about one's future, about one's ability to provide for one's family, about one's own prospects for an interesting and fruitful career, and so on. (Klein 2018, p. 239)

Marshall's case is not so different from the scenario we imagined above of a scientist deciding whether to accept a new job and hence inter alia what science to do. Choices like these seem at once to be matters of consequence (which carry with them moral and political responsibilities) and personal matters which the individual is entitled to decide for themself (in consultation with their family, perhaps, but not with all possible stakeholders). The space for scientists to believe in part on the basis of personal factors allows for cognitive diversity within the community, an important resource that is crucial for the development of science.[10]

---

[7] Brown himself notes the connection to Rudner (Brown 2020, p. 117).

[8] It is sometimes suggested that the distribution of cognitive labor creates a tension between individual rationality and group rationality. I have argued elsewhere that differences in personal values can resolve this tension, allowing individuals to be rational by their own lights while together forming a rational community (Magnus 2014). A similar suggestion is made by Will Fleisher (2018).

[9] I here echo my discussion of the case elsewhere (Magnus 2022). Kevin Zollman (2010) also discusses this case as one which illustrates the value of diversity.

[10] One might think that the extra space allows scientists to endorse unproven theories but not to believe them, where *endorsement* is an attitude related to the context of pursuit rather than to the context of justification (Fleisher 2018; Elliott and Willmes 2013). Perhaps Planck's early adoption of relativity just involved this more-guarded attitude. However, it looks as if Marshall genuinely

12.6 Resolutions

As I see it, someone committed to the ideal of moral imagination might respond to these considerations in one of three ways.

First: Accept Putnam's conclusion that existential choices are a limit to Deweyan democracy. There are cases where Jamesian considerations take over. In those cases, one has a right to believe as one will. On this view, the scientist in the hypothetical scenario has done something permissible.

The challenge for this first approach is to characterize which cases allow for such an exception. The boundary need not be perfectly precise, but we would need some answer to selfish jerks who ignore the interests of others and plead existentialism.

Second: Deny that such cases provide any exception. The demands of moral imagination and deliberation always apply. Failing to weigh the interests of stakeholders is wrong. On this view, the scientist in the hypothetical scenario has done something wrong.

This second approach simply accepts that the demands of morality can be unintuitive. If it seems to us that the scientist did nothing wrong, then it is a fault in us and not in the philosophical view.

However, one might worry that the obligation to consider stakeholders is too demanding. Every choice has small but genuine consequences for lots of different people, and trying to think through them all would lead to paralysis. If one took no action at all without considering the viewpoint of every possible stakeholder, then one would never get out of bed. Brown has an answer to this worry. For "second-order evaluation" of when the ideal of moral imagination should be *explicitly* invoked, he writes, the answer is still provided by the ideal of moral imagination (Brown 2020, p. 192). One need not self-consciously reflect on the ideal in order to act in accord with it, and sometimes reflecting too much will violate the ideal. In the scenario we are imagining, it might accord with the ideal for the scientist to only explicitly consider herself.

This suggests, third: Hold that the ideal of moral imagination is always the correct *standard of rightness* even when it is not the correct *rule for deliberation*. That is, the ideal of moral imagination is a standard which licenses some actions and not others, rather than a recipe which one must always consult in order to act rightly. On this view, the scientist's actions are permissible provided that one could have reached the same result while weighing the impact on stakeholders.

This third approach might be seen as a *via media* between the first two. However, there is a limit to this maneuver. Insofar as deliberation is not actually conducted, it may be indeterminate what the outcome of it would have been. Brown raises a similar worry about Philip Kitcher's ideal of well-ordered science. Brown writes, "Kitcher is wrong to privilege ideal, hypothetical democratic engagement over actual engagement in a variety of cases. Democratic ethicists like Jane Addams and

---

believed that peptic ulcer disease was caused by bacteria. He wanted to treat patients on that basis and failed to do so only because he was overruled. For more on this point, see Magnus (2022).

John Dewey would argue that in many cases it is better for the public to be involved and do badly…. Self-determination is itself an important value" (Brown 2020, p. 212). Dewey himself suggests that value cannot be something that exists apart from actual deliberation. He writes, "[t]o judge value is to engage in instituting a determinate value where none is given" (Dewey 1916, p. 368).

There are several ways of unpacking this concern. Consider three of them. First, it might be that deliberation has a value of its own in addition to the conclusion reached by the deliberation. So arriving at the same conclusion by other means would miss something important. Second, it might be that the outcome of deliberation is sensitive to particular, concrete details. Moral imagination and democratic deliberation *in the abstract* are at no particular time and in no particular circumstances, so they would not pick out one determinate outcome over another. Third, it might be that the value of a choice is constituted by its being the product of deliberation, so that no choice is right apart from explicit deliberation actually deciding on it. As Dewey says, there is only a determinate value when there is a judgement. Any of these possibilities would undo the third approach.

12.7 Coda

None of what I have said should be taken as a refutation of the ideal of moral imagination. Rather, the considerations from James and Putnam highlight a gap in how the ideal has been articulated so far. What matters is how proponents of the moral imagination framework like Brown decide to fill the gap. I think that all of the three responses discussed in the previous section are viable options. Each faces challenges, but so does every philosophical position.

References

Brown, M. 2020. *Science and Moral Imagination*. Pittsburgh: University of Pittsburgh Press.

Dewey, J. 1916. The Logic of Judgments of Practice. In *Essays in Experimental Logic*, ed. J. Dewey, 335–442. Chicago: University of Chicago Press.

Dewey, J. 1925/1998. The Development of American Pragmatism. In *The Essential Dewey*, ed. L. Hickman and T. Alexander, vol. 1, 3–13. Bloomington: Indiana University Press.

Douglas, H. 2000. Inductive Risk and Values in Science. *Philosophy of Science* 67 (4): 559–579.

Douglas, H. 2009. *Science, Policy, and the Value-Free Ideal*. Pittsburgh: University of Pittsburgh Press.

Elliott, K., and D. Willmes. 2013. Cognitive Attitudes and Values in Science. *Philosophy of Science* *80*(5): 807–817.

Fleisher, W. 2018. Rational Endorsement. *Philosophical Studies* 175 (10): 2649–2675.

James, W. 1896/1912. The Will to Believe. In *The Will to Believe and Other Essays in Popular Philosophy*, ed. W. James, 1–31. New York: Longmans, Green, & Co.

Klein, A. 2018. In Defense of Wishful Thinking: James, Quine, Emotions, and the Web of Belief. In *Pragmatism and the European Traditions: Encounters with Analytic Philosophy and*

Phenomenology Before the Great Divide, ed. M. Baghramian and S. Marchetti, 228–250. London: Routledge.

Magnus, P. D. 2013. What Scientists Know is Not a Function of What Scientists Know. *Philosophy of Science* 80 (5): 840–849.

Magnus, P. D. 2014. Science and Rationality for One and All. *Ergo* 1 (5): 129–138.

Magnus, P. D. 2022. William James on Risk, Efficacy, and Evidentialism. *Episteme* 19 (1): 146–158.

Marshall, B. 2006. Helicobacter Connections. In *Les Prix Nobel*, ed. K. Grandin, 250–277. Stockholm: Nobel Foundation.

Misak, C. 2004. Making Disagreement Matter: Pragmatism and Deliberative Democracy. *Journal of Speculative Philosophy* 18 (1): 9–22.

Peirce, C. S. 1878/1992. The Doctrine of Chances. In *The Essential Peirce*, ed. N. Houser and C. Kloesel, vol. 1, 142–154. Bloomington: Indiana University Press.

Putnam, H. 1990. A Reconsideration of Deweyan Democracy. *Southern California Law Review* 63 (6): 1671–1697.

Rudner, R. 1953. The Scientist qua Scientist Makes Value Judgments. *Philosophy of Science* 20 (1): 1–6.

Zollman, K. 2010. The Epistemic Benefit of Transient Diversity. *Erkenntnis* 72 (1): 17–35.

Chapter 13
Brown's Pragmatic Theory of Values and the Challenges of Commercial Science

Manuela Fernández Pinto

Abstract Matt Brown's *Science and Moral Imagination* (2020) offers a groundbreaking pragmatic pluralist theory of values, challenging the traditional value-free ideal in scientific inquiry. Drawing from American pragmatism and feminist philosophy, Brown argues for the cognitive status and evidential warrant of values in science. Despite its many virtues, in this chapter, I highlight some issues faced by this account, stemming from Brown's focus on individual scientists and the significant impact of the commercialization of science on research practices. The transition to privately funded research has influenced methodological choices and the integrity of scientific values, leading us to a dilemma. We could acknowledge that scientific practice has notably evolved to align with the commercial objectives of private companies, potentially necessitating a redefinition of scientific values. Alternatively, we could assert that commercialized science is dysfunctional, questioning the integrity of a vast quantity of contemporary scientific research. I conclude by pressing for the inclusion of the broader social context of science and its commercial influences in order to address the dilemma and develop a comprehensive theory of values in science.

Keywords Commercial interests · Privatization of science · Scientific values · Sponsorship bias · Pragmatic pluralism

---

M. Fernández Pinto (✉)
Department of Philosophy, Universidad de los Andes, Bogotá, Colombia

African Centre for Epistemology and Philosophy of Science, University of Johannesburg, Johannesburg, South Africa
e-mail: m.fernandezp@uniandes.edu.co

J. Y. Tsou et al. (eds.), *Values, Pluralism, and Pragmatism: Themes from the Work of Matthew J. Brown*, Boston Studies in the Philosophy and History of Science 347, https://doi.org/10.1007/978-3-031-92958-8_13

13.1 A Pragmatic Theory of Values

Matt Brown's book, *Science and Moral Imagination: A New Ideal for Values in Science* (2020), presents an enticing prospect for philosophers of science working on science and values. Drawing from American pragmatism and feminist philosophy, Brown introduces an innovative alternative to the value-free ideal. This alternative not only acknowledges the legitimacy of value judgments within scientific inquiry but also confronts the limitations posed by influential predecessors, notably feminist radical empiricism (Hankinson Nelson 1990; Clough 2003), social constructive empiricism (Longino 1990), the ideal of well-ordered science (Kitcher 2001), and the ideal of socially responsible science (Kourany 2010).

A significant breakthrough in Brown's work lies in its formulation of a theory of values. Recognizing a crucial gap in the philosophy of science, Brown addresses the inherent confusions and ambiguities resulting from the absence of such a theory in the science and values literature. The book's *pragmatic pluralist theory of values* emerges as a much-needed contribution. It aims to rectify the noncognitive stance on values perpetuated by the ideal of value-freedom, advocating instead for a pragmatic cognitivist understanding. This approach allows values to possess cognitive status and evidential warrant, acknowledging their diverse sources and functions.

Partially inspired by Anderson's seminal paper, "Uses of Value Judgments in Science" (2004), the pragmatic pluralist theory of values expounded in *Science and Moral Imagination* stands out as one of Brown's pivotal contributions to the philosophy of science. This theory represents a crucial milestone, finally establishing a comprehensive framework for values that meticulously challenges the fact/value distinction and provides compelling arguments in favor of a cognitivist understanding of value judgments. Without such a theory, no contender of the value-free ideal would be complete.

The second notable contribution is the *ideal of moral imagination*. Early in his work, Brown develops a general argument against the value-free ideal and in favor of the value-ladenness of science which he calls "the contingency argument." Central to this argument is the recognition that scientists make value-laden decisions throughout the research process and that scientists are morally responsible for such decisions. Brown emphatically concludes that "scientists have a responsibility to make value judgments about scientific contingencies, and this science is value-laden through and through" (2020, p. 19). This assertion encapsulates his thesis of science being inherently imbued with values at every stage.

Given the acknowledgment of science as inherently value-laden and the attribution of cognitive status to value judgments, the concept of moral imagination becomes fundamental. It serves as a crucial tool for comprehending the multifaceted consequences of decision-making and for creatively exploring diverse courses of action. Within this framework, Brown underscores three methods for exercising moral imagination: empathy, dramatic rehearsal, and creative problem-solving.

What I find particularly noteworthy in Brown's work is his explicit effort to outline practical methodologies for implementing the ideal of moral imagination. For instance, he offers practical strategies to address conflicts of values (2020, pp. 153–154) and illustrates instances of its application, such as within his Responsible Conduct of Research (RCR) training seminars (pp. 225–227). Brown's comprehensive approach not only conceptualizes moral imagination as an ideal but also delineates actionable strategies for its practical realization.

Considering these and various other reasons, *Science and Moral Imagination* stands as a fundamental contribution to the science and values literature, arguably representing one of the most conceptually groundbreaking works in recent years. Having said that, and without undermining the book's achievements, let me present some critical remarks about the book and its main argument.

13.2 Pragmatism and the Global Privatization of Science

Central to Brown's approach is his "heuristic focus" on the individual scientist, a stance he maintains throughout the book. Brown justifies this emphasis by asserting that numerous issues encompassing values, ethics, and politics manifest at a broader societal level, which, in his view, require substantial social, cultural, and political-economic transformations to be addressed effectively. Consequently, he opts to concentrate on the realm where he perceives potential for tangible advancement: the midstream of the research process, i.e., "the course of inquiry itself" (Brown 2020, p. 186), wherein he believes substantial progress can be made in refining science (p. 18).

Here, I would like to challenge the idea that we can make "real progress in ameliorating science" without taking into account "the larger social level" (pp. 12–13). I argue that the current social organization of science, and in particular the commercialization of scientific research, has a direct impact on the midstream of the research process. Therefore, when deliberating on the value-laden decisions made by scientists, it becomes imperative to consider how the broader social landscape shapes the research process.

As I have previously argued (Fernández Pinto 2014), the transition from publicly funded research to a commercially driven and privately funded research regime since the end of the Cold War has instigated a series of transformations across the research continuum. These alterations have notably influenced decisions not only concerning research priorities and the dissemination of findings but have impacted methodological choices, experimental design, data collection and interpretation, the framing of conclusions, and various other decisions within the midstream of the research process.

A prime example illustrating this impact can be observed in clinical trials. Commercial values have induced shifts in practices pertaining to patient recruitment (Fernández Pinto 2019a), endeavors to broaden the diversity of research subjects (Fernández Pinto 2018), and the selection of comparators, endpoints, and risk reduction measures (Fernández Pinto 2019b), among other critical aspects (a concise overview can be found in Bero and Rennie 1996).

Consequently, seemingly external conditions or broader social factors have gradually but consistently instigated shifts in scientific practice. Crucially pertinent to Brown's argument is the inescapable integration of commercial values into the decision-making processes of individual scientists and research groups. This assimilation of commercial values significantly shapes scientific practices. For instance, meta-analyses investigating sponsorship bias in clinical research highlight this influence by concluding that studies funded by industries are inclined to yield results favorable to their sponsors, despite the apparent superior quality of industry-funded studies (as evidenced by studies such as Lundh et al. 2017).

This issue gains even greater significance when one acknowledges that the overwhelming majority of contemporary scientific research is both financed by and conducted within the private sector. Recent statistics from the National Scientific Board illuminate this trend: in the United States, the private sector funds 71% and conducts 74% of Research and Development (R&D), while federal government funding accounts for only 21%, with less than 10% of R&D performed (NSB 2022). Similarly, the European Union exhibits analogous tendencies and figures (Eurostat 2023).

Consequently, it is no longer accurate to assert, as Brown posits, that "science is largely a publicly supported enterprise" (Brown 2020, p. 77).[1] The dominance of private sector involvement in both funding and execution underscores a substantial departure from the prior regime of publicly-funded research with national interests in mind.

Therefore, I consider that taking into account the broader globalized privatization of science (Mirowski 2011) is especially important for a pragmatist account of values in science, i.e., one that aims to understand and ameliorate the practice of science as it is. Even while focusing on the individual scientist's responsibilities, akin to Brown's stance, neglecting the influence of this expansive context on scientists' decision-making would render the account partial, if not altogether unrealistic.

13.3 The Core Values and the Challenges of Commercialization

To exemplify potential challenges for Brown's pragmatic pluralist theory of values when the broader commercialization of science isn't considered, I focus on his pragmatist assertion that "science is a complex and adaptable set of practices" (Brown 2020, p. 127). He contends that values evolve historically, while also defending the existence of "core values" essential for functional science (p. 129). Notably, in Chap. 4, Brown hesitates to compile a definitive list of "candidates for

[1] It is still true though that science receives large amounts of funding from the federal government, e.g., 139 billion in 2019, and that the federal government continues to be the main funder of basic research, circa 41% (NSB 2022).

core scientific values" (p. 127). This hesitation aligns with a pragmatist approach, as he seeks to avoid scientism and the assertion of a universally applicable value set. Instead, he acknowledges the adaptability of science and the potential for scientific values to shift over time.

Subsequently, however, Brown proceeds to outline a roster of potential basic scientific values, encompassing consistency, empirical control, systematicity, universalism, communalism, fallibilism, peer review, due credit, and whole-heartedness (pp. 127–129). Despite acknowledging that certain scientific practices may contravene some of these values, he concludes:

> Still, while we can recognize practices that follow the preponderance of these norms, even if they violate one or two, we should recognize inquiries that violate them as suspect and dysfunctional. *Healthy scientific inquiries abide by all of these norms* and cherish these values at least at a minimal level. (pp. 129-130, emphasis mine)

The problem is that contemporary commercially driven science appears to contravene not only one of these values (e.g., just the value of communalism, as he acknowledges), but also several other values on the list, at various junctures.

For instance, financial conflicts of interest pervade the peer-review process to such an extent that the essential critical scrutiny of scientific findings is significantly compromised. This distortion is evident in both conference and journal peer reviews, where private sponsorship can skew evaluations in favor of commercial interests (Resnik and Elmore 2018). Compromises in due credit also prevail, facilitated by publication planning firms and the use of ghostwriters in scientific publication production (Sismondo 2009). The value of whole-heartedness is similarly jeopardized due to stringent intellectual property rights governing research outcomes and publications, leading private companies to bury results incongruent with their interests (Turner et al. 2008; Every-Palmer and Howick 2014).

Furthermore, private companies have also co-opted many of these values to argue in favor of scientifically questionable views and to the detriment of what we would consider "legitimate" research. Notably, universalism, or the tempered equality of intellectual authority, has been co-opted by private corporations to demand equal attention and validation for scientifically unfounded views. This tactic, famously employed by tobacco companies, has proven successful and has transcended into other industries (Fernández Pinto 2014). Similarly, fallibilism and empirical control have been manipulated; consider the strategy employed by the tobacco industry, emphasizing the uncertainty in epidemiological studies linking tobacco smoking to lung cancer. This tactic of demanding more refined and comprehensive research was instrumental in the industry's successful delay of regulatory measures (Oreskes and Conway 2010). Similar approaches have been adopted by the petrochemical and pharmaceutical sectors, exploiting doubts to their advantage (Fernández Pinto 2017).

Moreover, systematicity has served as a core defense for industries facing accusations of bias due to financial conflicts of interest. Notably, meta-analyses revealing sponsorship bias also indicate that industry-backed research is often deemed "of better quality," according to existing quality assessment tools. This narrow interpretation of "trustworthy research" privileges commercially driven studies despite sponsorship bias (Fernández Pinto 2020).

Two primary concerns surface here. Firstly, contemporary scientific research driven by commercial interests often appears to contravene several of Brown's fundamental scientific values. Yet, despite these discrepancies, we continue to categorize these endeavors as science. Brown acknowledges this tension to an extent, noting, "It is important that we be able to point to the variety of ways in which this 'closed' and for-profit science is dysfunctional. But one could reasonably object that it goes too far to say that such inquiries are not scientific practices as such" (Brown 2020, p. 129). In reality, given that a significant portion of scientific research is funded by and conducted within the private sector, this "closed" and profit-driven science arguably represents the predominant form of scientific practice today.

This presents us with a dilemma—one that perhaps moral imagination can help unravel. We could acknowledge that scientific practice has notably evolved to align with the commercial objectives of private companies, potentially necessitating a redefinition of scientific values. For instance, communalism may have been supplanted by "closedness" or "secrecy." This position would imply that such a shift is acceptable, given science's adaptability to thrive in diverse contexts, but it would not help us evaluate the problems in commercialized science.

Or, alternatively, we could assert that commercialized science is dysfunctional, as we have good reasons to do in my opinion. However, this path would prompt us to question the integrity of the vast majority of contemporary scientific research. Such a questioning would create a discrepancy between the philosophical understanding of science and its actual practice, which might pose a problem for a pragmatist account of science, which wants to provide normative guidance to the practice as such.[2]

To develop a pragmatic pluralist theory of values for science, it is imperative not to overlook this dilemma. What is needed is a normative pluralist perspective on values in science, which is non-relativist yet sensitive to the diverse contexts and multifaceted purposes guiding scientific practice. Acknowledging the substantial influence of commercial values on science today is essential. However, equally crucial is establishing frameworks that allow evaluation of such practices without divorcing ourselves from their intricate social contexts.

Secondly, the co-optation of scientific values by private companies to serve their commercial interests not only raises concerns but also muddles the fundamental purpose of delineating core values. Case studies, particularly those scrutinizing the manufacturing of doubt, exemplify how private enterprises have exploited several of the core values outlined in Brown's list. These companies have utilized these values to cast doubt on legitimate research and researchers while validating their own research as the epitome of "legitimate science" (Fernández Pinto 2014, 2017).

---

[2]This might not necessarily undermine Brown's pragmatism insofar as he might not concede, following Feyerabend, that our philosophies of science should reflect scientific practice. But it would certainly undermine an account that aimed to follow the more naturalist pragmatists, such as Dewey. I thank Jamie Shaw for clarifying this point.

Consequently, constructing a list of core scientific values intended to differentiate between legitimate and illegitimate, or functional and dysfunctional science, might prove inadequate. Private entities adeptly discern and emulate how scientific values operate in practice, accommodating—or perhaps manipulating—these values to serve their commercial interests.

And while one might be inclined to label this as an illegitimate appropriation or interpretation of the core values, such a stance would merely lead us to establish second-order criteria for evaluating the legitimacy of value usage, which could lead in turn to an illegitimate appropriation of this second-order criteria, and so on to an infinite regress. This, rather than resolving the issue, aggravates it further. Once again, I contend that formulating a pragmatic theory of values necessitates addressing practical challenges such as those arising from commercialization. These challenges include the co-optation of values for commercial gains, which significantly influences the application of values in guiding scientific practice.

13.4 Conclusion

I do not necessarily dismiss Brown's emphasis on the individual scientist. I concur with his assertion that philosophies of science after the social turn may have neglected the moral responsibilities of individual scientists in favor of a (much-needed) social epistemology of science. However, as previously suggested, concentrating on the value judgments made by individual scientists does not imply disregarding the wider social context. Specifically, it is crucial to acknowledge the diverse constraints imposed by this context on scientists' decision-making processes.

For instance, in the commercial-private sphere, scientific autonomy differs significantly from research conducted within university laboratories. Value conflicts emerge not only within the decision-making process of scientists but also in the interactions between scientists and other overseeing entities, such as marketing departments. Narrowing our focus solely to the decision-making of individual scientists or research groups overlooks these distinct value conflicts—pragmatic perplexities—that characterize commercially driven science today.

Consequently, I believe that a robust theory of values in science needs a social orientation. For it to genuinely embody pragmatism, this theory should encompass the social dimensions inherent in scientific practices, particularly acknowledging the social constraints shaping the decisions of individual scientists today. Fortunately, within Brown's pragmatic pluralist theory of values, there exist foundational elements that can serve as a starting point for such considerations, but it would require to expand Brown's proposal to consider the relevant social aspects. Accordingly, further exploration is warranted on how social values, notably commercial ones, intersect with scientific values. Additionally, there is a need to elucidate how value judgments can facilitate the assessment and improvement of research within this framework. For instance, by offering pertinent evaluations of questionable practices often encouraged within commercially driven research settings.

References

Anderson, E. 2004. Uses of value judgments in science: A general argument, with lessons from a case study of feminist research on divorce. *Hypatia* 19:1–24.

Bero, L., and D. Rennie. 1996. Influences on the quality of published drug studies. *International Journal of Technology Assessment in Health Care* 12 (2): 209–237.

Brown, M. J. 2020. *Science and Moral Imagination: A New Ideal for Values in Science*. University of Pittsburgh Press.

Clough, S. 2003. *Beyond epistemology: A pragmatist approach to feminist science studies*. Lanham: Rowman & Littlefield.

Eurostat. 2023. R&D expenditure. Available at: https://ec.europa.eu/eurostat/statistics-explained/index.php?title=R%26D_expenditure&oldid=551418#R.26D_expenditure_by_sector_of_performance

Every-Palmer, S., and J. Howick. 2014. How evidence-based medicine is failing due to biased trials and selective publication. *Journal of Evaluation of Clinical Practice* 20:908–914.

Fernández Pinto, M. 2014. Philosophy of science for globalized privatization: Uncovering the limitations of critical contextual empiricism. *Studies in History and Philosophy of Science* 47:10–17.

Fernández Pinto, M. 2017. To know or better not to: The social construction of ignorance in commercial research. *Science & Technology Studies* 30 (2): 53–72.

Fernández Pinto, M. 2018. Democratizing strategies for industry-funded medical research: A cautionary tale. *Philosophy of Science* 85 (5): 882–894.

Fernández Pinto, M. 2019a. Doubly disadvantaged: The recruitment of diverse subjects for clinical trials in Latin America. Tapuya*:* Latin American Science, Technology and Society 2(1): 391–407.

Fernández Pinto, M. 2019b. Scientific ignorance: Probing the limits of scientific research and knowledge production. *Theoria* 32 (2): 195–211.

Fernández Pinto, M. 2020. Commercial interests and the erosion of Trust in Science. *Philosophy of Science* 87 (5): 1003–1013.

Hankinson Nelson, L. 1990. *Who knows?: From quine to feminist empiricism*. Philadelphia: Temple University Press.

Kitcher, P. 2001. *Science, truth, and democracy*. New York: Oxford University Press.

Kourany, J. 2010. *Philosophy of science after feminism*. Oxford: Oxford University Press.

Longino, H. 1990. *Science as social knowledge: Values and objectivity in scientific inquiry*. Princeton: Princeton University Press.

Lundh, A., J. Lexchin, B. Mintzes, J. Schroll, and L. Bero. 2017. Industry sponsorship and research outcome. *Cochrane Database of Systematic Reviews* 2:MR000033.

Mirowski, P. 2011. *Science-Mart: Privatizing American science*. Cambridge: Harvard University Press.

National Science Board. 2022. *Science and Engineering Indicators 2022: The State of U.S. Science and Engineering*. NSB-2022-1. Alexandria. Available at: https://ncses.nsf.gov/pubs/nsb20221

Oreskes, N., and E. Conway. 2010. *Merchants of doubt: How a handful of scientists obscured the truth on issues from tobacco smoke to global warming*. New York: Bloomsbury Press.

Resnik, D., and S. Elmore. 2018. Conflict of interest in journal peer review. *Toxicologic Pathology* 46 (2): 112–114.

Sismondo, S. 2009. Ghost in the machine: Publication planning in the medical sciences. *Social Studies of Science* 39 (2): 171–198.

Turner, E., A. Matthews, E. Linardatos, R. Tell, and R. Rosenthal. 2008. Selective publication of antidepressant trials and its influence on apparent efficacy. *New England Journal of Medicine* 358 (3): 252–256.

Part V
The Value-Free Ideal, Inductive Risk, and Epistemic Priority

Chapter 14
A History of Metaethics and Values in Science

Paul L. Franco

Abstract I pursue Matt Brown's (*Science and moral imagination: A new ideal for values in science*. Pittsburgh: Pittsburgh University Press, 2020) and James Leach's (*The British Journal for Philosophy of Science* 19 (2): 93–108, 1968a) historical suggestions that there is a relationship between metaethical noncognitivism and certain arguments for the value neutrality of the sciences. In particular, I relate C.L. Stevenson's emotivism to arguments by Herbert Feigl, Carl Hempel, and Ernest Nagel for the sciences' value neutrality. I also consider whether these arguments can be disentangled from their controversial metaethical claims by looking at Robert Alexander's (*Philosophical Studies* 25 (6): 391–401, 1974) account of value neutrality based on the view that a scientist's aims are discharged by making empirical statements. Drawing upon Leach's (*The British Journal for Philosophy of Science* 19 (2): 93–108, 1968a, *Inquiry: An Interdisciplinary Journal of Philosophy* 11 (1-4): 349–367, 1968b) defense of a form of the argument from inductive risk, I argue Alexander fails to offer a metaethically neutral version of the value neutrality of the sciences. Though I do not explicitly explore this, I think the history I sketch is relevant to recent calls for philosophers of science to more fully characterize the 'values' in 'values and science.'

Keywords Value-free ideal · Values and science · History of philosophy of science · Value judgments · Metaethics

---

P. L. Franco (✉)
Department of Philosophy, University of Washington, Seattle, WA, USA
e-mail: pfranco@uw.edu

J. Y. Tsou et al. (eds.), *Values, Pluralism, and Pragmatism: Themes from the Work of Matthew J. Brown*, Boston Studies in the Philosophy and History of Science 347, https://doi.org/10.1007/978-3-031-92958-8_14

14.1 Introduction

In his recent book, Matt Brown claims the worry that "if we let values into science, it will lead to bias, subjectivism, and wishful thinking… is based on an implicit view that values themselves have no cognitive status or somehow systematically inferior cognitive status to other components of inquiry, such as data, experiments, hypotheses, logic, and reasoning" (Brown 2020, pp. 89–90). He further claims, "[t]he implicit model of values that most philosophers of science work with, whether or not they think that science ought to be value-free, is generally an emotivist or at least noncognitivist one" (p. 103). Roughly, metaethical noncognitivism holds that value judgments are not truth-apt. One way to be a noncognitivist is to be an emotivist who holds that value judgments *express* one's attitudes or emotions in ways aimed at influencing others, but do not state anything true or false. Brown suggests this implicit model philosophers of science work with has roots in logical empiricism's outsized influence in the development of philosophy of science (p. 101).

Reaching slightly further back in history, James Leach says, "[f]ollowing the progressive elimination of teleology from the province of the scientist in the seventeenth century, the domains of science and values seemed finally isolated and immune from the other" (Leach 1968a, p. 93). As a result, "the very notion of scientifically supported values [was] exploded as a 'naturalistic fallacy'" (p. 93). Finally, on Leach's telling, "the convergence in value theory of… emotivists and oxonians, conjoined with the success and influence of a positivist philosophy of science, consummated the separation of science and value" (p. 93). To unpack the last of Leach's claims: Emotivists, like C.L. Stevenson, and Oxonians, like R.M. Hare, converged on noncognitivism, as did (some of) the logical empiricists.[1] Eliding important differences, on these accounts, value judgments are distinct from empirical statements in that they express emotions, commend or condemn, or issue imperatives rather than describe the world. For Leach, these developments "enthroned… the value neutrality thesis of science… which prohibits the scientist from making value judgments" (ibid).

In this paper, I pursue Brown's and Leach's historical suggestions that there is a relationship between metaethical noncognitivism and certain arguments for the value neutrality of the sciences. In particular, I look at Stevenson's influential noncognitivist view developed contemporaneous with logical empiricism and relate it to Herbert Feigl's, Carl Hempel's, and Ernest Nagel's respective arguments in the mid-twentieth-century for the value neutrality of science. I also consider whether these arguments can be disentangled from their controversial metaethical claims by looking at Robert Alexander's (1974) account of value neutrality based on the view that a scientist's aims are discharged by making empirical statements. Drawing upon Leach's (1968a, 1968b) defense of a version of the argument from inductive risk, I argue Alexander fails to offer a metaethically neutral version of the value neutrality of the sciences.

[1] See Anne Siegetsleitner (2022) for an overview of the noncognitivism of some logical empiricists.

I sketch this history for two reasons. First, it highlights one facet of the conceptual development of the value-free ideal in parts of philosophy of science that turns on the view that value judgments are neither true nor false.[2] Second, though I do not explore it here, I think the history is potentially relevant to calls for philosophers of science to more fully characterize the 'values' in 'values and science' rather than relying on pre-theoretical notions of values and their roles in science (Scriven 1972; Brown 2020; Ward 2021; Wieten 2021).

14.1.1 Plan for the Paper

In 14.2, I review early analytic metaethical arguments for a strong distinction between empirical statements and value judgments, focusing especially on Stevenson's emotivism. In 14.3, I show how metaethical commitments similar to Stevenson's inform accounts of the value neutrality of the sciences by Feigl, Hempel, and Nagel. In 14.4, I detail Alexander's (1974) argument that the value neutrality of the sciences can be formulated independent from the controversial metaethical commitments taken on by Feigl, Hempel, and Nagel. In 14.5 in response to Alexander, I use Leach (1968a, 1968b) to draw connections between metaethics and the argument from inductive risk. I conclude in 14.6.

14.2 C.L. Stevenson's Noncognitivist Emotivism

While it is well-known that A.J. Ayer (1946/1952, ch. VI) outlines an emotivist ethics and some logical empiricists develop related accounts (Brown 2020, pp. 101–103; Siegetsleitner 2022), I foreground Stevenson's emotivism, as his version was more developed and influential in analytic metaethics.[3] Moreover, Feigl endorses Stevenson's 1944 *Ethics and Language* as "one of the very few and exceptional books on ethics in which a sound scientific attitude combined with a logical analysis of great subtlety and thorough penetration is applied toward the clarification of moral discourse" (Feigl 1946, p. 80). Here, I focus on a 1937 article that provides an overview of Stevenson's emotivism, "The Emotive Theory of Ethical Terms." Despite Stevenson not being a logical empiricist, this article was reprinted in Ayer (1959) alongside canonical logical empiricist texts.

Some brief background for Stevenson's view will be helpful. Much of twentieth-century analytic metaethics is framed in relation to G.E. Moore's *Principia Ethica.*

[2] For other aspects of this development, some related to the themes of this paper, see Don Howard (2003), Heather Douglas (2009, ch. 3), and Krist Vaesen and Joel Katzav (2019).

[3] In the second edition of *Language, Truth and Logic*, Ayer says his emotive theory "needs to be supported by a more detailed analysis" (1946/1952, p. 20). In a footnote, he says, "this deficiency has been made good by C.L. Stevenson in…*Ethics and Language*" (p. 20, fn. 3).

For Moore, the primary subject matter of ethics is the property good. Moore understands good to be a simple, unanalyzable, and nonnatural property. He claims, "'good' is a simple notion, just as 'yellow' is a simple notion… just as you cannot, by any manner of means, explain to any one who does not already know it, what yellow is, so you cannot explain what good is" (Moore 1903/1959, p. 7). Accordingly, any account of good that defines it in terms of or reduces it to other properties misunderstands its very nature. Moore is particularly concerned with avoiding the so-called naturalistic fallacy. Ethical views make this mistake when they aim at "substituting for 'good' some one property of a natural object or of a collection of natural objects; and in thus replacing Ethics by some one of the natural sciences" (p. 40). Ethical judgments stand independent of empirical statements, and cannot be reduced to them, because for any natural property put forward as identical with good it always remains an open question whether that property is actually good. This is Moore's open-question argument. Consider hedonists who analyze goodness in terms of pleasure. Such views commit the naturalistic fallacy by reducing good to a natural property, namely the psychological state of pleasure, of which it remains an open question to ask, without contradiction, if pleasure is actually good. As Stevenson understands Moore, "[n]o matter what set of scientifically knowable properties a thing may have (says Moore, in effect), you will find, on careful introspection, that it is an open question to ask whether anything having these properties is good… That is, the scientific method is not sufficient for ethics" (Stevenson 1937, p. 18). If good is a nonnatural property not reducible to empirically discoverable natural properties, then science and ethics are separate domains.

It is important to note that two of Moore's reasons for separating the domain of ethics from the domain of the natural sciences would not be acceptable to philosophers of science who adopt an empiricist criterion of meaning even if they similarly claim that the ethical and the scientific are independent. First, Moore takes the subject matter of ethics to be an unanalyzable, nonnatural property irreducible to any natural property. Second, he thinks ethical statements are truth-evaluable and known intuitively, rather than empirically.

However, two decades later, C.K. Ogden and I.A. Richards offer a linguistic formulation of Moore's separation of science and ethics that resonates with Stevenson and later accounts from philosophers of science. In *The Meaning of Meaning*, Ogden and Richards divide "the functions of language into two groups, the symbolic and the emotive" (Ogden and Richards 1923/1945, p. viii). They understand Moore's claim that "'good' is…a unique unanalysable concept" to suggest the "peculiar ethical use of 'good' is…a purely emotive use" (p. 125). For Ogden and Richards, on one hand, "[t]he symbolic use of words is *statement*; the recording, the support, the organization and the communication of references" (p. 149). The scientific use of words is restricted to the statement (p. 124). On the other hand, "[t]he emotive use of words is a more simple matter, it is the use of words to express or excite feelings and attitudes" (p. 149).[4] Ogden and Richards think 'good' has no referential function; it does not pick out any property. Instead,

[4] Susan Stebbing, another contemporary and interlocutor of the logical empiricists, makes a similar distinction (Stebbing 1930/1942, pp. 16–17).

"it serves only as an emotive sign expressing our attitude...and perhaps evoking similar attitudes in other persons, or inciting them to actions" (p. 125). So, on this view, 'good' is unanalyzable because it is used to express attitudes and influence others rather than to refer to and make statements about properties in the world.

Stevenson credits Ogden and Richards's interpretation of Moore with inspiring his emotivist metaethics (Stevenson 1937, p. 23 fn. 1). Following them, Stevenson claims, "Broadly speaking, there are two different *purposes* which lead us to use language" (p. 21). The descriptive use of language is "(as in science) to record, clarify and communicate *beliefs*"; the dynamic use of language is "to give vent to our feelings (interjections), or to create moods (poetry), or to incite people to actions or attitudes (oratory)" (ibid).[5, 6] Stevenson understands the distinction between the descriptive and dynamic use of words pragmatically, in terms of the aims of the speaker, which the audience infers with the help of a speaker's "tone of voice... gestures, the general circumstances... and so on" (p. 22).

The emotive meaning of some words, understood semantically, makes them especially suited for dynamic uses. On Stevenson's view, "[t]he emotive meaning of a word is a tendency of a word, arising through the history of its usage, to produce (result from) *affective* responses in people. It is the immediate aura of feeling which hovers about a word" (p. 23). For him, "there is an important contingent relation between emotive meaning and dynamic purpose: the former assists the latter" (p. 24). Ethical discourse is aimed at directing people's interests to get them to act using words with strong emotive meaning that make them well-suited for dynamic purposes. For example, Stevenson says, "[t]he word 'good' has a pleasing emotive meaning which fits it especially for the dynamic use of suggesting favourable interest" (p. 25). When I call some action good, I express my favorable attitude towards it and suggest my audience take the same attitude. Given the strong feelings and interests associated with terms expressing moral approval and disapproval (p. 26), moral terms like 'good' are especially useful for influencing others via the dynamic use of language.

Stevenson thinks his emotive theory satisfies three requirements a metaethical theory must meet to capture the "vital" sense of 'good' (p. 16). Here, I highlight his first and third requirements. The first requirement is that "we must be able sensibly to *disagree* about whether something is 'good'"; the third requirement is related to the naturalistic fallacy and Moore's open-question argument, namely, that "the 'goodness' of anything must not be verifiable solely by use of the scientific method" (ibid). Consider his first requirement. Stevenson argues, "[w]e must

---

[5] Stevenson recognizes a speaker might use words with both purposes (1937, p. 21).

[6] An anonymous reviewer points out Rudolf Carnap also distinguishes descriptions of facts and expressions of attitudes in his criticism of metaphysics. Carnap: "the (pseudo) statements of metaphysics do not serve for the description of states of affairs... They serve for the expression of the general attitude of a person towards life" (1932/1959, p. 78). Such expressions are not truth-evaluable. Carnap (1935) argues moral claims are imperatives, which cannot be true or false and have no empirical consequences. Hans Reichenbach (1951, p. 280) makes a similar point. More on Carnap's noncognitivism in note 9.

distinguish 'disagreement in belief' (typical of the sciences) and 'disagreement in interest'" (p. 27).[7] For Stevenson, the latter "occurs when A has a favourable interest in X, when B has an unfavourable one in it, and when neither is content to let the other's interest remain unchanged" (ibid). Disagreement in interest is the type of disagreement characteristic of ethics. Resolving ethical disagreement starts with "the suggestive force of [an] ethical sentence," i.e., a sentence with emotive terms used dynamically (p. 28). This use "exerts enough pressure to start… trains of reasons" aimed at giving "a more thorough account of the object of interest" (ibid). This more thorough account may be descriptive, e.g., it might consist in specifying the likely consequences of some course of action relative to another. In this way, "the empirical method is relevant to ethics simply because our knowledge of the world is a determining factor to our interests. But," Stevenson continues, "note that empirical facts are not inductive grounds from which the ethical judgment problematically follows" (p. 28–29). Empirical statements made in attempts to resolve disagreement in interest are in the service of influence and persuasion; they are not logical grounds from which ethical conclusions follow. This relates to Stevenson's third requirement. Recalling Moore, Stevenson says that a factual description of a thing's natural properties leaves open the question of whether those properties are good; this is because "to ask whether it is good is to ask for influence. And whatever I may know about an object, I can still ask, quite pertinently, to be influenced with regard to my interest in it" (p. 30). Factual statements aimed at saying something true do not entail ethical judgments aimed at influence.

I turn now to showing how this claim about the logical relationship between ethics and the sciences, and the related distinction between the purposes of scientific language—to state facts—and ethical language—to express approval or disapproval and influence others—is central to arguments for the value neutrality of the sciences advanced by Feigl, Hempel, and Nagel.[8, 9]

[7] For more on this distinction, see Stevenson (1948/1963).

[8] See Douglas (2009, ch. 3) for more on Hempel's and Nagel's arguments as they relate to inductive risk. Both Douglas and Vaesen and Katzav (2019) explore the relationship between Reichenbach's (1951) noncognitivism and claims about the value neutrality of the sciences, complementing themes in 14.3.

[9] The same anonymous reviewer in note 7 helpfully suggests Carnap as another source of influence for the views in 14.3. I agree, but focus on Stevenson given Feigl's praise of Stevenson's book and because Stevenson develops a fuller account than Carnap. Christian Damböck notes Carnap did not publish a developed account of noncognitivism, which makes it difficult to trace the reception of his views (Damböck 2022, p. 498). See Damböck for more on connections between Reichenbach's and Carnap's noncognitivism, and Carnap's appreciation for Stevenson. Thanks to the same reviewer for suggesting Damböck.

14.3 Arguments for the Value Neutrality of the Sciences

I start with Feigl's 1950 paper, "The Difference Between Knowledge and Valuation."[10] Of interest for our purposes is his characterization of disagreement in the sciences and disagreement in ethics. Regarding disagreement in the sciences, Feigl says, "[o]ne of the characteristic features of science… is the possibility of removing disagreements concerning matters of fact. Observation, measurement, experimentation and statistical analysis, together with generally well agreed upon methods of interpretation and inference, provide at least in principle the means for the adjudication of scientific disputes" (Feigl 1950, p. 40). In principle, "[e]veryone, equipped with the requisite intelligence and instruments, is in a position to examine the validity of assertions or hypotheses regarding matters of fact. This is the essential meaning of the *objectivity* (better perhaps, 'intersubjectivity') of the scientific method" (p. 40). It is important to note that claims about the relationships between means and ends are straightforwardly empirical for Feigl. With regards to practical reasoning about the means necessary to achieve ends, the objectivity characteristic of the sciences can be achieved. So long as there is agreement on ends or shared purposes and interests, the empirical methods of the sciences are relevant to successful means-end reasoning.

Note, though, that agreement on ends is needed to "evaluate means (actions and instrumentalities)" (p. 41). Someone might accept an empirical claim that a particular course of action achieves some end while rejecting the end itself. Indeed, Feigl thinks this type of disagreement characterizes ethical discourse. When it comes to the interests and purposes that set our ends, "we cannot ignore the actuality (or at least the possibility) of cognitively irremediable disagreements in regard to purposes" (ibid). When choosing ends themselves, "it remains possible that no matter how complete the information of each party to a dispute over *ends*, no agreement on these ends may be attainable" (ibid). Unlike disagreement in the sciences, there is no agreed-upon method for settling disagreement in ethics.[11] For Feigl, "[t]his implies that there is nothing in the nature of evaluation that ensures its intersubjectivity;" like Stevenson, he thinks ethical disagreement is settled by "persuasion, compromise, separation or coercion" rather than scientifically (ibid).

In part, this is due to a more fundamental difference Feigl sees between the sciences and ethics. He argues, "scientific knowledge consists of *information*…Scientific statements can by their very nature not contain such words as 'advisable,' 'desirable,' 'should,' 'ought,' or 'must'" (pp. 41–42). Here, again, is a linguistic gloss on the differences between the sciences and ethics. The sciences

[10] See Matthias Neuber (2022, §5) for an overview of Feigl on value judgments. See also Feigl (1952).

[11] An anonymous reviewer asks if Feigl considers the possibility of scientific disagreement that cannot be settled. Feigl notes "the notorious practical difficulties of the decision between rival hypotheses," but seems to think these difficulties are in-principle solvable by shared methods (1950, p. 40).

describe, whereas ethics provides "[t]he stamp of approval, the accent of endorsement, the force of encouragement, the factor of exhortation" in light of "the needs and interests of… agents" (p. 41). Scientific statements are silent with regards to approval, endorsement, encouragement, or exhortation. As such, "[s]cientific knowledge, even if it concerns the conditions and consequences of valuations, is in a perfectly clear sense value-free or value-neutral" (p. 41). Empirical information does not entail ethical judgments about purposes, interests, or ends.

In "Science and Human Values," Hempel also distinguishes two types of value judgments when considering the relationship between the scientific method and ethics. The first type, "a *relative*, or *instrumental, judgment of value…* states that a certain kind of action, *M*, is good (or that it is better than a given alternative M_1) if a specified goal is to be attained; or more accurately, that *M* is good, or appropriate, for the attainment of goal *G*" (Hempel 1960/1965, p. 84). Like Feigl, Hempel thinks the empirical methods of the sciences are relevant to instrumental judgments of value. Indeed, "a relative, or instrumental, judgment of value can be reformulated as a statement which expresses a universal or a probabilistic kind of means-end relationship, and which contains no terms of moral discourse… at all" (p. 85). Still, recalling Feigl's view, such hypothetical judgments do not require anyone adopt the particular ends in question.

Hempel's second type of value judgment relates to ends themselves, i.e., "an *absolute*, or *categorical, judgment of value* to the effect that a certain state of affairs (which may have been approved as a goal or end) is good, or that it is better than some specified alternative" (ibid). Consider a categorical judgment like 'Killing is evil'; Hempel claims "it serves to express a standard for moral appraisal or a norm of conduct" (ibid). Recalling noncognitivist accounts of ethical language, categorical judgments of value express approval or commitment to certain standards and norms, but they do not state anything like instrumental judgments do (ibid). Thus, Hempel claims, "[c]ategorical judgments of value, then, are not amenable to scientific test and confirmation or disconfirmation; for they do not express assertions but standards or norms for conduct" (p. 86). For Hempel, judgments about ends do not describe, and so do not have observational consequences and are not truth-apt (p. 85).

Still, when it comes to an instrumental judgment of value, "[s]cience can render an indispensable service by providing us with increasingly extensive and reliable information relevant to our purpose" (p. 89). However, the methods and results of science do not speak for or against categorical judgments of values typical of ethical discourse. For Hempel, "no set of scientific statements logically implies an unconditional judgment of value" (p. 91). Given the lack of logical connection between descriptive statements and categorical judgments of value, "it remains for us to *evaluate* the various probable sets of consequences of the alternative choices under consideration. And this requires the adoption of pertinent valuational standards which are not objectively determined by the empirical facts" (p. 89). Moreover, like Feigl and Stevenson, Hempel countenances the possibility of cognitively irremediable disagreement over the standards needed to evaluate the choices and consequences relevant to ethical decisions.

Themes and distinctions similar to Feigl's and Hempel's account of the value neutrality of the sciences appear in Nagel's discussion of the social sciences in *The Structure of Science*.[12] There, Nagel distinguishes two types of value judgments. The first is "the sense in which a value judgment expresses *approval* or *disapproval* either of some moral (or social) ideal, or of some action (or institution) because of a commitment to such an ideal" (Nagel 1961, p. 492). This recalls Hempel's characterization of categorical value judgment, the function of which is to express commitment to some value or approval or disapproval of an action. Nagel calls these "'appraising value judgments'" which hold that "some envisaged or actual state of affairs is worthy of approval or disapproval" (p. 493).

Nagel's second sense of value judgment is related to Feigl's and Hempel's views on means-end judgments. This is "the sense in which a value judgment expresses an *estimate* of the degree to which some commonly recognized (and more or less clearly defined) type of action, object, or institution is embodied in a given instance" (p. 492). Nagel uses 'anemia' as an example. He says, "although the meaning…can be made quite clear, it is not in fact defined with complete precision; for example, the notion of a 'normal' number of red corpuscles that enters into the definition of the term is itself somewhat vague, since this number varies with the individual members of a species as well as with the state of a given individual at different times" (ibid). On Nagel's account, given the vagueness associated with the term 'anemia,' a scientist "must judge whether the available evidence *warrants* the conclusion that the specimen is anemic" by using "the measure provided by… [an] assumed standard" (ibid). He calls these "characterizing value judgments" and they consist in "evaluations of the evidence, which conclude that a given characteristic is in some degree present (or absent) in a given instance" (p. 492). On my reading, characterizing value judgments single out a property normally presumed to be desirable or related to something like normal functioning, and are made when that property is determined to be present relative to an antecedently accepted standard.

Having distinguished two senses of value judgments, Nagel asks whether the social sciences can be value-neutral in light of arguments that the social scientist uses thick terms in the explanation of human behavior, e.g., in characterizing survey responses as "uninformed, deceitful, or irrational" (p. 494). Nagel grants the social scientist uses thick terms, which "as commonly used have a widely recognized pejorative overtone" (ibid); or, to draw upon Stevenson, they use words with emotive meanings. Nagel even grants that "anyone who employs such terms to characterize human behavior can normally be assumed to be stating his disapprobation of that behavior…and not simply characterizing it" (ibid). Nevertheless, he maintains social scientists using such terms need only assume the standard necessary to make a characterizing value judgment, and need not take the additional step of making an appraising value judgment.

This is because of an important difference between the functions of the types of value judgments. Appraising value judgments express approval or disapproval or

[12] See Anna Alexandrova (2018) for recent work critically engaging Nagel.

commitment to some norm, whereas characterizing value judgments make an empirical claim relative to an assumed standard, which someone may or may not approve or disapprove. For Nagel, "an investigator making a characterizing value judgment is not…logically bound to affirm or deny a corresponding appraising evaluation" (p. 493). While "it is not always easy to make the distinction formally explicit in the social sciences" (p. 494), keeping the two judgments separate in the practice of science and in a philosophical account of science "are essentially practical rather than theoretical problems" (p. 495). In other words, the use of thick terms by the social scientist does not show fact and value are logically intertwined in a way that makes the value neutrality of the social sciences impossible. We can draw a logical distinction between the two types of value judgments, even if it is practically difficult to do so in practice.

Yet again, like Feigl and Hempel, Nagel advances an argument for the value neutrality of science turning on the claim that value judgments about ends are not logically entailed by factual statements, even if means-end value judgments are empirical. One reason Nagel offers in support of this conclusion concerns a difference in the linguistic functions of scientific statements and ethical judgments. As I have been suggesting, these claims are central to the metaethical noncognitivism advanced by Stevenson decades earlier.

14.4 Is Value Neutrality Metaethics Free?

Nagel (1961, p. 493) maintains his formulation of the value neutrality of the sciences does not assume any particular metaethical view. Alexander says this is "clearly incorrect" (Alexander 1974, p. 394). As he makes clear, one way to understand Nagel's view—and *mutatis mutandis* Feigl's and Hempel's views—is that "*characterizations do not entail any appraisals*" (ibid). This holds only if empirical statements are not inductive or deductive grounds from which appraising value judgments expressing approval or disapproval follow. However, as Alexander argues, "[i]f one is a naturalist…the meaning of appraisals is rendered transparent by virtue of definitions composed entirely of characterizations. Thus, since the appraisal is ultimately nothing more than a characterization for naturalism, it is an obvious truth that the appraisals are entailed by characterizations on this metaethical view" (ibid). An example will illustrate Alexander's point.

Take Philippa Foot, who advances the sort of naturalist thesis Alexander has in mind. Foot is concerned with whether an evaluative conclusion can follow from nonevaluative premises (Foot 1958, pp. 506–509). To tackle this question, she considers the thick term 'rude.' According to her, 'rude' "expresses disapproval, is meant to be used when action is to be discouraged, implies that other things being equal the behavior to which it is applied will be avoided by the speaker, and so on" (p. 507). So, the judgment, e.g., 'Interrupting someone who is speaking is rude,' expresses one's disapproval of interrupting and is meant to discourage that behavior. In this way, uses of 'rude' have the same features as Stevenson's dynamic uses of words with emotive meaning, Hempel's categorical judgments of value, and Nagel's appraising value judgments. Yet, any application of 'rude' must meet factual criteria to be used correctly. In calling an action 'rude,' we are not only expressing a personal attitude, but claiming certain facts hold. We do not call just any action rude if we disapprove of it; 'rude' is applicable in cases where "behavior causes offence by indicating lack of respect" and inapplicable otherwise (ibid). Now, whether an instance of interrupting someone causes offense in a way that indicates a lack of respect is an empirical question. For Foot, one cannot make such a factual determination and deny the action was rude (p. 509). This holds for all competent speakers of a language; one cannot unilaterally reject such a definition simply because they do not have the requisite attitude.[13] In Alexander's terms, the meaning of the appraisal, 'That act of interrupting was rude,' is rendered transparent by the characterization, 'That act of interrupting caused offense in a way indicating lack of respect.' Where the latter empirical statement holds, the former value judgment follows.[14]

As such, arguments for the value neutrality of the sciences that claim empirical statements do not entail value judgments make metaethical assumptions a naturalist would reject. However, according to Alexander, there is a formulation of the value neutrality thesis that does not beg the question against naturalistic metaethics. In considering whether or not the sciences are value-neutral, Alexander thinks "we also have to consider whether the scientist is trying to achieve something with his descriptions or explanations which would make appraisals relevant and necessary to his job, and consequently, something he should include if his product is to be adequate and complete" (Alexander 1974, pp. 394–395). So, he asks: "what is the guiding conception of the empirical scientist's job?" (ibid). For Alexander, the answer to this question will tell us if the scientist is simply out to make descriptive statements, or if they must also make something like appraising value judgments to achieve their aims (ibid).

Alexander's answer is that the job of the scientist is "'to explore, to describe, to explain, and to predict the occurrences of the world we live in'" (p. 395; quoting Hempel 1966, p. 1). In light of this, he thinks we can grant metaethical naturalism, and still hold fast to value neutrality. Even if appraising value judgments follow directly from facts, given the task of the scientist, "the entailed appraisals are not *independently* relevant for scientific confirmation...since they are either epistemologically posterior to characterizations or strictly redundant" (Alexander 1974, p. 397). On one hand, if an appraising value judgment adds something on top of the characterizing judgments it logically follows from, e.g., some expression of approval, then a scientist need not make it; their job of describing, explaining, or predicting is discharged by the characterizing, or factual judgment since approval is not relevant to confirmation. On the other hand, if, as the naturalist claims, "an appraisal is simply some characterization...the appraisal can be ignored as redundant in favor

[13] On Foot's view, "whether a man is speaking of behaviour as rude or not rude, he must use the same criteria as anyone else" (Foot 1958, p. 509).

[14] Foot argues value judgments are derivable from empirical statements. Steven Diggin (2022) engages literature in science and values and argues empirical statements are derivable from ethical judgments.

of its definitionally equivalent characterization" (p. 398). Alexander argues, then, that no matter one's metaethical view, "[n]ot all of the logical consequences of an acceptable scientific proposition are scientifically relevant" in the sense of confirming or disconfirming the scientist's descriptions, explanations, or predictions (p. 399). Even when it comes to the act of publication, the scientist need not make explicit any appraisals that follow from characterizations since that would be taking on "the role of appraiser, and no longer simply… the role of a scientist" (p. 399).

Alexander's formulation of science's value neutrality does initially appear independent of controversial, noncognitivist glosses on ethical judgments. However, drawing upon Leach (1968a, 1968b), I argue in the next section that Alexander assumes a divide between the realms of theoretical and practical reason that a version of the argument from inductive risk challenges.

14.5 Inductive Risk, and the Practical and Theoretical

Alexander is right that securing the value neutrality of the sciences on the grounds that value judgments are not logically entailed by scientific statements because value judgments merely express approval or disapproval and scientific statements are factual is metaethically controversial. His attempt to formulate a metaethically neutral version of value neutrality appeals to the aims of the scientist, namely, to make well-supported, reliable claims and predictions about the world. Alexander's argument also relies on a distinction "between the *process* of science (the scientist's *activities* of choosing problems, forming concepts, validating hypotheses, etc.) and the *product* of that process (the *propositions* accepted into the body of scientific knowledge)" (Alexander 1974, p. 391). He restricts his focus to the product of knowledge, arguing "the scientific product, at least in principle, is value neutral in the sense that these propositions need only contain factual claims, and not also value claims" (ibid). He thinks even if value judgments are logically entailed by scientific statements, as ethical naturalists maintain, the scientist's job is discharged by making factual claims. Were the scientist to "use his scientific results for purposes of making practical appraisals," they would no longer be acting "qua scientist…but qua governmental consultant, concerned citizen, moral agent, or human being" (p. 400).

Importantly, Alexander thinks questions raised by versions of the argument from inductive risk about the relationship between theoretical reason and practical reason are independent of his argument for the value neutrality of the product of science; considerations of inductive risk apply only to what he identifies as the process and not the product of scientific practice (p. 392). I think this is mistaken because his conception of the confirmation of empirical statements makes assumptions about the relationship between belief and action that some would find controversial.

Consider Leach's (1968a, 1968b) defense of a particular form of the argument from inductive risk. Leach says arguments about the value neutrality of the sciences fail to specify "what sorts of value the scientist should avoid and hence what type of neutrality to maintain" (Leach 1968a, p. 94). Leach prefigures Alexander's focus on the scientist's role when he claims that one way to make value neutrality more specific is to look at "the goals of scientific inquiry" given that "the values ones advocates, seeks or avoids are correlated with goals or objectives" (ibid). Leach considers three different ways to formulate the goals of science corresponding to stronger or weaker forms of value neutrality. The strongest version—"the 'odds maker' view of science" (p. 95)—holds that the scientist only "assigns degrees of confirmation or probability to hypotheses so that others, decision makers, can decide as to the truth or falsity of the beliefs" (94). A less strong version—"the 'truth seeker' view"—holds "that the scientist, though himself a decision maker, seeks only one goal, has only one overriding value: truth at any cost" (p. 95). Finally, the weakest version of value neutrality—the "'cognitive decision maker' version" (ibid)—views "science as a quest not only for true hypotheses but also for ones with additional cognitive, theoretical, or epistemic values" (ibid).

Leach thinks the cognitive decision maker version of the value neutrality thesis is the "most cogent" (p. 95), and it appears to be the version Alexander endorses insofar as he includes activities like exploration and explanation in his account of the aims of science. To argue against this view, Leach defends "the 'practical decision maker' view of science ... [:] the scientist in the logic of his inquiry also makes pragmatic value judgments in pursuit of practical goals" (ibid). Leach's defense of the practical decision maker view turns on a reformulation of Richard Rudner's (1953) inductive risk argument in light of criticisms from Isaac Levi (1960). Levi argues Rudner's argument from inductive risk assumes too strong a connection between belief and action. Namely, Levi thinks the argument works only if accepting a hypothesis "is equivalent to choosing to act on the basis of H relative to some specific objective P" (Levi 1960, p. 348). He questions this equivalence by providing cases where one might accept the truth of a hypothesis, but be unwilling to act on it. Importantly, his argument parallels arguments for value neutrality that assume value judgments are not entailed by empirical statements. In this case, Levi denies a connection between theoretical belief (or acceptance) and practical action.

However, Leach thinks his version of the argument from inductive risk need not assume *equivalence* between accepting a hypothesis and acting on the basis of that hypothesis. Instead, he claims, "[t]he only conditions required for the argument are that belief be logically tied to action in a such a way that beliefs can be *partially justified* by practical considerations of action, and that different mistakes in action be taken with different degrees of seriousness" (Leach 1968a, pp. 100–101). Making this logical connection involves positing a relationship between accepting a hypothesis, having a disposition to act on the basis of that hypothesis relative to some practical goal, and acting in the appropriate circumstances. For Leach, practical considerations related to whether one is warranted, in an ethical sense, in acting in the appropriate circumstances on the basis of a hypothesis are relevant to assessing whether that hypothesis is rationally acceptable. On his view, "the reasonableness or rational acceptability of H relative to P [rests] not only on whether H is true or has high epistemic utility, but also upon its pragmatic value, upon whether or not action based upon H is warranted" (p. 102). Thus, Leach posits an "epistemic interdependence of acts, dispositions and beliefs" that need not be as strong as the equivalence relation Levi thinks Rudner's inductive risk argument needs (ibid).

How is this related to metaethics and Alexander's argument? First, it pushes back against Alexander's claim that value judgments "are not *independently* relevant for scientific confirmation" (Alexander 1974, p. 397). If there are connections or an interdependence between beliefs, dispositions to act, and actions, then value judgments are relevant to confirmation. Second, for Leach, inductive risk "forces a reconsideration...of the relations between theoretical, technological and policy making aspects of rational inquiry" (Leach 1968a, p. 107). This raises questions about the connection between belief and action related to metaethical disputes about the nature of fact, value, and the scopes of theoretical reason and practical reason. When it comes to the picture of rational inquiry advanced by proponents of the various versions of the argument from inductive risk, Leach thinks "No less at issue are the embarrassingly complex conceptual interrelationships between 'belief', 'value', 'action', 'accepting a belief or hypothesis', 'acting upon a belief' and 'reasonable belief'" (Leach 1968b, p. 366). And for him, it is "[a]lmost inconceivable... that an adequate analysis of 'belief' could be provided which is logically unrelated to 'action'. For there would then seem to be no way to account for the important influence of beliefs on action" (Leach 1968a, p. 102). Thus, Leach posits a connection between belief in a hypothesis and acting on that hypothesis that Alexander rejects. Value judgments are relevant to the aims of the scientist to describe and explain because descriptions and explanations are connected to action, and value judgments are needed to determine whether or not an action is reasonable. In turn, judging an action based on some hypothesis reasonable or unreasonable is relevant to confirming said hypothesis, e.g., in determining the amount of evidence we demand to accept it.

In effect, Leach's defense of something like Rudner's argument from inductive risk suggests Alexander cannot focus on the *product* of scientific investigation without also looking at the *process*. If Leach's argument from inductive risk goes through, value judgments relative to the costs of acting on an erroneous hypothesis are relevant to whether one's acceptance of that hypothesis is rational. I follow Leach in thinking that rejecting this connection requires substantive metaethical assumptions about the connection between the domains of theoretical reason (belief) and practical reason (action). So, while Alexander's argument for the value neutrality of the sciences does not follow Hempel, Feigl, and Nagel in taking a controversial stance on the cognitive status of value judgments that recalls Stevenson's emotivism, it is not completely metaethically innocent.

14.6 Conclusion

The history I sketched supports Brown's (2020) and Leach's (1968a) historical suggestions about the relationship between metaethical noncognitivism—especially as formulated by Stevenson—and certain arguments for the value neutrality of the sciences—especially Feigl's, Hempel's, and Nagel's. I also think Alexander's (1974) attempt to formulate a role-based, value neutrality thesis independent of metaethical claims prompts us to consider connections between the various arguments from inductive risk and metaethical questions. Drawing upon Leach (1968a, 1968b), I argued that Alexander's intervention raises further questions about the connections between theoretical reason and the generation of belief and practical reason and action. Importantly, Brown pursues related questions about the empirical status of value judgments, as well as the relationship between epistemic values, belief formation, and acts of assertion. In doing this, he explicitly eschews any "deep metaethics" (Brown 2020, p. 90 fn. 1). Yet, I think the metaethical issues touched on in the history I sketched, as well as other related issues, merit revisiting in light of recent calls and attempts to more fully characterize the nature and role of values in 'values in science.'[15]

In closing, I note the narrow focus of my historical sketch. Metaethical noncognitivism stands in stark contrast to the pragmatist tradition Leach (1968a, pp. 107–108) hints at adopting and Brown (2020) explicitly works within. In this tradition, given the tight connection between knowledge, valuation, and action, there is no hard distinction drawn between fact and value like there is for Stevenson. For many pragmatists, value judgments are a type of empirical statement. This is also true of some work in the feminist, pragmatist tradition Brown draws upon in developing his view, e.g., Lynn Hankinson Nelson (1990), Elizabeth Anderson (2004), and Sharyn Clough (2011). Note, too, that the metaethical characterizations of value judgments I examined are distinct from Thomas Kuhn's discussion of value judgments and theory choice. Kuhn (1977) glosses a value judgment as a nonalgorithmic decision made on the basis of commitments to certain theoretical virtues.[16] While scientists share the same commitments, different scientists might reasonably disagree on how to bring them to bear on theory choice. Given Kuhn's influence, as well as recent interest in pragmatist philosophies of science, a more complete history of the 'values' in 'values in science' would compare and contrast the noncognitivist, pragmatist, feminist, and Kuhnian treatments of value judgments.[17]

[15] See also Michael Scriven (1972), Dan Hicks (2014), Brown (2020), Zina Ward (2021), and Sarah Wieten (2021). An anonymous reviewer also mentions historical antecedents and related issues beyond the scope of this paper. These include David Hume's claim that passions, rather than belief, are needed to motivate action, and whether actions, which are nonpropositional, can ever be conclusions of arguments. I agree it could be fruitful to look into how these issues relate to arguments for and against the value-free ideal.

[16] Thanks to P.D. Magnus for a discussion of this point.

[17] Of particular interest to a more complete history is the robust debate in feminist philosophy concerning the empirical status of value judgments. For a recent overview of this debate as it relates to science and values, see Christopher ChoGlueck and Elisabeth Lloyd (2023). See Audrey Yap (2016) for critical engagement with Nelson, Anderson, and Clough on the empirical status of value judgments. Kristen Intemann (2001) is also of interest since she pushes back against those who would conflate 'empirical' and 'descriptive.' For Intemann, value judgments might be empirical even if not descriptive. Thanks to Jer Steeger for helpful conversations related to points in this note.

Acknowledgments Versions of this paper were presented at the Philosophy of Science Association's 2022 poster session, Cal State-Long Beach's Philosophy Day!, and UW Philosophy's works-in-progress series. Thanks to those audiences for discussion and encouragement. Thanks also to two anonymous reviewers for helpful suggestions, and to the students in my Autumn 2023 seminar for working through some of the history of science and values with me. Special thanks to Matt Brown for his support over the years.

References

Alexander, J. 1974. Metaethics and value neutrality in science. *Philosophical Studies* 25 (6): 391–401.

Alexandrova, A. 2018. Can the science of Well-being be objective? *British Journal for the Philosophy of Science* 69:421–445.

Anderson, E. 2004. Uses of value judgments in science: A general argument, with lessons from a case study of feminist research on divorce. *Hypatia* 19 (1): 1–24.

Ayer, A. J. 1946/1952. *Language, truth and logic*. New York: Dover.

Ayer, A. J., ed. 1959. *Logical Positivism*. New York: The Free Press.

Brown, M. 2020. *Science and moral imagination: A new ideal for values in science*. Pittsburgh: Pittsburgh University Press.

Carnap, R. 1932/1959. The elimination of metaphysics through the logical analysis of language. In *Logical Positivism*, ed. A. J. Ayer, 60–82. New York: The Free Press.

Carnap, R. 1935. *Philosophy and logical syntax*. London: Kegan Paul.

ChoGlueck, C., and E. Lloyd. 2023. Values as heuristics: A contextual empiricist account of assessing values scientifically. *Synthese* 201:220.

Clough, S. 2011. Radical interpretation, feminism, and science. In *Dialogues with Davidson*, ed. J. Malpas, 405–426. Cambridge: MIT Press.

Damböck, C. 2022. The politics of Carnap's non-cognitivism and the scientific world-conception of left-wing logical empiricism. *Perspectives on Science* 30 (4): 493–524.

Diggin, S. 2022. Ethical evidence. *Synthese* 200 (329): 1–24.

Douglas, H. 2009. *Science, policy, and the value-free ideal*. Pittsburgh: Pittsburgh University Press.

Feigl, H. 1946. Review of ethics and language by C.L Stevenson. *Philosophy of Science* 13 (1): 80–80.

Feigl, H. 1950. The difference between knowledge and valuation. *Journal of Social Issues* 6 (4): 39–44.

Feigl, H. 1952. Validation and vindication: An analysis of the nature and the limits of ethical arguments. In *Readings in ethical theory*, ed. W. Sellars and J. Hospers, 667–680. New York: Appleton-Century-Crofts.

Foot, P. 1958. Moral arguments. *Mind* 67 (268): 502–513.

Hempel, C. 1960/1965. Science and human values. In *Aspects of scientific explanation and other essays in the philosophy of science*, ed. C. Hempel, 81–96. New York: The Free Press.

Hempel, C. 1966. *Philosophy of natural science*. Upper Saddle River: Prentice-Hall.

Howard, D. 2003. Two left turns make a right: On the curious political career of north American philosophy of science at midcentury. In *Minnesota studies in the philosophy of science volume XVIII: Logical empiricism in North America*, ed. G. Hardcastle and A. W. Richardson, 29–93. Minneapolis: University of Minnesota Press.

Hicks, D. 2014. A new direction for science and values. *Synthese* 191:3271–3295.

Intemann, K. 2001. Science and values: Are value judgments always irrelevant to the justification of scientific claims? *Philosophy of Science* 68 (3): S506–S518.

Kuhn, T. 1977. Objectivity, value judgment, and theory choice. In *The essential tension: Selected studies in scientific tradition and change*, ed. T. Kuhn, 320–339. Chicago: University of Chicago Press.
Leach, J. 1968a. Explanation and value neutrality. *The British Journal for Philosophy of Science* 19 (2): 93–108.
Leach, J. 1968b. Historical objectivity and value neutrality. *Inquiry: An Interdisciplinary Journal of Philosophy* 11 (1-4): 349–367.
Levi, I. 1960. Must the scientist make value judgments? *The Journal of Philosophy* 57 (11): 345–357.
Moore, G. E. 1903/1959. *Principia Ethica*. Cambridge: Cambridge University Press.
Nagel, E. 1961. *The structure of science: Problems in the logic of explanation*. Vol. 29, 716. New York: Harcourt, Brace, and World.
Nelson, L. 1990. *Who knows: From quine to a feminist empiricism*. Philadelphia: Temple University Press.
Neuber, M. 2022. Herbert Feigl. *The Stanford encyclopedia of philosophy* (Winter 2022 Edition), Edward N. Zalta & Uri Nodelman (eds.): URL: https://plato.stanford.edu/archives/win2022/entries/feigl/
Ogden, C., and I. Richards. 1923/1945. *The meaning of meaning*. 7th ed. New York: Harcourt, Brace and Company.
Reichenbach, H. 1951. *The rise of scientific philosophy*. Berkeley: University of California Press.
Rudner, R. 1953. The scientist qua scientist makes value judgments. *Philosophy of Science* 20 (1): 1–6.
Scriven, M. 1972. The exact role of value judgments in science. *PSA: Proceedings of the biennial meeting of the Philosophy of Science Association*: 219–247.
Siegetsleitner, A. 2022. Noncognitivism. In *The Routledge handbook of logical empiricism*, ed. T. Uebel and C. Limbeck-Lilienau, 168–175. London: Routledge.
Stevenson, C. 1937. The emotive meaning of ethical terms. *Mind* 46 (181): 14–31.
Stevenson, C. L. 1948/1963. "*The Nature of Ethical Disagreement*." Facts and Values: Studies in Ethical Analysis. New Haven: Yale University Press: 1–9.
Stebbing, S. 1930/1942. *A Modern Introduction to Logic*. London: Methuen & Co.
Vaesen, K., and J. Katzav. 2019. The National Science Foundation and philosophy of science's withdrawal from social concerns. *Studies in History and Philosophy of Science* 78:73–82.
Ward, Z. 2021. On Value-Laden Science. *Studies in History and Philosophy of Science, Part A* 85:54–62.
Wieten, S. 2021. Review of *science and moral imagination* by Matthew J. Brown: Implications for values in medicine. *Studies in History and Philosophy of Science Part A* 89:299–300.
Yap, A. 2016. Feminist radical empiricism, values, and evidence. *Hypatia* 31 (1): 58–73.

Chapter 15
Science, Values, and Action Guidance: Can we Stop Talking about the Value Free Ideal?

Greg Lusk

Abstract There has recently been a resurgence of interest in defending the value-free ideal (VFI). Many of these defenses are similar in style: the VFI is supposed to play the role of a regulative ideal. In this chapter, I examine what it would mean for the VFI to play the role of a regulative ideal in light of the argument from inductive risk and assess what kind of guidance the VFI could provide. I argue that the VFI (on its own) is not very determinate: it gives too little guidance on how to make decisions in which values are required. In fact, we can almost always expect better guidance to be available via appeals to more fundamental normative goods, the same normative goods that might motivate value free idealists. These normative goods, however, are not synonymous with the VFI, are conceptually richer than the VFI, and do not uniquely recommend the VFI. I suggest there are good reasons to appeal to these goods rather than the VFI, and doing so places philosophical discussions on richer ground. Thus, we have good reasons to stop talking about the value free ideal.

Keywords Value free ideal · Non-epistemic values · Argument from inductive risk · Normative ideals · Regulative ideals · Values in science

15.1 Introduction

> Philosophers of science and scientists over the past century at least have worried about the influence of values over science as a destructive, biasing factor. Many have sought to defend the "ideal of value-free science," according to which scientists ought to strive to eliminate the influence of values over science. Of course, scientists are human, and thus imperfect, but the scientific process includes checks and balances that limit the impact of their biases.

G. Lusk (✉)
Department of Philosophy, Durham University, Durham, UK
e-mail: greg.lusk@durham.ac.uk

J. Y. Tsou et al. (eds.), *Values, Pluralism, and Pragmatism: Themes from the Work of Matthew J. Brown*, Boston Studies in the Philosophy and History of Science 347, https://doi.org/10.1007/978-3-031-92958-8_15

> These concerns, while understandable, are ultimately mistaken…. For one, they are based on a misunderstanding of the nature of values and value judgment…. For another, the need for value judgment is an inescapable feature of the scientific process…" (Brown 2020, p. 131).

It had once seemed that the position adopted by Matt Brown above was becoming the philosophical consensus: non-epistemic values (e.g., personal or political values) are (at minimum) legitimate to employ when genuine uncertainty arises in decisions concerning scientific methodology (e.g., Douglas 2000; Steel 2010; Brown 2013; Elliott 2017; ChoGlueck 2018). Some have even argued that such values should have wider purchase, perhaps even trumping epistemic considerations (e.g., Elliott and McKaughan 2014; Brown 2017; Lusk and Elliott 2022). However, there seems to be a resurgence of interest in maintaining the so-called "value-free ideal" which prohibits non-epistemic values from playing a role in the core of science. Several recent accounts argue that the ideal is something to strive for and propose ways of enacting its benefits (e.g., Hudson 2021; Carrier 2022; Parker 2024).

In response to this resurgence, I defend the view that we would be better off if we stopped discussing the value-free ideal. I examine what it would mean for the VFI to play the role of a regulative ideal in light of a strong challenge from the argument from inductive risk. That is, I assess what kind of guidance the VFI provides in the face of the conceptual ineliminability of values in the core of science. I argue that the VFI (on its own) is not very determinate: it gives too little guidance on how to make decisions in which values are required. In fact, we can almost always expect better guidance to be available via appeals to more fundamental normative goods, the same normative goods that might motivate value free idealists. These normative goods, however, are not synonymous with the VFI, are conceptually richer than the VFI, and do not uniquely recommend the VFI. Thus, I suggest there are good reasons to appeal to these goods rather than the VFI, and doing so places philosophical discussions on richer ground.

15.2 The Value Free Ideal(s)

The general thrust of the value-free ideal (VFI) is that it prohibits non-epistemic values (i.e., values that are not truth apt, which are typically "social", "political", "contextual" or "personal" values, like the value of life or preference for profit) from playing any substantial role in central aspects of scientific reasoning. The aspects over which the ideal is thought to govern are typically the epistemic or rational components of investigation; those that were historically thought of as part of the "context of justification" (see Brown 2020, p. 132). For example, those that defend the VFI are generally uncomfortable with non-epistemic values setting the standards of evidence for theory evaluation, only permitting so-called epistemic values (e.g., simplicity, accuracy) to play such roles. Value free idealists though, often take no issue with non-epistemic values helping to select which projects to pursue or how results are communicated. There is thus a distinction between the rational "core" of science, where non-epistemic value influence is impermissible, and aspects of science laying outside the core, that can be influenced by such values. Henceforth I will use "value" to refer exclusively to non-epistemic values.

The value free ideal is pithily described by Douglas as the "epistemic-values-only-in-scientific-inference ideal" (Douglas 2015, p. 5).[1] For our purposes here, it's worth distinguishing between two versions of the ideal. The first is a *normative* version, which claims that science *should* proceed in a value-free way, while the second is a *conceptual* version that claims there is a procedure by which science *can* proceed in a value-free way. We will see how the argument from inductive risk seeks to undermine value freedom in both ideals simultaneously.

15.3 The Strong Version of the Argument from Inductive Risk

Science, due to its overwhelming reliance on induction, is inherently uncertain: one may never make an evidence-based decision with certainty. Thus, there is typically a non-negligible risk of erring when advancing or adopting a scientific claim, that is, there exists inductive risk. As the inductive risk argument goes, successfully dealing with inductive risk purportedly involves the use of non-epistemic values in the core of scientific reasoning. While what is included in "the core of science reasoning" may be contentious, aspects that are commonly included are those that were once closely linked to the "context of justification" including hypothesis testing, model assessment, instrument selection, and data analysis. Inductive risk is widely acknowledged (e.g., see Brown 2020, p. 165) as one of the most influential and important counters to value freedom (though, even among VFI opponents, it has detractors, e.g., de Melo-Martín and Intemann 2016). Several variations of argument based on inductive risk appear in the literature, but here I only consider one of the most influential.[2]

[1] It is worth noting that there are other conceptions of the VFI, like that offered by Melo-Martín and Intemann (2016), which suggests that removing scientists' values in inductive risk contexts is sufficient to meet the demands of the VFI. I will concentrate on Douglas' formulation because of its prevalence in the literature, but also because it's questionable why the replacement of one set of values by another would be anymore permissible or legitimate (see Lusk 2020). See Elliott (2025) for a longer discussion of different conceptions of the VFI.

[2] Ward (2021) enumerates four forms of the argument that can be found in the literature though AIR-Must is singled out as the strongest.

15.4 AIR-Must

The AIR-Must argument is sometimes considered to be the strongest argument against value freedom (Ward 2021). The conclusion of the argument is that social values are ineliminable (in a conceptual sense) even in core aspects of scientific decision making. AIR-Must can be articulated as follows:

1. Core aspects of scientific reasoning force decisions that involve a non-negligible risk of error.
2. Some standards must be adopted to specify what degree of error would be tolerable when making core scientific decisions.
3. Such standards require consideration of the non-epistemic consequences of error and cannot be formulated using only epistemic considerations.
4. Therefore, non-epistemic considerations or values must play a role in setting standards for making such decisions, and thus are required to enter the core of scientific reasoning.

Essentially, the argument's first two premises establish that there are decisions within the core of scientific reasoning that are uncertain. Such decisions could involve hypothesis acceptance, choosing a model, a measurement procedure, or an analysis method, etc. (see Rudner 1953; Douglas 2009; Intemann 2015; Parker and Lusk 2019). The third premise establishes that these decisions are *epistemically unforced*, that is, how error should be handled cannot be justified with recourse to only epistemic factors or values. I take this premise for granted and refer the reader to the substantial literature for justification (e.g., see Winsberg 2012; Douglas 2017; Elliott and Richards 2017; Ward 2021; Brown 2020). In the conclusion (line four above), I use "enter the core of scientific reasoning" as to remain agnostic about whether the standards themselves are part of the core, or merely that the non-epistemic values they express apply to core decisions. Regardless, since core decisions in scientific reasoning cannot be made on purely epistemic grounds (and such decisions must be made), then non-epistemic factors must play a role in these decisions.

AIR-Must's strength seemingly comes from its ability to undermine both the normative and conceptual ideals. That it undermines the conceptual ideal is quite clear: if there are decisions in the core of science that require non-epistemic values, then such values play an essential role in scientific reasoning. The need for non-epistemic values demonstrates that they are essential for successful scientific reasoning. The conceptual ideal is defeated. I assume that the argument is successful in this respect.[3]

However, what makes the AIR-Must argument powerful is that ineliminability is thought by many to undermine the *normative* ideal. As Christopher ChoGlueck

[3] AIR-Must has its detractors, which include Levi (1960), Mitchell (2004), and John (2015); however, Ward (2021) argues convincingly that these authors use values in a motivational sense rather than the justificatory sense in which the argument is formulated. One might also think that there are other ways to epistemically force a decision—by convention, for example (Wilholt 2022). I leave the reader to enjoy the relevant literature.

articulates it, "[i]f, however, scientific reasoning ought to embody the value-free ideal, then it must be the case that scientific reasoning could be free from these extrascientific values" (ChoGlueck 2018, p. 705). Reiss and Sprenger express something similar: "[f]or value-freedom to be a reasonable ideal, it must not be a goal beyond reach and be attainable at least to some degree" (2020). An anonymous reviewer has perhaps put it most clearly:

> ...there is an implicit principle at work in the background of this argument that would permit it to undermine not just the [conceptual] ideal, but also the normative ideal. That principle is some version of "ought implies can." That is, if it ought to be the case that science is value free, as the normative ideal claims, then it ought to be the case that science can be value free. And since science cannot be value free, therefore, it ought not be value free. This line of thought would undermine the normative ideal.[4]

If indeed, one holds that ought implies can, then AIR-Must is a decisive argument against the VFI. However, that ought implies can (OIC)—or in this case, that cannot implies ought not—is not the truism that philosophers sometimes make it out to be. A popular intuition that supports the truism is that for an ought to provide guidance for action—which is essentially what any obligation seems to do—then it must be the case for one to do what they ought. Roughly then, if one ought to ϕ, then one must be able to ϕ; an unfulfillable obligation cannot be an obligation at all. If one cannot ϕ then it would seem as if the obligation to ϕ cannot provide any guidance for action.[5] In the terms that we are interested in: if one ought to keep the core of science value free, then it seems that they ought to be able to operate in a value-free way. Otherwise, what guidance does the VFI provide the scientist? Since, as many philosophers agree that AIR-Must shows that non-epistemic values are ineliminable, the VFI creates an unfulfillable obligation that cannot provide guidance for scientific decision making.

However, as Brian Talbot (2016) points out, there are many unfulfillable obligations that provide action guidance. Consider their example of a compulsive kleptomaniac. It seems clear—given social norms—that one ought not to steal, especially when the act of stealing is not for subsistence or livelihood. Yet, an untreated compulsive kleptomaniac is unable to stop stealing regardless of context, and thus has an unfulfillable obligation. Yet, though this obligation is unfulfillable, it still seems able to provide action guidance. For example, the compulsive kleptomaniac should recognize the wrongness of their actions and apologize for them if they have the ability to. The obligation also provides guidance for third parties: friends of the kleptomaniac have a duty to inform those that might become victims and have an obligation not to accept stolen goods themselves. Unfulfillable obligations are thus not irrelevant for guiding actions.

The more important counter to OIC for our purposes come from ideals: "There may be oughts which are neither fulfillable by the obligated agent nor by ordinary

[4] I realize directly quoting an anonymous reviewer, from a different piece of submitted and now published work, is somewhat atypical. However, the reviewer so nicely expressed the position that simply thanking them is insufficient (though it is still necessary). They deserve credit; their comments were thought provoking. Please accept my gratitude and appreciation, whoever you are.

[5] For discussions of OIC along these lines see Williams (1995) and Howard-Snyder (2006).

agents. Good examples come from the consideration of ideals" (Talbot 2016, 380). Ideals are often formulated in such a manner that they are unfulfillable; Talbot's examples are "the ideal to love all people as one loves oneself" or "the ideal to bring one's emotions and intellect into full accord" (ibid). Despite these ideals perhaps being unobtainable end states, the desire to achieve them may still provide guidance to individuals. For example, when faced with a choice to forgive a friend or completely unfriend them, the ideal to love all people would recommend forgiveness.

Value free idealists respond to the AIR-Must argument in exactly the way suggested above: by interpreting the VFI as a *regulative* ideal that points toward an obligation, but an obligation that is unfulfillable (in most instances, if not always). The role of values may be ineliminable, but we should see the VFI as a valuable guide. For example, Robert Hudson specifies that "Value- freedom, rather, is a normative ideal that scientists strive to attain" (Hudson 2016, p. 168). In later work, Hudson (2021) specifies how science can be made more value free.[6] Similarly, Martin Carrier advocates for value freedom while also asserting that there are "limits to upholding the value-free ideal in scientific research" (Carrier 2022, p. 5). These authors take the VFI qua regulative ideal to be action guiding, and thus, they can maintain their position even in the face of value ineliminability.[7] As long as value free idealists can demonstrate the action guiding nature of their position, then it remains viable. However, I suggest that herein lies the difficulty.

15.5 To What Extent Can the VFI Guide Action?

For the VFI to be viable, it must be shown that it can guide action in reasonable ways when inductive risk is present. Can it? Those sympathetic to the value free ideal have several suggestions about how to ensure science is consistent with the ideal. For example, one might be able to report results in coarse intervals rather than as precise predictions to reduce the influence of values on results—as is done with uncertainties connected with climate model projections (Parker 2014). Similarly, one might try to hedge results to eliminate the uncertainties that generate the need for values (Betz 2013). Still, others sympathetic to the VFI suggest strategies like adopting values specified by political bodies or other external groups upon which scientists conditionalize results (Carrier 2022; Parker 2024), probing the influence of different values on the analysis and reporting their impact, or being transparent about the values employed (e.g., by listing them alongside the results, see Carrier 2022). Carrier claims, "[p]olicy advice given along these conceptual lines could proceed in agreement with the value-free ideal" (Carrier 2022, p. 17). These are all

[6] While Hudson talks about striving to manage values, it is unclear whether they believe the obligations of the VFI are unfulfillable.

[7] For a particularly innovative way of treating the VFI as an ideal see Menon and Stegenga (2023).

strategies that might be employed to reduce the influence of values even in decision contexts where values are ineliminable.

Given the framing I've adopted here, for the VFI to be viable, it should be able to serve as a regulative ideal, even when values are ineliminable. That is, there should be some way to use the value free ideal in AIR-Must (i.e., inductive risk) situations to guide decisions towards value freedom. But what does that mean? Within the contemporary philosophy of science literature, there is very little discussion to help with this question. Perhaps more surprisingly, even in areas of philosophy where ideals are commonly discussed and play central theoretical roles (e.g., ideal vs. non-ideal theory in political philosophy), what is meant by action guidance is seldom made clear (see Chahboun 2019). To some extent then, we are treading on new ground within the philosophical debate on values in science. From here on out then, I will try to plot a fruitful path forward, fully acknowledging that it might be difficult to see other routes from within the thicket.

With any new journey, one must find a place to start. One of the only explicit lines of thinking is that action guidance is the degree to which an action is determined by an ideal or principle. For example, Richard North (2017) argues that action guidance is the extent to which something can function as a decision-making procedure. Thus, the more specific an ideal recommends which action should be taken the more determinant or action guiding the ideal is. Let us assume this definition and see where we end up.

The difficulty for proponents of the VFI is that the ideal is not very determinant in the cases where AIR-Must applies and values are ineliminable. For example, how could the VFI be used to decide between the value mitigating procedures above? For example, whether to employ values prescribed by a commissioning political body, or probe the influence of different sets of values and report the results, or both? Such a decision involves value judgements (Douglas 2009) and a different decision can be reasonably expected to have different justifications and outcomes. For example, using values prescribed by a political body is likely to be justified by appeal to a certain set of political or democratic values, whereas probing differential influence might be justified by appeal to a value of transparency or distributive justice. It's also not clear that doing both would create a dual justification rather than produce value tension (e.g., probing different sets of values and including those not held by the political body could be perceived as politically undermining). The VFI doesn't provide a procedure or justification for choosing one of these approaches over another, and the choice—rather than eliminating value influence—appears to maintain it. Perhaps even more importantly, if the choice of approach requires appeal to inherently contextual factors—as the example above suggestions—then the VFI can never be *fully* determinant.

One obvious response by the value idealist would be to claim the VFI does provide a clear decision procedure: minimize the influence of non-epistemic values as much as possible. But again, it's not clear what this kind of guidance this amounts to. Pretend one faces a decision between probing the outcomes of different values within an investigation or inserting a single set of values adopted by convention by a scientific association. It's unclear which of these decisions would minimize the value influence. Both reduce the influence of values held by individual investigators. But on one hand, probing the outcomes of different values inserts several different values into a scientific process and then reveals how the outcome depends on them. On the other hand, adopting conventional values allows a limited number of values to dominate the research process across many contexts. One gives little weight to a lot of values, the other lots of weight to a few values. Which is more in line with the VFI? It's not clear what the relevant "influence" to be minimized is and thus it's unclear what one should do when values are ineliminable. One needs some kind of measure or gauge to compare the value influence of certain decisions, but as of yet, there's no general measure on offer.

It is sometimes possible to give something like a rank order of value influence. Take Parker's (2014) example of coarse graining the uncertainty estimates associated with climate change projections. As many have noted, decisions made in the process of developing a climate model and producing model projections involve value decisions. If one were to have made slightly different decisions, based on the inclusion of slightly different values, then it's expected that one would get a different distribution for the projection. If one reports a very precise uncertainty estimate with that distribution, the chance that the values used to justify the climate modeling process influenced that estimate would be rather high. Thus, one can reduce the influence of values on the uncertainty estimates by coarse graining them and making them less precise: "the influence of those values on estimates of uncertainty will be reduced when coarser estimates are given. The influence will be reduced insofar as choices in model development will less often make a difference to the uncertainty estimates produced. Whereas any change to the distribution of modeling results in an ensemble study will change the single-valued probabilities estimated as a function of that distribution (assuming no fortuitous cancelation in the effects of the changes to the distribution), coarser estimates will be less sensitive to such changes, often much less so" (Parker 2014, p. 28). In short, coarser estimates are less likely to be influenced by values than precise ones.

The above example might represent one of the stronger cases for the VFI providing action guidance: the ideal specifies that one should minimize value influence where possible, and one method of reporting results can be seen to minimize the influence of values relative to another; i.e., the VFI recommends coarse graining. However, that may be where its action guidance ends. Once one begins coarse graining results, they must decide how coarse grained the results should be. This is a decision that is also open to inductive risk considerations, as assigning the wrong uncertainty to results is likely to have social consequences. Again, it's somewhat unclear what action guidance the VFI recommends. One could select the most coarse-grained approach possible, and thus hopefully maximally curtail the influence of values, but such information may be too coarse to be epistemically useful in situations where such information is requested.[8] One could try to match the coarseness of the information to the level of indifference that users of the information have, but this would be permit user-values to guide action, and not the VFI. One could adopt a "reasonable doubt" stopping criterion, as Betz (2013) does in a similar proposal about how drastically to hedge hypotheses. Again, however, it's unclear what justifies that criterion in light of ineliminable inductive risk, nor how the VFI recommends it.

Still there is a strong intuition that the VFI isn't completely impotent in cases where values are ineliminable. For example, all things being equal, it seems more problematic for scientists to insert their own values into, for example, national climate policy or pandemic planning, than it is for them to rely on the values held by political representatives or the conventions of a particular discipline.[9] Where values cannot be eliminated, relying on values held by political representatives respects the division of labor in democracies, where representatives are typically the ones elected to make normative decisions. Relying on the conventions of academic associations similarly relieves individual scientists of making the burden of making value decisions and would likely reduce the number of different values employed in decisions.

However, it is worth examining what guides appeals to representatives or conventions (or other mitigating approaches). While the above may be positive ways to handle values in certain contexts, what justifies employing the values of political representatives, for example, is not the VFI, but rather fundamental principles of democracy that specify political neutrality or the division of political labor. Respect for democratic principles is a common motivation of those that promote the VFI (see Steele 2012; Betz 2013; Carrier 2022). However, appealing to a theory of democracy to guide action is very different from appealing to the VFI as a regulative ideal, because what is best for democracy may come apart from what the VFI recommends (see Lusk 2021). So even if promoting democratic values happened to be the right course of action, the proper locus of action guidance would be fundamental principles of democracy.

Herein lies my suggestion to move this conversation forward. It is clear to see that there are a variety of normative motivations for the VFI, like democratic legitimacy, trust in science, scientific authority, and scientific veracity or objectivity. While a desire to further these normative goods motivate VFI defenders, there is no reason to think that they will all align: they will likely recommend different, and often conflicting, actions when values are ineliminable. For example, it's easy to imagine that scientists employing the values of elected representatives within policy advice might decrease scientific authority (by deferring to politicians) but enhance political legitimacy (by deferring to elected bodies) or erode trust (by bucking disciplinary norms). But unless one thinks that the VFI is desirable in its own

[8] See the exchange between Betz (2013) and John (2015) for a similar concern when it comes to hedged hypotheses.

[9] See Wilholt (2022) for a particularly insightful analysis of epistemic interests and disciplinary conventions.

right—which seems rare (see Holman and Wilholt 2022)—striving for the VFI is not an equal means of promoting all the ends that typically motivate it.

I suggest that this conversation can be moved forward by dropping talk of the VFI and replacing it with discussion of the substantive goods that often motivate it. That is, in instances where values in science are ineliminable or legitimate, substantive normative principles—like those related to political legitimacy, trust, veracity or objectivity—seem much more capable of providing action guidance. For example, if one needs to decide how coarsely one should report results or uncertainties, political principles about divisions of labor can seemingly recommend a level of granularity that corresponds to the level of indifference of elected policy makers in a way that the VFI cannot. If one is deciding whether to present an ensemble of results that track different value influences, or follow a disciplinary convention, principles of trust are more likely to recommend one action over another, in a way that the VFI does not. They not only provide enhanced action guidance, but the potential consequences of decisions made for these reasons can be more easily measured and assessed. This suggests a form of VFI eliminativism, where the VFI could be reduced to more fruitful philosophical concepts and then dispensed with.[10]

This suggestion is not a significant loss or gain for either side of the VFI debate; it essentially asks both sides to more explicitly articulate the principles by which value laden decisions should be navigated. Opponents of the VFI may be more comfortable appealing to such normative motivations, however, proponents of the VFI self-describe as being motivated by these normative goods, even if the VFI itself does not necessarily promote them. After all, value free idealists do not want to pursue particular normative goods because of their commitment to the VFI, rather, it's the other way around.

15.6 Conclusion

Faced with AIR-Must, value free idealists often retreat to a position where the VFI serves as an unobtainable regulative ideal. However, the value-free idealist faces a challenge: to serve as a regulative ideal, the VFI must be action guiding. I've provided one influential notion of action guidance and argued that the VFI is not very determinate. Decisions are better determined by appeals to normative goods not mentioned in the VFI, but that often motivate its deployment. Furthermore, not all these normative goods are likely to uniquely recommend an action in line with the VFI, nor would they all agree on which action to take. The burden is on the idealist to show how the VFI can successfully function as a regulative ideal.

Rather than try to meet this burden, I suggest a shift in the way that philosophers discuss values in science. Rather than advocate or attack the value free ideal, the target of analysis should be the more fundamental normative goods that motivated

[10] See Elliott's (2025) contribution to this volume for another view on how we might complicate or dispense with the VFI.

proponents of the ideal and its detractors. These normative goods might in fact limit or promote certain values in certain cases, but regardless, the resulting conversation would be more rigorous and fruitful.

References

Betz, G. 2013. In Defence of the value free ideal. *European Journal for Philosophy of Science* 3 (2): 207–220.

Brown, M. 2013. Values in science beyond underdetermination and inductive risk. *Philosophy of Science* 80 (5): 829–839.

Brown, M. 2017. Values in science: Against epistemic priority. In *Current controversies in values and science*, ed. K. Elliott and D. Steel, 64–78. London: Routledge.

Brown, M. 2020. *Science and moral imagination: A new ideal for values in science*. Pittsburgh: University of Pittsburgh Press. (ebook pagination).

Carrier, M. 2022. What does good science-based advice to politics look like? *Journal for General Philosophy of Science* 53 (1): 5–21.

Chahboun, N. 2019. Ideal theory and action-guidance: Why we still disagree. *Social Theory and Practice* 45 (4): 549–578.

ChoGlueck, C. 2018. The error is in the gap: Synthesizing accounts for societal values in science. *Philosophy of Science* 85 (4): 704–725.

Douglas, H. 2000. Inductive risk and values in science. *Philosophy of Science* 67 (4): 559–579.

Douglas, H. 2009. *Science, policy, and the value-free ideal*. Pittsburgh: University of Pittsburgh Press.

Douglas, H. 2015. Values in Science. In *Oxford handbook in the philosophy of science*, ed. P. Humphreys, 609–631. Oxford: Oxford University Press.

Douglas, H. 2017. Why inductive risk requires values in science. In *Current controversies in values and science*, ed. K. Elliott and D. Steel, 81–93. London: Routledge.

Elliott, K., and D. McKaughan. 2014. Nonepistemic values and the multiple goals of science. *Philosophy of Science* 81 (1): 1–21.

Elliott, K. 2017. *A tapestry of values: An introduction to values in science*. Oxford: Oxford University Press.

Elliott, K. 2025. Characterizing the value-free ideal: From a dichotomy to a multiplicity? In *Values, pluralism, and pragmatism: Themes from the work of Matthew J. Brown*, ed. J. Tsou, J. Shaw, and C. Fehr. Cham: Boston Studies in the Philosophy and History of Science.

Elliott, K., and T. Richards. 2017. *Exploring inductive risk: Case studies of values in science*. Oxford: Oxford University Press.

Holman, B., and T. Wilholt. 2022. The new demarcation problem. *Studies in History and Philosophy of Science* 91:211–220.

Howard-Snyder, F. 2006. Cannot' implies 'not ought. *Philosophical Studies* 130 (2): 233–246.

Hudson, R. 2016. Why we should not reject the value-free ideal of science. *Perspectives on Science* 24 (2): 167–191.

Hudson, R. 2021. Should we strive to make science bias-free? A philosophical assessment of the reproducibility crisis. *Journal for General Philosophy of Science* 52 (3): 389–405.

Intemann, K. 2015. Distinguishing between legitimate and illegitimate values in climate modeling. *European Journal for Philosophy of Science* 5 (2): 217–232.

John, S. 2015. The example of the IPCC does not vindicate the value free ideal: A reply to Gregor Betz. *European Journal for Philosophy of Science* 5 (1): 1–13.

Levi, I. 1960. Must the scientist make value judgments? *The Journal of Philosophy* 57 (11): 345–357.

Lusk, G. 2020. Political legitimacy in the democratic view: The case of climate services. *Philosophy of Science* 87 (5): 991–1002.
Lusk, G. 2021. Does democracy require value-neutral science? Analyzing the legitimacy of scientific information in the political sphere. *Studies in History and Philosophy of Science Part A* 90:102–110.
Lusk, G., and K. Elliott. 2022. Non-epistemic values and scientific assessment: An adequacy-for-purpose view. *European Journal for Philosophy of Science* 12 (2): 35.
Melo-Martín, I., and K. Intemann. 2016. The risk of using inductive risk to challenge the value-free ideal. *Philosophy of Science* 83 (4): 500–520.
Menon, T., and J. Stegenga, 2023. Sisyphean science: Why value freedom is worth pursuing. *European Journal for Philosophy of Science* 13: 48
Mitchell, S. 2004. The prescribed and proscribed values in science policy. In *Science, values, and objectivity*, ed. P. Machamer and G. Wolters, 245–255. Pittsburgh: University of Pittsburgh Press.
North, R. 2017. Principles as guides: The action-guiding role of justice in politics. *The Journal of Politics* 79 (1): 75–88.
Parker, W. 2014. Values and uncertainties in climate prediction, revisited. *Studies in History and Philosophy of Science Part A* 46:24–30.
Parker, W. 2024. The epistemic projection approach to values in science. *Philosophy of Science* 91 (1): 18–36.
Parker, W., and G. Lusk. 2019. Incorporating user values into climate services. *Bulletin of the American Meteorological Society* 100 (9): 1643–1650.
Reiss, J., and J. Sprenger. 2020. Scientific objectivity. In *The stanford encyclopedia of philosophy*, ed. E. Zalta. https://plato.stanford.edu/archives/win2020/entries/scientific-objectivity/.
Rudner, R. 1953. The scientist qua scientist makes value judgments. *Philosophy of Science* 20 (1): 1–6.
Steel, D. 2010. Epistemic values and the argument from inductive risk. *Philosophy of Science* 77 (1): 14–34.
Steele, K. 2012. The scientist qua policy advisor makes value judgments. *Philosophy of Science* 79 (5): 893–904.
Talbot, B. 2016. The best argument for 'Ought Implies Can' is a better argument against 'Ought Implies Can'. *Ergo, an Open Access Journal of Philosophy* 3.
Ward, Z. 2021. On value-laden science. *Studies in History and Philosophy of Science Part A* 85:54–62.
Wilholt, T. 2022. Epistemic interests and the objectivity of inquiry. *Studies in History and Philosophy of Science* 91:86–93.
Williams, B. 1995. Internal reasons and the obscurity of blame. In *Making sense of humanity: And other philosophical papers 1982–1993*, ed. B. Williams, 35–45. Cambridge: Cambridge University Press.
Winsberg, E. 2012. Values and uncertainties in the predictions of global climate models. *Kennedy Institute of Ethics Journal* 22 (2): 111–137.

Chapter 16
Characterizing the Value-Free Ideal: From a Dichotomy to a Multiplicity

Kevin C. Elliott

Abstract The value-free ideal (VFI) has played a significant role in the recent philosophical literature on "values and science." The current dominant position among philosophers of science is that the VFI should be rejected. Despite the centrality of the VFI for the literature on values and science, I argue in this chapter that it has not been adequately scrutinized. Douglas's foundational characterization of the VFI focuses on excluding values from a particular group of scientific activities (namely, the "internal" judgments involved in evaluating and accepting scientific results), but one could characterize the VFI in a more or less stringent manner depending on which aspects of science one thought should be preserved as free from value influences and in what ways. I focus on three central concepts to construct a three-part framework for organizing different ways that it can be characterized. I suggest that one can arrive at different characterizations of the VFI depending on how one interprets values, what one means by freedom from value influences, and what kind of an ideal one is putting forward. I conclude that it would be better to formulate the

It is particularly fitting that I am writing this chapter for a book dedicated to Matt Brown because he started me down the path of thinking that it might be fruitful to distinguish different characterizations of the VFI. Matt participated in a workshop that I co-organized on "Cognitive Attitudes and Values in Science" at the University of Notre Dame in 2013. At the workshop, he pointed out that it wasn't clear whether I was truly abandoning the VFI, given my inclination to limit the influences of social and ethical values to the cognitive attitude of acceptance rather than belief. (See Sect. 3 of this chapter for further discussion about cognitive attitudes and the VFI.) After reflecting on his question, I concluded that I was in fact abandoning the VFI because I regarded science as a pragmatic enterprise in which scientists are virtually always engaged in accepting claims for the purposes of guiding action, and therefore I regarded social and ethical values as being relevant to the central activities of scientific reasoning. Nevertheless, his questioning helped me to realize that I might be rejecting one characterization of the VFI while holding onto other characterizations of it. This chapter is my attempt to flesh out my long-running thoughts about different ways of formulating the VFI.

K. C. Elliott (✉)
Lyman Briggs College, Department of Fisheries and Wildlife, and Department of Philosophy, Michigan State University, East Lansing, MI, USA
e-mail: kce@msu.edu

J. Y. Tsou et al. (eds.), *Values, Pluralism, and Pragmatism: Themes from the Work of Matthew J. Brown*, Boston Studies in the Philosophy and History of Science 347, https://doi.org/10.1007/978-3-031-92958-8_16

VFI not in terms of a single characterization but rather in terms of a multiplicity of distinct but related characterizations.

Keywords Value-free ideal · Idiosyncrasy-free ideal · Argument from inductive risk · Context of justification · Cognitive attitudes

16.1 Introduction

The value-free ideal (VFI) for science has been an important topic of debate for hundreds of years (Proctor 1991), and it has played a particularly significant role in the recent philosophical literature on "values and science."[1] In her groundbreaking book, *Science, Policy, and the Value-Free Ideal* (2009), Heather Douglas examined the history of philosophical debates about values in science during the latter half of the twentieth century. On the basis of that analysis, she defined the VFI as the view that "the value judgments internal to science, involving the evaluation and acceptance of scientific results at the heart of the research process, are to be as free as humanly possible of all social and ethical values" (2009, p. 45). Most introductions to the field have adopted roughly the same formulation as Douglas (see e.g., Biddle 2013; Brown 2020; Elliott 2017; Elliott 2022),[2] and at this point the dominant position among philosophers of science is that this VFI should be rejected (Brown 2024; Douglas and Branch 2024; Parker 2024). By adopting this conclusion, scholars have opened up a range of important questions about how to manage the influences of values in science and how to distinguish between appropriate and inappropriate roles for values (e.g., Holman and Wilholt 2022).

Despite the centrality of the VFI for the recent literature on values and science, I argue in this chapter that the nature of the VFI has not been adequately scrutinized. Although Douglas provided an influential historical analysis of the development of the VFI among philosophers of science during the mid- to late-twentieth century, that literature does not exhaust the range of options for thinking about and characterizing the VFI. As Proctor put it in his own historical analysis of this topic, "'Value-freedom' has meant different things to different people at different times. Slogans like 'sciences must be value-free' or 'all knowledge is political' must be understood in light of specific fears and goals that change over time" (1991, p. x). He emphasized that figures throughout history have promoted science's value freedom for many different reasons, including efforts to resist the suppression of science by states or religions, to guard against the intrusion of personal interests in

[1] For some of the history leading up to this recent literature on values and science, and especially for the important role that feminist scholarship played in that history, see e.g., Crasnow (2024), Douglas (2016), and Intemann (2021).

[2] Throughout the paper, I will often refer to "Douglas's" formulation of the VFI because she developed the canonical formulation of it, but it is important to recognize that she developed this account on the basis of her historical analysis of previous work in the philosophy of science.

science, to promote the professionalization or secularization of science, and to conceal the social aspects of scientific practice. Thus, different motivations and historical/scholarly vantage points can generate different ways of thinking about the VFI. Even within the philosophy of science, some scholars have responded to recent debates about the VFI by suggesting other plausible ways of characterizing it (see e.g., de Melo-Martín and Intemann 2016; Menon and Stegenga 2023; Stegenga and Menon 2023).[3]

I contend that ongoing philosophical debates over the cogency of the VFI (see e.g., Brown 2024; Stamenkovic 2024) could be improved by developing a more systematic assessment of the different options for characterizing it. Instead of providing a historical analysis, as figures like Douglas (2009) and Proctor (1991) have done, my focus will be on providing a preliminary taxonomy of different ways that value freedom could, in theory, be conceptualized. Ultimately, I think this analysis suggests that one could potentially maintain some formulations of the VFI, despite the arguments that have been raised against it (for an overview of these arguments, see Douglas 2016; Elliott 2022). Nevertheless, even if the VFI is not officially "dead," it is probably not the most fruitful concept for guiding future debates about values in science; philosophers are likely to communicate their insights more effectively to the scientific community and to society as a whole by talking about strategies for managing values instead. On my view, however, these value management strategies could encompass both the kinds of strategies advocated by proponents of the VFI and those promoted by opponents of the VFI, depending on a variety of local factors at play in different research contexts.

As a framework for guiding my analysis, I focus on the three central concepts associated with the VFI: values, freedom, and ideals. In other words, I suggest that one can arrive at different characterizations of the value-free ideal depending on how one interprets *values*, what one means by *freedom* from value influences, and what kind of an *ideal* one is putting forward. On the basis of this analysis, I conclude that it would be better not to think of the VFI in terms of a single characterization but rather in terms of a multiplicity of distinct but related characterizations. This conclusion has significant implications for the literature on values and science because it shifts the focus from asking a relatively simple yes/no question ("Should

---

[3] One might argue that there is no need to consider multiple characterizations of the VFI because (at least as Douglas envisioned it) contemporary accounts of the VFI are intended to state the specific view that a group of prominent thinkers held about science during a particular historical period (namely, the 1950s and 1960s). Thus, even if one could, in principle, formulate the VFI in different ways, the only formulation that matters is the one propounded by these influential figures during the mid-twentieth century. There are multiple problems with this argument, however. First, not everyone held the same views about the proper roles for values in science even during the mid-twentieth century. For example, as mentioned later in this paper, some figures held different views than others about the roles that ethical and social values should play in the context of discovery (see e.g., Polanyi 1962). Second, even if Douglas intended to focus on a particular historical characterization of the VFI, others are clearly not committed to interpreting the VFI in that particular way (see e.g., de Melo-Martín and Intemann 2016; Menon and Stegenga 2023); thus, it is important to consider the full variety of ways in which the VFI can be interpreted.

the VFI be rejected?") to asking much more complex questions (e.g., "Should all forms of the VFI be rejected?" or "Is it fruitful to try to develop a form of the VFI that could be maintained?").

This shift in focus has at least two significant benefits. First, it softens the opposition that has developed between proponents and opponents of the VFI. Rather than generating two "camps" with opposing positions, this shift draws attention to a more diverse array of positions that one might accept or reject. By considering all these potential positions, both the critics and the defenders of the VFI might develop more nuanced views and soften their opposition. Those uncomfortable with the current oppositional framing of the literature on values and science might also be encouraged to join the discussion. Second, this shift in focus can encourage new questions for scholars studying the topic of values and science. The past several decades of work on this topic have made it possible to develop much more nuanced views about many of the concepts under discussion. Attending to these nuances and working with multiple different characterizations of the VFI can highlight new opportunities for philosophical scholarship.

Nevertheless, attending to the many different possible ways of characterizing the VFI also raises questions about whether it is in fact the most fruitful principle for organizing the literature on values and science. On the positive side, framing scholarship on values and science in terms of opposition to the VFI has been an important strategy for drawing renewed attention to this area of inquiry and challenging earlier philosophical assumptions about the legitimate roles for values in science (see e.g., Douglas 2009). Also, as long as the VFI is characterized in a multiplicity of different ways, it can serve as a helpful heuristic tool for "mapping" the wide range of possible relationships between values and scientific practice.[4] However, when communicating their insights to the scientific community and to society as a whole, philosophers may find that talking about the VFI is more confusing than it is helpful. For example, as scholarship on values and science has progressed, it has become clear that many of the important ways in which values influence scientific practice involve elements of science that would typically be labeled as "outside" the core of scientific reasoning by the traditional VFI (see e.g., Elliott and McKaughan 2009; Holman and Bruner 2017; O'Connor and Freeborn 2025; Okruhlik 1994; Winsberg 2018). These include decisions about which research questions to pursue, which studies to perform, and how to "curate" and frame all the information stemming from the resulting studies (on information curation, see Elabbar 2023). Therefore, for audiences outside the philosophical community, talking about the VFI may already be confusing because they may not appreciate the nuanced senses in which it does and does not allow for values to influence science. Once one realizes that there are many other reasonable ways in which the VFI could be formulated (all of which allow values to play some roles in science), it becomes even more confusing. Thus, given that almost everyone involved in current debates (both proponents and opponents of the traditional VFI) can agree that values have appropriate roles to

[4] I'm grateful to Mousa Mohammadian for helping me think through this point.

play in science, it may be more fruitful to frame discussions about values in science in terms of value management rather than defending or rejecting the VFI.

16.2 Values

According to my three-part framework, the first way in which characterizations of the VFI can vary is by employing different notions of the "values" that are to be excluded. The basic idea of the VFI is that science is supposed to remain free of values or value influences, but one can arrive at different characterizations of the VFI depending on which *categories* or *kinds* of values it excludes. This point is not a new one. In her classic formulation of the VFI, Douglas (2009) noted that it does not exclude all values from scientific reasoning; it excludes ethical and social values but not "epistemic" or "cognitive" ones. Thus, one avenue for challenging that version of the VFI is to deny that it is even possible to distinguish between epistemic and non-epistemic values (see e.g., Longino 1996; Rooney 2017). However, even if one granted the defenders of the VFI that a general distinction could be made between epistemic and non-epistemic values, the precise contours of this distinction would still be subject to debate (see e.g., Laudan 1984; Douglas 2013; Lacey 2017; Rooney 2017). For example, aesthetic values have received relatively little attention in the literature on values and science (although see Morgan 2013; Ivanova 2017; Currie 2023), and it is unclear whether they should be regarded as epistemic or non-epistemic. Faced with these sorts of difficult distinctions between values that straddle the borderline between epistemic and non-epistemic ones (or cognitive and noncognitive ones, or constitutive and contextual ones), it seems most reasonable to acknowledge that one could formulate different characterizations of the VFI depending on which kinds of values it allowed and which kinds it excluded.[5]

Beyond the question of what *categories* of values to exclude, characterizations of the VFI could also vary based on what kinds of things they count as values, i.e., what *concept* of value they employ. Scholars working on the topic of values and science commonly worry that the word 'value' is used in a wide variety of different ways (e.g., Biddle 2013; Ward 2021; Elliott 2022). For example, according to Miriam Solomon:

[5] One might respond to the fuzziness of the epistemic/non-epistemic distinction by adopting Daniel Steel's (2010) approach, which defines epistemic values as those that promote the acquisition of true beliefs and non-epistemic values as those that do not. Although this approach creates a clear-cut distinction between epistemic and non-epistemic values, it can leave scientists with uncertainty about which values are genuinely epistemic or non-epistemic in a particular case until they have ultimately determined which beliefs are true or not. (After all, Steel emphasizes that whether or not a value is epistemic can be a contextual matter.) Thus, if the VFI is supposed to be action-guiding for scientists (a point I will discuss later in the paper), Steel's approach may not provide an adequate basis for formulating it.

> "Value" has been used to include political values, aesthetic preferences, psychological biases, cognitive goals, personal and societal goals, ideologies, and pre-theoretic intuitions. So "value" is, in practice, not restricted to ethical values or even aesthetic values. "Values" include pre-theoretic assumptions, ethical conduct of inquiry, and causes of preference for one theory over another. (2012, pp. 332-333)

Thus, efforts to distinguish different concepts of values (as well as other related phenomena that might be excluded by the VFI) can potentially generate different characterizations of the VFI (see e.g., Ward 2021; Hilligardt 2022).

Consider, for example, four concepts of values that Rebecca Korf and I recently proposed (Elliott and Korf 2024). First, many of those working on the topic of values and science have regarded values as criteria or standards for evaluating scientific theories or other phenomena, such as predictive accuracy, scope, or consistency (e.g., Kuhn 1977; McMullin 1983; Anderson 1993; Lacey 1999). Second, values are sometimes regarded as causal factors that influence scientific decision making, such as heuristics or ideologies or psychological inclinations (e.g., Longino 1990; Solomon 2012; Zhao 2022).[6] Third, values are sometimes treated as beliefs or attitudes about what is desirable, such as ethical beliefs about the importance of environmental sustainability or justice (e.g., Douglas 2009; Steel 2010; Brown 2020). Finally, values can also be regarded as desirable things themselves, such as public health or animal welfare (e.g., Elliott 2017; Brown 2020). Admittedly, the distinctions between these four categories are not entirely sharp. For example, criteria for choice often involve beliefs about what is desirable, and those beliefs or attitudes may serve as causal factors that influence scientific decision making. Nevertheless, although there are likely to be considerable overlaps among these concepts, they do not overlap completely, and so something can serve as a value in one sense without serving as a value in another sense.[7]

These different concepts of values can give rise to different characterizations of the VFI. For example, there are important differences between saying that values *in the sense of causal factors* (or particular types of causal factors, such as non-epistemic ones) should be excluded from scientific reasoning versus saying that values *in the sense of beliefs or attitudes about what is desirable* should be excluded from scientific reasoning. In fact, this difference sheds light on a recent disagreement between Robert Hudson (2021) and Heather Douglas and myself (Douglas

[6] As Rebecca Korf and I note in our paper, this definition may be too broad as it stands because there are such a wide array of causal factors that could influence scientists' decision making. One way to limit this concept would be to focus only on *psychological* causal factors and not on other sorts of causal factors. For the purposes of this chapter, though, I think it is best not to prejudge precisely which causal factors should count as values.

[7] It is also worth emphasizing that the nature of the overlaps between these concepts could vary depending on how exactly one interprets them. For example, if one treated criteria for choice and beliefs about what is desirable as causal factors, then our first and third concepts of values would be subsumed as a subset under our second concept of values. However, if one treated criteria for choice and beliefs about what is desirable as reasons, and if one held the metaphysical view that reasons are distinct from causal factors, then our first and third categories of values would not overlap with the second.

and Elliott 2022). Hudson (2021) claimed that those who reject the VFI open the door for scientific reasoning to be corrupted by biases, and he worried that this would exacerbate the lack of reproducibility that afflicts some areas of scientific research. In contrast, Douglas and Elliott (2022) argued that there are important distinctions to be made between biases and values, such that the rejection of the VFI does not mean that biases should be accepted as legitimate elements of scientific reasoning. Once one begins to consider different concepts of values, it becomes clear that this dispute arises partly because Hudson is employing a different concept of values when he thinks about the VFI.[8] Hudson insists that "scientists are affected on a daily basis by all sorts of non-epistemic factors: social forces, moral obligations, political interests, and so on" (Hudson 2021, p. 12). He then contends that, because Douglas and I hold that non-epistemic values can legitimately influence science, we are accepting the influences of all these potentially biasing factors. In contrast, Douglas typically conceptualizes values as beliefs about what is desirable (e.g., the ethical belief that public health tends to be more important than short-term corporate profits). Thus, Douglas and I would agree with Hudson in accepting a characterization of the VFI that focuses on excluding biases from scientific reasoning (see Douglas and Elliott 2022), but we would insist that this does not justify accepting characterizations of the VFI that exclude values conceptualized in other ways.

Another source of variation when characterizing the VFI is *whose* values should be excluded from scientific reasoning. Douglas's (2009) formulation of the VFI does not specify whose values are at play, but one could characterize the VFI in such a way that this matters. For example, de Melo-Martín and Intemann (2016) appear to characterize the VFI as the view that scientists should not allow their *own idiosyncratic values* to influence their reasoning. They imagine a situation in which scientists address value-laden decisions in their research by employing the values of stakeholders rather than the scientists' own personal values. They claim that such an approach would not violate the VFI: "This seems perfectly consistent with the claim made by proponents of the VFI that scientists, qua scientists, ought to refrain from allowing their personal value judgments to influence their decision making" (de Melo-Martín and Intemann 2016, p. 513). This interpretation of the VFI seems importantly different from the way Douglas (2009) conceives of it, insofar as Douglas characterizes the VFI as the view that scientists should refrain from allowing any social or ethical values from influencing their reasoning, not merely that they should block their *personal* social or ethical values. Nevertheless, recent commentators have suggested that Isaac Levi (1960) may have thought of the VFI in roughly the same manner as de Melo-Martín and Intemann (see e.g., Staley 2017; Boulicault and Schroeder 2021). Similarly, Marion Boulicault has recently proposed an "idiosyncrasy-free ideal," according to which scientists should not allow

[8] I say that this is dispute is "partly" because different concepts of values are at play. Another crucial part of the dispute is that Hudson seems to think that whenever values affect the statistical standards of evidence for accepting a claim, those values play a biasing role. Douglas and Elliott (2022) strongly resist this claim.

their unique individual features to influence their decision making (see Boulicault and Schroeder 2021). Extending these ideas, one could adopt different characterizations of the VFI depending on whose values one excluded from influencing science. For example, one could handle the values of individual scientists, communities of scientists, policy makers, institutional structures, or broad communities of "interested and affected parties" differently (see e.g., Levy 2025).

16.3 Value Freedom

Turning next to the notion of value *freedom*, it turns out that one can also generate different characterizations of the VFI depending on how one handles this idea. First, one has to specify *what elements* of science are supposed to remain free of values. As noted in the introduction, Douglas's (2009) formulation of the VFI focuses on the "internal" aspects of science, which she specifies as involving the evaluation and acceptance of hypotheses and theories. She acknowledges that even most proponents of the VFI would accept that values can appropriately influence more "external" aspects of science, such as decisions about what projects are undertaken and how scientific findings are applied. Matt Brown (2020) also emphasizes this point. He notes that most proponents of the VFI rely on a rough distinction between the "context of discovery" and the "context of justification," and they formulate the VFI so that it excludes values from the context of justification while allowing values in the context of discovery. (For a sophisticated contemporary development of an approach along these lines, see Parker 2024). Nevertheless, one could hold an even stricter interpretation of the VFI that restricts value influences even in the context of discovery; Percy Bridgman and Michael Polanyi may have held something like this view, insofar as they thought the pursuit of "basic" or "pure" scientific research projects ought to be based solely on "internal" considerations about which ideas were most scientifically promising (see Nye 2011; Shaw 2021; Douglas and Branch 2024). Those concerned to prevent values from influencing the context of justification might find this restrictive view to be all the more tempting in light of recent scholarship showing that the influences of values in the context of discovery are likely to bleed over into the context of justification (see e.g., Okruhlik 1994; Elliott and McKaughan 2009; Holman and Bruner 2017; Winsberg 2018; O'Connor and Freeborn 2025), and value influences in the context of justification could potentially be shifted into the context of discovery (Parker 2024). Thus, those who are sympathetic to the VFI might want to develop characterizations of it that limit the influences of values on at least some aspects of study design or question formulation.[9]

[9] For example, as noted in an earlier footnote, Dan Hicks characterizes the VFI such that it prohibits values not only from influencing hypothesis acceptance but also from influencing the process of "gathering evidence" (Hicks 2018, p. 166)—although one should keep in mind that Hicks is an opponent rather than a proponent of the VFI.

Even if one focuses solely on excluding values from the "internal" aspects of scientific reasoning, there is room for different views about precisely which aspects of scientific reasoning should be classified as internal. Menon and Stegenga (2023) recently brought this issue to the fore because they insisted that the VFI should focus only narrowly on excluding values from the activity of *scientific inference* and not on other activities, such as the interpretation of scientific concepts. This is significant because prominent critics of the VFI have sometimes appealed to the value-ladenness of scientific concepts as an argument against the VFI (e.g., Dupré 2007; Elliott 2022). If one characterizes the VFI so that it focuses only on keeping values out of scientific inference, however, the value-ladenness of scientific concepts may not challenge this characterization of the VFI.[10] Once one starts distinguishing different aspects of scientific reasoning, one could argue that even Douglas (2009) maintains a form of the VFI because she insists that the activity of assessing the degree of evidence in favor of a hypothesis or theory should not be influenced by social or ethical values. On her view, values become relevant once that level of evidence has been determined and scientists turn to the question of whether that level of evidence is sufficient to accept the hypothesis or theory. Others would challenge this view and argue that values cannot and should not be excluded from assessing the amount of evidence in favor of a hypothesis or theory (see e.g., Bluhm 2017; Brown 2020). In addition, Harvard and Winsberg (2022) have recently emphasized that scholars working on the topic of science and values should pay attention not only to inference but also to representation; thus, their work raises the question of whether values should play a role in various activities involving scientific modeling (see also Intemann 2015). Finally, some scholars working on the topic of values and science suggest that their primary concern is to keep scientists from engaging in wishful thinking, which would involve driving inquiry toward pre-determined conclusions (e.g., Anderson 2004; Brown 2013). Thus, one might argue that these scholars are accepting a characterization of the VFI that excludes values from influencing science in ways that pre-determine the results of inquiry.[11]

In addition to this variation regarding the *elements of science* that should be kept free of values, characterizations of the VFI can also vary in terms of the *form of value-ladenness* that is meant to be avoided. Zina Ward (2021) has helped to clarify this point because she has identified four different ways in which judgments in science can be value-laden; values can relate to scientific judgments as: (1) motivating

[10] One might think that when scientific concepts presuppose or incorporate value judgments, inferences involving those concepts would also sometimes presuppose or incorporate value judgments, and thus those values would make a difference to inference. However, Menon and Stegenga (2023) dispute this conclusion and argue that scientific concepts could presuppose values without making a difference to inference.

[11] One might think that there is no room for debate about whether values should be excluded from pre-determining the results of inquiry, but Peters (2021) has argued that allowing values to pre-determine inquiry at the level of the individual scientist could still allow inquiry to advance effectively at a social level. Thus, there would be room for genuine debate about whether or not to defend a characterization of the VFI that focused on excluding values from pre-determining the outcome of a scientist's inquiry.

reasons; (2) justifying reasons; (3) causes; or (4) effects.[12] It is not entirely clear whether the VFI as presented by Douglas (2009) was designed to prevent values from playing a role as *motivating reasons* for judgments involved in scientific reasoning or whether it was intended to exclude values from being treated as *justifying reasons*. Some of the debates over whether scientists should defer value-laden judgments to policy makers (thereby maintaining the VFI) seem to presuppose that ethical or social values could legitimately serve as *justifying* reasons for these kinds of choices but that they should not serve as *motivating* reasons for scientists (e.g., Elliott 2011; Betz 2013; Betz 2017; Havstad and Brown 2017). One could formulate different characterizations of the VFI depending on which form of value-ladenness one decided to focus on. In principle, one could also develop a characterization of the VFI that focused on preventing the judgments involved in scientific reasoning from generating value-laden *effects*. Given the pervasive way in which different scientific judgments generate differing conclusions and consequences for society, however, one would probably have to abandon this characterization of the VFI as untenable (Elliott 2017; Ratti and Russo 2024).

Finally, characterizations of the VFI can differ depending on which *cognitive attitudes* are supposed to remain free of values. For example, Elliott and Willmes (2013) have argued that it is important to maintain a distinction between the cognitive attitudes of *belief* and *acceptance*. By employing this distinction, one could maintain a version of the VFI that focuses on keeping scientists' beliefs free of values while at the same time affirming that scientists should allow their values to influence what they accept for various practical purposes (see also Fleisher 2018).[13] Along these lines, Hugh Lacey makes cognitive attitudes central to his views about the proper roles for values in science:

> Cognitive but not social values play essential roles in making the judgment that a theory or hypothesis is impartially held of a set of phenomena. However, social values have proper and ineliminable roles in other aspects of science, for example, when adopting a theory for the sake of giving direction to a research project, or endorsing a theoretically articulated hypothesis for the sake of informing practical action (Lacey 2017, p. 15).

Lacey is difficult to categorize straightforwardly as a proponent or an opponent of the VFI if one does not distinguish different characterizations of it. On one hand, he explicitly prohibits incorporating ethical and social values from influencing scientific reasoning if that reasoning involves impartially holding theories; this appears to make him a proponent of the VFI. Nevertheless, if one were to treat science as a pragmatic enterprise in which scientists are almost always adopting or endorsing

[12] Observant readers might note that Ward's list of four forms of value-ladenness has similarities to the list that Korf and I developed of four concepts of values. However, these lists should not simply be equated. Korf and I intended to develop concepts of what values *are*, whereas Ward focuses on what values can *do* or how they can *relate* to scientific judgments. Perhaps one could argue that different concepts of values arose in part based on the different relationships that can arise between values and other things, but that is a question that goes beyond the scope of this chapter.

[13] Importantly, Matt Brown (2015) has rejected the belief/acceptance distinction on Deweyian grounds in order to reject this approach to maintaining a form of the VFI.

conclusions for the sake of giving direction to research projects or informing practical action, then he would appear to be an opponent of the VFI. DiMarco and Khalifa (2019) provide another example of appealing to cognitive attitudes in order to clarify roles for values in science; they argue that one can maintain some forms of the VFI while denying that it applies to judgments of the *pursuit worthiness* of scientific hypotheses, theories, or models (see also Shaw 2022).

16.4 The Notion of "an Ideal"

Characterizations of the VFI can also vary based on different notions of the *ideal* toward which they are aiming. One potential source of variation is the relationship between descriptive and normative elements in the ideal. Given that the very notion of an ideal seems to be inherently normative, it probably does not make sense to interpret the VFI as being solely descriptive. However, this still leaves room for varying positions on whether and how the ideal might include some descriptive elements along with normative ones. For example, one could characterize the VFI in a sociological way, as the claim that scientists typically regard the exclusion of values from their reasoning as a feature of good science. Claims like this would constitute descriptions of individuals' or communities' normative views about science. Or one could characterize the VFI as the view that scientists, when they are operating at their best or doing proper science, do not incorporate values in their reasoning. Lacey (1999) may be thinking along these lines when he structures his book around the claim that "science is value-free." He is clearly not intending his analysis to be merely descriptive; he is interested in the characteristics of "good" or "proper" science. Justin Biddle's account of the VFI (or, as he puts it, the ideal of epistemic purity), has a similar mixture of descriptive and normative elements: "this view maintains that (1) the proper application of scientific methods will, as a matter of fact, always screen out all contextual factors, and (2) scientists ought to apply scientific methods properly, thereby screening out all contextual factors" (Biddle 2013, p. 125). Douglas (2009) seems to be almost completely normative when she characterizes the VFI as the view that scientists *ought* to exclude values from the core of scientific reasoning. However, her normative claims still pertain to what scientists should *actually do* under ordinary circumstances; one could focus instead on normative claims about what scientists *should strive for or pursue* under ideal circumstances (de Melo-Martín and Intemann 2016).

Building on this point, Menon and Stegenga (2023) have shown that one could distinguish between thinking of ideals in science as *states to be achieved* versus thinking of ideals as *pursuits* that scientists should try to engage in. This distinction is important because Menon and Stegenga contend that the VFI can be a worthy ideal in the sense of a pursuit even if it is not a worthy ideal in the sense of an end state. In other words, they acknowledge that scientists should not (and cannot) actually achieve the end state of keeping their judgments completely free of value influences. Nevertheless, they argue that it is beneficial for scientists to try to minimize the extent to which their conclusions are value-laden, and thus they contend that the VFI is a worthy ideal if it is regarded as a form of pursuit. Stamenkovic (2024) also appears to sympathize with this way of understanding the VFI.

One could develop a closely related distinction between characterizations that treat the VFI as a state to be achieved versus characterizations that treat it as a practical guideline for working scientists. (For more on ideals as states vs guidelines, see Philippi 2020). One might initially think that if scientists understood the state that they should be trying to achieve (e.g., excluding non-epistemic values from specific aspects of scientific reasoning), they would automatically be able to guide their decision making by pursuing that state. However, this idea might be too simple. For example, as noted in an earlier footnote, it might be difficult for scientists to determine which values are genuinely epistemic vs non-epistemic. Thus, even if one employed the language of "epistemic" vs "non-epistemic" values in a characterization of the VFI designed to specify the state to be achieved, one might need to describe values in more easily and immediately recognizable ways in order to characterize the VFI as a guideline for working scientists.

Finally, de Melo-Martín and Intemann (2016) have pointed out that one needs to clarify the *circumstances* under which the VFI is actually a meaningful ideal. In other words, one must consider whether it applies only under special circumstances or whether it applies under the real-life circumstances in which scientists typically find themselves. De Melo-Martín and Intemann contend that this distinction is important for understanding Douglas's (2009) critique of the VFI. In their view, she is critiquing the VFI only under non-ideal circumstances, and they claim that she would still accept the VFI as a claim that applies to ideal circumstances. They draw this conclusion because Douglas contends that values have less of a role to play as uncertainty decreases and as the evidence for a hypothesis increases (Douglas 2009). Therefore, under conditions of perfect evidence, de Melo-Martín and Intemann conclude that Douglas would actually accept the VFI. They use this observation as a source of motivation for exploring stronger arguments that would challenge characterizations of the VFI that apply even under ideal evidential circumstances.

One could extend de Melo-Martín and Intemann's point by claiming that even under non-ideal circumstances, there might be some conditions under which a VFI holds and other conditions under which it can be overridden. As Lorraine Daston (2022) has recently emphasized, rules do not have to be regarded as exceptionless; one could regard them instead as generalizations or models that admit of exceptions. One might think about ideals in a similar way, as generalizations or models that still have some exceptions (see e.g., Sheykh-Rezaee and Bikaraan-Behesht 2023). In fact, Stegenga and Menon appear to hold this view, insofar as they propound a particular characterization of the VFI while acknowledging that it could legitimately be violated under a narrow range of conditions (typically involving the application of science for decision making; Stegenga and Menon 2023, p. 438). This notion that the VFI could be a viable ideal even while having exceptions is highly significant because it makes the VFI more difficult to challenge; to reject a VFI of this sort requires not merely showing that it should be violated in some cases but rather showing that it is problematic in a deeper or more systematic fashion.

16.5 The Upshot

I have argued that it is possible to characterize the VFI in a surprisingly wide variety of ways, and I have organized these different characterizations based on how one interprets *values*, what one means by *freedom* from value influences, and what kind of an *ideal* one is discussing. Table 16.1 summarizes how variations in the VFI can be organized based on this three-part framework.

To further clarify how different characterizations of the VFI relate to each other, one could employ visualization techniques like those developed by Palider et al. (2021). As a brief example, Fig. 16.1 illustrates how Douglas's (2009) characterization of the VFI and de Melo-Martín and Intemann's (2016) characterization of the VFI handle three of the nine questions provided in Table 16.1. The box on the left captures how Douglas's characterization of the VFI answers the questions, the box on the right captures how de Melo-Martín and Intemann's characterization of the VFI answers the questions, and the darker area shared by the two boxes represents the overlap where their characterizations provide the same answers. The three questions provided in the figure were chosen to illustrate three different kinds of relationships that can be represented using these sorts of visualizations. For the first question, the figure demonstrates that neither Douglas nor de Melo-Martín and Intemann are very explicit about the specific concept of values that they are employing. For the second question, the figure shows that both Douglas's characterization of the VFI and de Melo-Martín and Intemann's characterization of the VFI exclude the values of scientists from influencing scientific reasoning. However, as discussed above, Douglas's characterization of the VFI also excludes the values of non-scientists (e.g., stakeholders) from influencing scientific reasoning, whereas de Melo-Martín and Intemann do not appear to think that the VFI excludes such influences. For the third question, the figure shows that Douglas and de Melo-Martín and Intemann appear to answer the question in the same way. Thus, Fig. 16.1 illustrates

Table 16.1 Different ways of characterizing the VFI, organized in terms of three categories

Values	Value-freedom	Ideal
• Which kinds of values should be excluded? • What concept of values are we employing (i.e., what are we talking about when we refer to values)? • Whose values should be excluded?	• What elements of science should be kept free of values? • What form of value-ladenness should be prevented? • Which cognitive attitudes should be kept free of values?	• How does the ideal incorporate descriptive vs normative elements? • Does the ideal refer to a state, a pursuit, or a guideline? • Under which circumstances does the ideal hold?

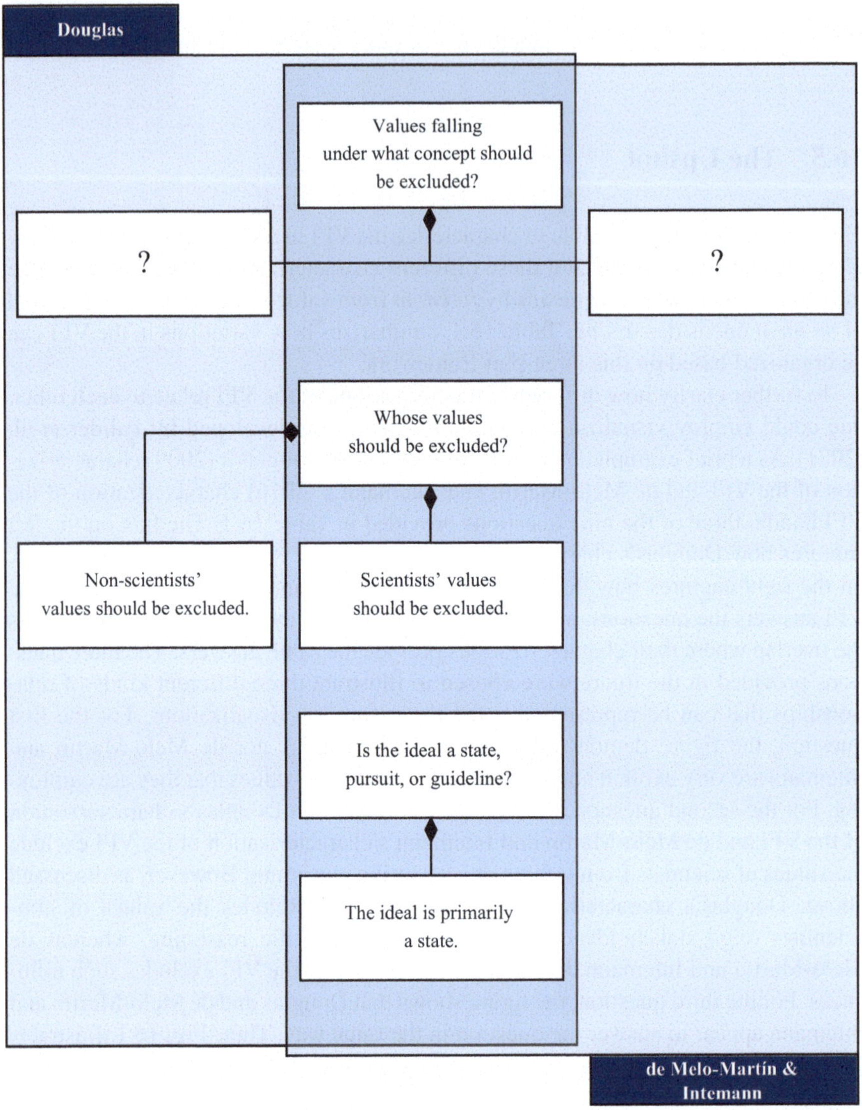

Fig. 16.1 An example of how different characterizations of the VFI could be visualized using the diagramming techniques developed in Palider et al. (2021). The figure compares the characterizations of the VFI provided by Douglas (2009) and de Melo-Martín and Intemann (2016) based on three of the nine questions provided in Table 16.1

the heuristic role that diagrams like these could play in clarifying the relationships between different characterizations of the VFI and highlighting issues that merit further clarification.

The remainder of this section highlights two benefits that emerge from distinguishing all these different characterizations of the VFI: (1) it frames the literature on values and science in a more irenic and nuanced fashion; and (2) it suggests promising questions for future investigation.

First, consider how attention to different ways of characterizing the VFI could foster a more irenic approach to values and science. At present, discussions about the role of values in science tend to be framed in a dichotomous manner, such that scholars are divided into proponents and opponents of the VFI. There is a tendency to portray the proponents (e.g., Betz 2013; Hudson 2016) as being in a small minority that holds an implausible view while the opponents of the VFI are portrayed as the victors (see e.g., Elliott 2022; Brown 2024). Thus, the dichotomous approach to characterizing the VFI encourages a framing of the literature on values and science as an agonistic conflict that needs to be settled or a debate that needs to be won.

Once one recognizes that the VFI can be characterized in different ways, however, it seems more plausible to think in terms of a multiplicity of different positions about the roles that values should play in science. A benefit of this approach is that it shifts the debate from a two-sided battle between the proponents and the opponents of the VFI and frames it as a multi-faceted discussion about possible roles for values in science. For example, even opponents of the "traditional" VFI might be willing to accept characterizations of the VFI that preclude values from serving as a form of evidence in favor of a theory or hypothesis (for a defense of this alternative form of the VFI—although not expressed as such—see Douglas 2009; for criticisms of it, see e.g., de Melo-Martín and Intemann 2016; Bluhm 2017; Brown 2020; Lewens 2025). Some opponents of the traditional VFI might also be willing to consider accepting characterizations of it that strive to exclude scientists' idiosyncratic values from scientific reasoning (for discussions of this issue, see e.g., Wilholt 2009; John 2015; Elliott 2017; Boulicault and Schroeder 2021; Levy 2025). On the other side of the issue, some proponents of the traditional VFI might be willing to reconsider their position if the VFI were distinguished into different characterizations. For example, some proponents might reject forms of the VFI that go so far as to exclude values from playing a role in assessing mixed claims (see e.g., Alexandrova 2018; Lewens 2025), but they might still hold onto forms of the VFI that exclude values from playing other roles in science. Similarly, proponents of the traditional VFI might be willing to reject alternative characterizations of the VFI that exclude values from playing a role in the conclusions that scientists "endorse" for the purposes of regulatory decision making (see e.g., Lacey 2017; Levy 2025). Once one recognizes the potential for these debates, it seems somewhat arbitrary to choose a particular characterization of the VFI and to divide everyone into two camps based on those who accept it and those who reject it. Instead, it might be more fruitful to explore the wide variety of positions that one might hold regarding the proper roles for values in science. This approach might soften the seemingly hardened positions of proponents and critics of the VFI.

This leads us to the second benefit of clarifying different ways of characterizing the VFI, namely, that it highlights promising questions for further investigation. Consider three examples drawn from the preceding analysis. First, by highlighting all the different aspects of scientific reasoning that values could potentially influence, this analysis can spur closer attention to the relationships between these elements of science and the varying roles for values among them. For example, as noted previously, philosophers have been exploring how values operating in what seem to be "external" aspects of science (e.g., steering the details of the research questions that scientists investigate) can have a profound influence on the conclusions that scientists draw (which seems to be an "internal" aspect of scientific reasoning; see e.g., Okruhlik 1994; Lacey 1999; Elliott and McKaughan 2009; Holman and Bruner 2017; Winsberg 2018). There also continue to be debates (even among critics of the "traditional" VFI) about the extent to which values should influence what counts as evidence in the first place, so the analysis in this paper encourages further efforts to distinguish the appropriate roles for values among different activities like this (for discussion of this point, see Bluhm 2017 and Lewens 2025). A second example of questions opened up by this analysis comes from the way it highlights different groups (e.g., individual scientists, communities of scientists, policy makers) that could strive to keep their values out of scientific reasoning. This analysis encourages further research to consider the feasibility and merits of different options along these lines (see e.g., John 2015, 2025; Boulicault and Schroeder 2021; Levy 2025). Third, figures like de Melo-Martín and Intemann (2016) and Stegenga and Menon (2023) have very fruitfully drawn attention to the possibility of thinking about the concept of an "ideal" in different ways. This raises very important questions about whether, even if scientists cannot or should not always exclude values from their reasoning, it could still make sense to hold up value freedom as a kind of ideal in some research contexts. The analysis in this paper highlights this issue as a crucial ongoing research question.

16.6 An Objection

There is an obvious objection that critics are likely to emphasize in response to the arguments in this chapter. This objection is that even if, *in theory*, there could be variations in how the VFI is characterized, ultimately there is only one main question at stake *in practice*, and that is the one that Douglas (2009) herself emphasized on the basis of her historical analysis of the VFI.[14] This fundamental question is whether non-epistemic values should be excluded from scientific reasoning. As Brown recently put it, "[T]he VFI is an all-or-nothing affair—either social and ethical values should play a role in the internal phases of scientific reasoning, or they should not" (Brown 2024, p. 111). Thus, even if there is some potential for creating different formulations of the VFI, critics will insist that none of the variations ultimately matter very much compared to the core question of whether to incorporate non-epistemic values in scientific reasoning.

[14] I thank Jacob Stegenga for helping me to clarify this objection.

I have taken steps throughout the chapter to forestall this objection, but it may be helpful to reiterate my response here. First, I would return to Proctor's (1991) point that the VFI has taken different forms throughout history, depending on the interests and concerns at the forefront of people's attention. Thus, even if philosophers of science during the late twentieth century were particularly concerned about whether social and ethical values should play a role in scientific reasoning, a new historical moment could justify characterizing the VFI in a different way. For example, the state of the debate over values in science has shifted in recent decades (largely in response to the arguments in Douglas 2009), so it is now difficult to argue that social and ethical values should never play any role in the "internal" stages of scientific reasoning. Nevertheless, one might still think that the influences of these values should be "minimized" or lessened in various ways (see e.g., Menon and Stegenga 2023; Stamenkovic 2024). Given Proctor's point that the definition of the VFI is not set in stone, I do not see any reason why this goal of minimizing value influences could not be called an "ideal" of value freedom. This would be a different way of thinking about the notion of an ideal than philosophers have employed in the past when debating the VFI (after all, this would be the kind of ideal that can never be fully achieved), but that does not mean it is incoherent. Thus, I do not think that accepting or rejecting the VFI is an "all-or-nothing" matter; one could reject the characterization of the VFI that has been dominant in the recent literature while accepting another characterization that treats value freedom as something worth pursuing to the extent feasible. It is crucial, however, for the proponents of alternative characterizations of the VFI to be clear about how they are defining it so they can avoid talking past defenders of the "traditional" VFI.

My other response to critics is that the development of the literature on values and science in recent years has helped to sharpen our concepts, which makes it possible for us to develop more nuanced characterizations of the VFI than we were able to develop in the past. As a result, we are finding that the traditional VFI was somewhat ambiguous between different characterizations that are meaningfully different and that should not be glossed over. For example, as discussed above, there has been increasing attention to the different cognitive attitudes that scientists can take toward their hypotheses and representations. It is difficult to maintain that there is only one meaningful way of characterizing the VFI when one considers these cognitive attitudes. If my critics insist that there is only one fundamental question at issue in debates about the VFI, my response is that I am not sure which cognitive attitude this question involves. Is the crucial question whether scientists should prevent non-epistemic values from influencing their beliefs? Or is the question whether they should prevent values from influencing what they accept? Or does this question have to do with what they adopt, endorse, or impartially hold (see Lacey 2017)? Surely it makes more sense to acknowledge that there is not a single question at issue; there are multiple interesting questions that correspond to these different cognitive attitudes, and reducing them all to a single characterization of the VFI obscures this complexity.

This paper has also emphasized that it is becoming increasingly difficult to specify what counts as the "core" of scientific reasoning for the purposes of the VFI.[15] Thus, there is ample room for disagreement about which of the following activities a proponent of the VFI should be keeping insulated from values: choosing research questions, choosing methodologies, formulating categories and concepts, developing study designs, collecting data, developing models, assessing models, interpreting data, assessing the amount of evidence in support of a hypothesis, determining whether there is sufficient evidence to accept a hypothesis, and deciding how to frame or communicate the outcomes of inquiry.[16] Why should we accept that there is a clean distinction between the activities in this list that are "internal" or "core" to scientific reasoning and those that are not? And once one rejects, or at least problematizes, this internal/external distinction, then it seems relatively clear that the VFI can be formulated in a number of meaningful ways.

These are only two of the nine forms of variation described in Table 16.1, but they should be sufficient to show that different characterizations of the VFI are genuinely interesting and are not merely trivial variations on a single question. The other forms of variation in Table 16.1 add even more richness to the discussion. For example, as noted previously, deciding whose values should be excluded from science raises the question of whether the VFI should be focused on removing *everyone's* values from science or only *individual scientists' idiosyncratic values* (Boulicault and Schroeder 2021). This is a live question that may go back to Levi (see Staley 2017; Boulicault and Schroeder 2021) and that has been attracting new attention (Levy 2025). Therefore, to insist that there is only one legitimate definition of the VFI and that all other potential variations on it constitute misunderstandings seems like a missed opportunity to engage in fruitful conceptual clarification of this research area.

---

[15] Although I am blurring the distinction between the "internal" and "external" aspects of scientific reasoning by focusing on different scientific activities that could be either internal or external, one could also draw on the cognitive attitudes mentioned in the preceding paragraph to blur the internal/external distinction. For example, one might think that the attitude of acceptance counts as an internal aspect of scientific reasoning, whereas the attitude of pursuit counts as an external aspect of scientific reasoning. However, it is not clear that these attitudes are cleanly distinguished in published journal articles, thereby generating confusion about whether the activities described in those articles count as internal or external aspects of scientific reasoning (see e.g., Davis-Stober et al. 2025; Shaw and Barseghyan 2019). I am grateful to Jamie Shaw for sharing this insight with me.

[16] For example, Dan Hicks appears to interpret the "traditional" VFI such that it prohibits values from influencing the process of "gathering evidence" (2018, p. 166), but it is not clear that this activity should actually be classified as "internal" to scientific reasoning, given that gathering evidence seems similar to selecting research questions or deciding what to study.

16.7 Conclusion

I have argued that even though philosophers developed a particular way of thinking about the VFI during the latter part of the twentieth century, it is conceptually possible to characterize it in a number of different ways. I have organized this variety of characterizations using a three-part framework that focuses on the values to be excluded, the ways in which science is supposed to be free of those values or their influences, and the nature of the ideal. Thinking about the VFI in terms of a variety of different potential characterizations rather than a single characterization is significant because it shifts discussions of the VFI from the dichotomous question "Should the VFI be rejected?" to more complex questions, like "What form(s) of the VFI should be rejected?" or "Is it fruitful to try to develop a form of the VFI that can still be maintained?"

I have argued that this shift in focus has at least two potential benefits. First, it has the potential to soften the oppositional framing that has emerged in the literature on science and values. Rather than creating an "in-group" (those who oppose the VFI) and an "out-group" (those who defend the VFI), my proposed shift portrays everyone as holding one among a multiplicity of different views about the appropriate roles for values in science and the ways they should be limited. Softening the oppositional framing might open up the critics and the defenders of the VFI to consider more nuanced positions among this multiplicity, and it might encourage those uncomfortable with the current framing of the literature to engage more fully in the discussion. Second, this shift in perspective could potentially help to broaden and nuance research on the topic of values and science. By highlighting the ambiguity associated with key concepts, such as what counts as an "internal" component of scientific reasoning or what counts as an "ideal," this shift encourages new research questions and different ways of thinking about the relationships between values and science.

This chapter should also encourage those working on the topic of values and science to consider the contexts in which the VFI serves as a fruitful way of framing the literature on values and science as well as the contexts in which the VFI is more likely to limit or obscure important issues and discussions. Once one begins to think about different ways of characterizing the VFI, it becomes clear that the VFI itself may be more confusing than it is helpful, especially for those outside the philosophical community. The dominant form of the VFI that emerged from philosophical discussions throughout the latter part of the twentieth century was already fairly nuanced and allowed for values to play important roles in science. Given the potential ambiguities associated with the VFI that this chapter has brought to the fore (e.g., what is meant by the "internal" or "core" elements of scientific reasoning, what cognitive attitudes the VFI involves, questions about whether it focuses on individual scientists or on some other group), it is unclear why ongoing discussions of values in science should revolve around this concept. Why not simply acknowledge that values have important roles to play in science—which even the proponents of the traditional VFI should be willing to accept—and proceed to discuss how best to manage values in science? Thus, although I am less likely than Brown (2024) to dismiss the VFI as indefensible, we ultimately come to a fairly similar conclusion, namely, that future work on values and science would do well to focus instead on questions about how best to manage values in science.

Acknowledgments I am extremely grateful to Hakob Barseghyan for his generous assistance in developing Fig. 16.1 and to Jon Tsou for inviting me to contribute to this volume. I also thank Heather Douglas, Carla Fehr, Jamie Shaw, Jacob Stegenga, Jon Tsou, an anonymous reviewer, and the members of a reading group led by Arnon Levy for their excellent suggestions on earlier versions of this chapter. Jamie Shaw was particularly gracious in providing extensive comments throughout the development of the chapter. Finally, I am so thankful for Matt Brown—for our friendship through the years, for his impact on the field of values and science, and for his influence on this chapter. My conversations with him a number of years ago helped to inspire this chapter, and my recent conversations with him improved the chapter while I was working on it.

References

Alexandrova, A. 2018. Can the science of Well-being be objective? *British Journal for the Philosophy of Science* 69:421–445.

Anderson, E. 1993. *Values in ethics and economics*. Cambridge: Harvard University Press.

Anderson, E. 2004. Uses of value judgments in science: A general argument, with lessons from a case study of feminist research on divorce. *Hypatia* 19:1–24.

Betz, G. 2013. In Defence of the value free ideal. *European Journal for Philosophy of Science* 3:207–220.

Betz, G. 2017. Why the argument from inductive risk doesn't justify incorporating non-epistemic values in scientific reasoning. In *Current controversies in values and science*, ed. K. Elliott and D. Steel, 94–110. New York: Routledge.

Biddle, J. 2013. State of the Field: Transient underdetermination and values in science. *Studies in History and Philosophy of Science* 44:124–133.

Bluhm, R. 2017. Inductive risk and the role of values in clinical trials. In *Exploring inductive risk: Case studies of values in science*, ed. K. Elliott and T. Richards, 193–212. New York: Oxford University Press.

Boulicault, M., and S. Schroeder. 2021. Public trust in science: Exploring the idiosyncrasy-free ideal. In *Social trust: Foundational and philosophical issues*, ed. K. Vallier and M. Weber, 102–121. London: Routledge.

Brown, M. 2013. Values in science beyond Underdetermination and inductive risk. *Philosophy of Science* 80 (5): 829–839.

Brown, M. 2015. John Dewey's pragmatist alternative to the belief-acceptance dichotomy. *Studies in History and Philosophy of Science* 53:62–70.

Brown, M. 2020. *Science and moral imagination: A new ideal for values in science*. Pittsburgh: University of Pittsburgh Press.

Brown, M. 2024. For values in science: Assessing recent arguments for the ideal of value-free science. *Synthese* 204:112.

Crasnow, S. 2024. Feminist Perspectives on Science. In *The Stanford Encyclopedia of Philosophy*, ed. E. N. Zalta and U. Nodelman. (Spring 2024 Edition), forthcoming URL = https://plato.stanford.edu/archives/spr2024/entries/feminist-science/.

Currie, A. 2023. Epistemic engagement, aesthetic value, and scientific practice. *The British Journal for the Philosophy of Science* 74 (2): 313–334.

Daston, L. 2022. *Rules: A short history of what we live by*. Princeton: Princeton University Press.

Davis-Stober, C., A. Sarafoglou, B. Aczel, S. Chandramouli, T. Errington, S. Field, A. Fishbach, J. Freire, J. Ioannidis, K. Oberauer, and F. Pestilli. 2025. How can we make sound replication decisions? *Proceedings of the National Academy of Sciences* 122 (5): e2401236121.
de Melo-Martín, I., and K. Intemann. 2016. The risk of using inductive risk to challenge the value-free ideal. *Philosophy of Science* 83 (4): 500–520.
DiMarco, M., and K. Khalifa. 2019. Inquiry tickets: Values, pursuit, and Underdetermination. *Philosophy of Science* 86 (5): 1016–1028.
Douglas, H. 2009. *Science, policy, and the value-free ideal.* Pittsburgh: University of Pittsburgh Press.
Douglas, H. 2013. The value of cognitive values. *Philosophy of Science* 80:796–806.
Douglas, H. 2016. Values in Science. In *The oxford handbook of philosophy of science*, ed. P. Humphreys, 609–632. New York: Oxford University Press.
Douglas, H., and T. Y. Branch. 2024. The social contract for science and the value-free ideal. *Synthese* 203.
Douglas, H., and K. Elliott. 2022. Addressing the reproducibility crisis: A response to Hudson. *Journal of General Philosophy of Science* 53 (2): 201–209.
Dupré, J. 2007. Fact and value. In *Value-free science? Ideals and illusions*, ed. H. Kincaid, A. Wylie, and J. Dupré, 27–41. New York: Oxford University Press.
Elabbar, A. 2023. The curatorial view of assessment and the ethics of scientific advice: Beyond decisional autonomy towards distributive epistemic justice. *International Union for the History and Philosophy of science and technology* (IUHPST)(essay prize in the history and philosophy of science). Available online at IUHPST. Org/media/pdf/Elabbar_IUHPST_Prize_2023. Pdf.
Elliott, K. 2011. *Is a little pollution good for you? Incorporating societal values in environmental research.* New York: Oxford University Press.
Elliott, K. 2017. *A tapestry of values: An introduction to values in science.* New York: Oxford University Press.
Elliott, K. 2022. *Values in Science.* Cambridge: Cambridge University Press.
Elliott, K., and R. Korf. 2024. Values in science: What are values, anyway? *European Journal for Philosophy of Science* 14:53.
Elliott, K., and D. McKaughan. 2009. How values in scientific discovery and pursuit Alter theory appraisal. *Philosophy of Science* 76:598–611.
Elliott, K., and D. Willmes. 2013. Cognitive attitudes and values in science. *Philosophy of Science* 80 (Proceedings): 807–817.
Fleisher, W. 2018. Rational endorsement. *Philosophical Studies* 175 (10): 2649–2675.
Harvard, S., and E. Winsberg. 2022. The epistemic risk in representation. *Kennedy Institute of Ethics Journal* 32 (1): 1–31.
Havstad, J., and M. Brown. 2017. Inductive risk, deferred decisions, and climate science advising. In *Exploring inductive risk: Case studies of values in science*, ed. K. Elliott and T. Richards, 101–124. New York: Oxford University Press.
Hicks, D. 2018. Inductive risk and values in science: A comment on de Melo-Martín and Intemann. *Philosophy of Science* 85:164–174.
Hilligardt, H. 2022. Looking beyond values: The legitimacy of social perspectives, opinions and interests in science. *European Journal for Philosophy of Science* 12 (4): 58.
Holman, B., and J. Bruner. 2017. Experimentation by industrial selection. *Philosophy of Science* 84:1008–1019.
Holman, B., and T. Wilholt. 2022. The new demarcation problem. *Studies in History and Philosophy of Science* 91:211–220.
Hudson, R. 2016. Why we should not reject the value-free ideal of science. *Perspectives on Science* 24 (2): 167–191.
Hudson, R. 2021. Should we strive to make science bias-free? A philosophical assessment of the reproducibility crisis. *Journal for General Philosophy of Science* 52:389–405.
Intemann, K. 2015. Distinguishing between legitimate and illegitimate values in climate modeling. *European Journal for Philosophy of Science* 5:217–232.

Intemann, K. 2021. Feminist perspectives on values in science. In *The Routledge handbook of feminist philosophy of science*. ed. Crasnow and K. Intemann, 201–215. New York: Routledge.
Ivanova, M. 2017. Aesthetic values in science. *Philosophy Compass* 12 (10): e12433.
John, S. 2015. Inductive risk and the contexts of communication. *Synthese* 192:79–96.
John, S. 2025. Weber's elephant: Rethinking science advice. *British Journal for the Philosophy of Science.*
Kuhn, T. 1977. Objectivity, value judgement, and theory choice. In *The essential tension*, ed. T. Kuhn, 320–329. Chicago: University of Chicago Press.
Lacey, H. 1999. *Is science value free? Values and scientific understanding*. New York: Routledge.
Lacey, H. 2017. Distinguishing between cognitive and social values. In *Current controversies in values and science*, ed. K. Elliott and D. Steel, 15–30. New York: Routledge.
Laudan, L. 1984. *Science and values: The aims of science and their role in scientific debate*. Berkeley: University of California Press.
Levi, I. 1960. Must the scientist make value judgments? *The Journal of Philosophy* 57 (11): 345–357.
Levy, A. 2025. *A (consequence-oriented) critique of the argument from inductive risk*. Forthcoming: Philosophical Studies.
Lewens, T. 2025. Towards a direct role for values in the heart of science. *Philosophy of Science* 1–25.
Longino, H. 1990. *Science as social knowledge*. Princeton: Princeton University Press.
Longino, H. 1996. Cognitive and non-cognitive values in science: Rethinking the dichotomy. In *Feminism, science, and the philosophy of science*, ed. L. Hankinson Nelson and J. Nelson, 39–58. Boston: Kluwer.
McMullin, E. 1983. Values in Science. In: PSA 1982. In *East Lansing*, ed. P. Asquith and T. Nickles, vol. 1982, 3–28. Philosophy of Science Association.
Menon, T., and J. Stegenga. 2023. Sisyphean science: Why value freedom is worth pursuing. *European Journal for Philosophy of Science* 13 (4): 48.
Morgan, G. 2013. The value of beauty in theory pursuit: Kuhn, Duhem, and decision theory. *Open Journal of Philosophy* 3 (1): 9–14.
Nye, M. 2011. *Michael Polanyi and his generation: Origins of the social construction of science*. Chicago: University of Chicago Press.
O'Connor, C., and D. Freeborn. 2025. *Industrial distraction*, 1–22. Philosophy of Science.
Okruhlik, K. 1994. Gender and the biological sciences. *Canadian Journal of Philosophy Supplementary* 20:21–42.
Palider, K., A. Sarwar, H. Barseghyan, P. Patton, J. Da Silva, T. Doppelt, N. Levesley, J. Rapson, J. Shaw, Y. Zhang, and A. Zulfiqar. 2021. A diagrammatic notation for visualizing epistemic entities and relations. *Scientonomy: Journal for the Science of Science* 4:87–139.
Parker, W. 2024. The epistemic projection approach to values in science. *Philosophy of Science* 91 (1): 18–36.
Peters, U. 2021. Illegitimate values, confirmation bias, and Mandevillian cognition in science. *The British Journal for the Philosophy of Science.* 72 (4): 1061–1081.
Philippi, C. L. 2020. Well-ordered science's basic problem. *Philosophy of Science* 87:365–375.
Polanyi, M. 1962. The republic of science. *Minerva* 1:54–73.
Proctor, R. 1991. *Value-free science?* Cambridge: Harvard University Press.
Ratti, E., and F. Russo. 2024. Science and values: A two-way direction. *European Journal for Philosophy of Science* 14 (1): 6.
Rooney, P. 2017. The borderlands between epistemic and non-epistemic values. In *Current controversies in values and science*, ed. K. Elliott and D. Steel, 31–45. New York: Routledge.
Shaw, J. 2021. Feyerabend, funding, and the freedom of science: The case of traditional Chinese medicine. *European Journal for Philosophy of Science* 11 (2): 37.
Shaw, J. 2022. On the very idea of Pursuitworthiness. *Studies in History and Philosophy of Science* 91:103–112.
Shaw, J., and H. Barseghyan. 2019. Problems and prospects with the Scientonomic workflow. *Scientonomy: Journal for the Science of Science* 3:1–14.

Sheykh-Rezaee, H., and H. Bikaraan-Behesht. 2023. Value-free ideal is an epistemic ideal: An objection to the argument from inductive risk. *Principia: An International Journal of Epistemology* 27 (1): 137–163.

Solomon, M. 2012. Socially responsible science and the Unity of values. *Perspectives on Science* 20:331–338.

Staley, K. 2017. Decisions, decisions: Inductive risk and the Higgs boson. In *Exploring inductive risk: Case studies of values in science*, ed. K. Elliott and T. Richards, 37–55.

Stamenkovic, P. 2024. Straightening the 'value-laden turn': Minimising the influence of extra-scientific values in science. *Synthese* 203:20.

Steel, D. 2010. Epistemic values and the argument from inductive risk. *Philosophy of Science* 77:14–34.

Stegenga, J., and T. Menon. 2023. The difference-to-inference model for values in science. *Res Philosophica* 100:423–447.

Ward, Z. 2021. On value-laden science. *Studies in History and Philosophy of Science* 85:54–62.

Wilholt, T. 2009. Bias and values in scientific research. *Studies in History and Philosophy of Science* 40 (1): 92–101.

Winsberg, E. 2018. *Philosophy and climate science*. Cambridge: Cambridge University Press.

Zhao, H. 2022. *The nature of values in science: What they are and how they guide*. Unpublished Manuscript.

Chapter 17
Inquiry and Epistemic Priority: Lessons from Segregation Research

Kareem Khalifa, Jared Millson, and Mark Risjord

Abstract In this paper, we offer a novel account of epistemic priority, which we dub "inquisitive due diligence." We then show how our account both outperforms the two prominent species of epistemic priority—Douglas's inductive risk account and Steel's values-in-science account—while also rebutting objections from the leading critic of epistemic priority, the infamous Matthew J. Brown. We illustrate these points using examples from segregation research in the 1980s.

Keywords Inquiry · Epistemic priority · Inductive risk · Science and values · Defeasible justification

17.1 Introduction

Despite the elimination of Jim Crow laws and the institution of Civil Rights legislation in the 1960s, US residential segregation by race has persisted. In the 1970s and 1980s, explaining the persistence of Black residential segregation was central to the politics of segregation. On the left, racial discrimination was seen as a chief driver of entrenched segregation. As such, segregation was regarded as a form of oppression and a legitimate target for political intervention. This narrative was challenged

K. Khalifa (✉)
Department of Philosophy, University of California, Los Angeles, CA, USA
e-mail: kkhalifa@humnet.ucla.edu

J. Millson
Department of Philosophy, Rhodes College, Memphis, TN, USA
e-mail: millsonj@rhodes.edu

M. Risjord
Department of Philosophy and Institute for the Liberal Arts, Emory University, Atlanta, GA, USA
e-mail: mrisjor@emory.edu

J. Y. Tsou et al. (eds.), *Values, Pluralism, and Pragmatism: Themes from the Work of Matthew J. Brown*, Boston Studies in the Philosophy and History of Science 347, https://doi.org/10.1007/978-3-031-92958-8_17

by newly formed and privately funded conservative think tanks, such as the Manhattan Institute and the Heritage Foundation, whose class-based explanations of segregation appealed to individuals' preferences to live among those with a similar socioeconomic status and to enjoy the benefits of better neighborhoods associated with higher socioeconomic status. Insofar as a persistent "underclass" was unable to escape, this was attributable to a "lower class culture or lifestyle" where a person "lived from moment to moment" and was "either unable or unwilling to take account of the future and control his impulses" (Banfield 1974, p. 54). Thus, on the conservatives' analysis, segregation was due to individuals' class preferences and moral character, which militated against political intervention by a liberal-democratic state. In the 1980s, Douglas Massey and his coauthors were at the forefront of empirical research that showed that race affects residential segregation independently of class (Massey and Mullan 1984; Massey and Bitterman 1985; Massey and Denton 1985). In doing so, Massey's results called the conservative explanation into question, and thereby undermined a central rationale for a *laissez-faire* political approach.

Predictably, each side of the debate over segregation accused the other of biased, value-laden inquiry. To distinguish themselves as scientists—as opposed to mere political commentators—sociologists of residential segregation valorized evidence and empirically grounded models over non-epistemic values. Some philosophers of science would applaud this embrace of so-called *epistemic priority* (Douglas 2009; Steel 2010, 2017). Others, however, see epistemic priority as a hangover from the much-reviled value-free ideal (Brown 2013a, b, 2017; Hicks 2014).

Given the salience of residential segregation's moral and political implications, epistemic priority's viability comes into sharp relief. Massey and his collaborators were not chasing a pipe dream in their attempts to rise above the political fray through sober empirical analysis. They were exercising their *inquisitive due diligence*, which we will argue provides a novel and promising account of epistemic priority. Before doing so, we clear the ground (Sect. 17.2) by giving a more precise characterization of epistemic priority than has previously been articulated and by situating leading epistemic prioritarians within it. In Sect. 17.3, we introduce and articulate the concept of inquisitive due diligence and illustrates its core features with Massey's segregation research. In the remaining sections, we then show that inquisitive due diligence has several advantages over other prominent accounts of epistemic priority.

17.2 What Is Epistemic Priority?

As already noted, debates about epistemic priority have gained steam in recent years. But what exactly is all the *Sturm und Drang* about? That is not always so clear. To that end, we approach the literature with a guiding idea: that epistemic priority theses form a genus with multiple species. After characterizing that genus

(Sect. 17.2.1), we describe the two most developed species of that genus, both of which place inductive risk at the forefront (Sect. 17.2.2).

17.2.1 Generic Epistemic Priority

In slogan form, epistemic prioritarians claim that epistemic considerations "trump" non-epistemic considerations in science's central inferential tasks. However, specifying this "trumping" relation is no simple task. As we see it, epistemic priority is characterized by a pair of commitments:

(EP1) There are certain "basic" epistemic standards, i.e., standards that all/most correctly accepted hypotheses must satisfy, but no value that all/most correctly accepted hypotheses must satisfy.

(EP2) Values can only legitimately influence acceptance of hypotheses by modifying non-basic epistemic standards in prescribed ways; some epistemic standards legitimately influence acceptance independently of any values.

While we use the term "hypothesis," we take our points to apply to claims about theories, models, methodologies, data, etc. We will leave the ways in which a hypothesis is "correctly accepted" unanalyzed, relying on scientific practice to fix the meaning of this term.[1] A standard, *E*, is epistemic just in case it is truth-conducive and, at a minimum, there is at least one hypothesis, *h*, such that *E* is either necessary or sufficient for correctly accepting *h*. Basic epistemic standards have greater scope, being necessary for correctly accepting all or most hypotheses. A value is any moral, political, social, or pragmatic consideration. Finally, we'll say that *x* influences acceptance of a hypothesis *h* just in case scientists would have been less likely to accept *h* in the absence of *x*, and that this influence is legitimate just in case it's correct for the scientists to accept *h* given *x*'s role in the process that led to scientists' acceptance of *h*.

We hope that it's clear why EP1 and EP2 each prioritize the epistemic. First, EP1 claims that science has a core of basic epistemic standards but lacks an analogous non-epistemic core. In the sociology of segregation, this means that any acceptable explanation of why Black segregation persisted would have to satisfy these basic epistemic standards no matter what the political consequences. Second, EP2 claims that the only legitimate influences of non-epistemic values on acceptance are mediated by epistemic standards, but denies the analogous requirement that epistemic standards be filtered through non-epistemic values. Epistemic priority thus counsels segregation researchers to alter non-basic epistemic standards to reflect the political ramifications of their research. While the rhetoric of scientific writing in this period forbade explicit appeal to values, EP1 and EP2 suggest that segregation researchers

[1] Crucially, we take correct acceptance to be fallible. If one likes, we are interested in "correct acceptance at time *t*" or "*prima facie* correct acceptance." In order not to prejudice the conclusion, we assume that correct acceptance entails neither epistemic priority nor its denial.

should use more demanding statistical tests, higher evidential thresholds, and so on, than their colleagues who worked in less fraught domains.

17.2.2 Risky Epistemic Priority

More detailed characterizations of epistemic priority can camp out under EP1 and EP2's big-tent definition. Prioritarians will differ with respect to the epistemic standards they deem basic, whether such standards must apply to all or only most correctly accepted hypotheses, and the legitimate influences values may exert on epistemic standards. We do not deign to enumerate all of these variants.[2] However, it will be instructive to unpack the most developed species of prioritarianism, which comes from those who take the primary function of values in hypothesis acceptance to be the mitigation of *inductive risk*. We characterize this *risky epistemic priority* thesis as follows:

(REP) For all hypotheses *h*, it's correct to accept *h* if and only if

(a) *h* satisfies all of the basic epistemic standards and
(b) *h* satisfies any other epistemic standards as modified by values to suitably mitigate risk.

Douglas (2009) most readily fits REP's profile. Douglas (2009, p. 94) takes internal consistency and predictive competency to be basic epistemic standards in the sense of minimal "criteria that all theories must succeed in meeting," thereby satisfying the first condition. For her, values may only influence correct acceptance "indirectly," i.e., by lowering or raising non-basic epistemic standards proportionally to risk. Roughly stated, when the non-epistemic costs of errantly accepting a hypothesis are high, these epistemic standards are raised; when the non-epistemic costs of errantly not accepting a hypothesis are high, these epistemic standards are lowered (though never below the basic epistemic standards).[3] Moreover, critics of epistemic priority clearly regard these as potential sticking points (Brown 2013a, pp. 69–71; b, p. 836; 2017, p. 67).

Steel differs from Douglas in how he interprets both (a) and (b). With respect to (a), while Steel sometimes uses the word "basic," his more precise formulations typically refer to "robust" epistemic standards, which he defines as values that are truth-conducive in *most* scientific contexts (Steel 2010, p. 20; Steel and Whyte 2012, p. 168). Contrast this with Douglas's characterization of minimal criteria that apply to *all* correctly accepted hypotheses. Closely related, whereas Douglas takes internal consistency and predictive competency as her only basic epistemic

[2] In particular, we do not discuss underdetermination-based accounts of epistemic priority, e.g., Longino (1990).

[3] Insofar as no risk is to be mitigated, EP entails that satisfying basic epistemic standards is sufficient for correct acceptance. We will not broach the vexed issue as to whether such cases ever arise.

standards, Steel's (2017, p. 58) preferred "basic epistemic value" is the "weak severity criterion, according to which a test counts as evidence for a hypothesis only if it was likely to issue a negative result if the hypothesis were indeed false." However, he appears to countenance other epistemic standards as basic, including Douglas's.

With respect to REP's second condition (b), Steel "allows nonepistemic values to influence the design, interpretation, or dissemination of scientific research that is practically feasible and ethically permissible when, and only when, epistemic values fail to indicate a unique best option" (Steel and Whyte 2012, p. 170). Importantly, this allows values to influence acceptance in non-risky ways, so we will occasionally consider *Steel's variant* of thesis (b):

(b*) no values interfere with *h* satisfying as many extant epistemic standards as is scientifically feasible.

While a hypothesis might or might not satisfy numerous epistemic standards, Steel's variant only concerns those that are *extant* and *feasibly satisfied*.[4] Both of these characteristics are indexed to particular scientific communities at particular times. A standard is *extant* insofar as scientists are responsive to it in their practices of constructing, considering, and evaluating hypotheses. For instance, prior to the advent of probability theory, only the simplest of epistemic standards for evaluating stochastic hypotheses were extant. Additionally, *feasibility* concerns the ability of scientists to ascertain whether a hypothesis satisfies an epistemic standard. For example, it was not scientifically feasible for Bohr and his contemporaries to evaluate their hypotheses using evidence from the Large Hadron Collider. Combined, these provide a non-utopian account of what it means for the scientific community to do its "epistemic best."

While Steel's variant makes no explicit reference to risk, his most detailed discussions of how values bear on epistemic standards are developed through the lens of inductive risk (e.g., Steel 2010, 2013; Kaivanto and Steel 2019). Through that lens, Steel allows non-epistemic values to mitigate risk in much the same way Douglas does, though he "insists that values in this role must not conflict with epistemic values" and especially the weak severity principle (Steel and Whyte 2012, p. 171).

17.3 Inquisitive Due Diligence

Epistemic priority need not be risky. In what follows, we carve out an alternative niche in the prioritarian ecosystem. Its centerpiece is the concept of *inquisitive due diligence*. Roughly stated, inquisitive due diligence is the requirement that scientists

[4] Note that we are using these as terms of art, and not trying to capture "ordinary language" senses with much fidelity.

ask themselves "tough" or "pointed" questions, do their damnedest to answer those questions correctly, and revise their commitments in light of these efforts. More precisely:

(IDD) For all hypotheses *h*, it's correct to accept *h* if and only if ***inquisitive due diligence*** has been exercised, i.e.,
(i) all of the relevant pointed questions against *h* have been asked,[5]
(ii) those questions have been correctly answered, and
(iii) none of those answers defeat *h*.

Since it will prove useful below, we contrast *pointed* questions with *focal* questions. While the former challenge a hypothesis, the latter are the inquiry-guiding questions to which *h* is a potential answer. By a pointed question, we mean a question that has a *defeater* as one of its potential answers. To a first approximation, defeaters downgrade an accepted statement's positive epistemic status. While defeaters can assume myriad forms,[6] we focus on two: *rebutting* and *undercutting*. Rebutting questions aim to defeat hypotheses by showing them to be false. To ask about an anomalous phenomenon is to challenge a generalization with a rebutting question. By contrast, undercutting questions aim to defeat the purported justification for a hypothesis. In scientific inquiry, undercutting questions typically target aspects of a study's methodology.

In our case study, the focal question was why Black segregation persisted in the US after the introduction of anti-segregation and anti-discrimination policies. Racial discrimination and class-based assimilation competed as answers to this question. To distinguish the effects of race and class, Massey and colleagues developed models that contrasted trajectories of Black and Hispanic urban populations as their socioeconomic status (SES) increased (Massey and Mullan 1984; Massey and Bitterman 1985; Massey and Denton 1985). The models treated SES as causally determining two aspects of spatial assimilation: the residential distance from highly segregated Black and Hispanic neighborhoods and the probability of encountering either non-Hispanic Whites (what Massey and Mullan called "Anglos") or individuals of the other racial/ethnic group. The strength of these causal associations was estimated using 1970 Census data, with one model for Hispanics and another for Blacks.

These models demonstrated clear differences in the mechanisms by which changes in SES were translated into spatial assimilation. Hispanics were able to translate improvements in education, income, and occupational status directly into residential assimilation with Anglos, and the effects were quite strong. On the other

[5] This condition is not nearly as demanding as its universal quantification might suggest. The relevance of a question is highly context sensitive—being determined, *inter alia*, by scientists' background assumptions, personal interests, and (disciplinary and institutional) role-responsibilities. If, as seems plausible, the latter are subject to an ought-implies-can rider, then asking all of the relevant questions in a given inquiry should be achievable by most well-trained scientists.

[6] For a standard taxonomy of defeaters, see Sudduth (2023). Dutant and Littlejohn (2021, p. 223) enumerate eight different kinds of defeaters—and footnote two more for good measure.

hand, while Black residents were able to improve their neighborhoods as their SES rose, there was no direct effect on contact with Hispanics, and a very small effect on contact with Anglos. Consequently, given identical educational starting points, Hispanics were able to move significantly farther from highly segregated neighborhoods than Blacks. Massey and Mullan (1984, p. 870) conclude that:

> If there is an underclass in the United States, Hispanics cannot be considered "permanent" members of it in the same way as blacks. Our results point consistently to an ongoing process of assimilation among Hispanics circa 1970. Unlike blacks, they are able to translate social mobility into residential mobility. Hispanics are simply not trapped in the barrio in the same way that blacks are trapped in the ghetto.

These results suggest that race and racial discrimination affect segregation independently of SES.

Massey and Mullan addressed important "technical problems" in their research (1984, p. 859), which IDD captures as potential undercutting questions. The issues arise because the numeric values of the model variables were estimated from mean values for census tracts. Census tracts are relatively permanent subdivisions of a geographic region (in this case, a metropolitan area) typically consisting of 1200 to 8000 people. The first problem is that census tracts are spatially autocorrelated, meaning that the demographic properties of one tract are correlated with those of adjacent tracts. Since the assumption of independence between observations is thereby violated, this poses problems for statistical testing.[7] Indeed, Massey and Mullan chose not to report any p-values for their estimation because of this autocorrelation. The models' estimation of causal power was thus imprecise, and this imprecision threatened to undermine the justification for their conclusions.

In response to the pointed question "Could the actual values of the coefficients be different than those represented in the model?", Massey and Mullan admit that the model values could be (and probably were) biased. They argued, however, that "we do not center our interpretation on specific values of coefficients but on the relative strength and direction of relationships" (Massey and Mullan 1984, p. 860). The inferences about the effects of race are drawn from a comparison of the two models, and insofar as the values are biased, the bias did not affect Massey and Mullan's coarse-grained contrast between Black and Hispanic causal mechanisms. In IDD's parlance, Massey and Mullan were arguing that the pointed question regarding the coefficients' precise values was not *relevant* to their inquiry.

In doing so, Massey and Mullan do not draw any strong conclusions that the coefficient values in the model are accurate, only committing to qualitative claims (e.g., that Hispanics can translate social mobility into residential mobility more easily than Black Americans) while remaining agnostic about the specific magnitudes. This highlights how "bracketing" pointed questions (i.e., treating them as irrelevant) often comes at a cost: the answers one can justifiably offer to the focal question are correspondingly constrained. More generally, the informativeness of an answer to a

[7] Specifically, the potential variance around the true value could be substantially higher than what the standard tests of statistical significance would show.

focal question tends to be inversely proportional to its defeasibility. So, scientists must often sacrifice some of the scope, precision, or boldness of their hypotheses when they bracket pointed questions.

Massey and Mullan's strategy highlights that not *all* questions pointed at a hypothesis are relevant. If they were, inquisitive due diligence would be impossible to exercise, as scientists would have to rule out evil demon hypotheses, answer objections that nobody can reasonably be expected to entertain, run an inexhaustible number of robustness checks, and so on. Many things can make a question relevant—personal interests, professional role-responsibilities, and other questions in the inquiry (Khalifa and Millson 2020). Here, we focus on just two. First, *non-epistemic values* can make certain focal and pointed questions relevant. According to IDD, this is the mechanism by which values ought to influence scientific inquiry, not—in contrast to REP—by modifying epistemic standards to manage risk.

Second, we claim that *role-responsibilities* often determine the relevance of pointed questions. Here, we adopt Hart's (1968, p. 212) account of role-responsibilities:

> ...whenever a person occupies a distinctive place or office in a social organization, to which specific duties are attached to provide for the welfare of others or to advance in some specific way the aims or purposes of the organization, he is properly said to be responsible for the performance of these duties, or for performing what is necessary to fulfill them.

Scientists frequently occupy distinctive roles in their fields—as theorists, experimenters, etc. Moreover, results in one field contribute to others, e.g., segregation research bears on educational psychology, environmental science, and epidemiology. As such, individual scientists play roles in the broader scientific community that go beyond their immediate circle of fellow specialists. In short, other scientists (and in many cases, laypersons) rely on scientists' results. Because of this, scientists must address a host of pointed questions before it is correct for them to accept hypotheses.[8]

Massey and Mullan's defense against another undercutting question about their methodology clearly arises from both their role-responsibilities as scientists and from non-epistemic values. In a follow-up essay, Massey and Denton (1985) considered a further potential defeater: that census tracts are arbitrary constructs of the Census process; they do not reflect neighborhoods. One might ask, then, "[a]re the differences between the models of Black and Hispanic segregation the result of how the census-tract boundaries were drawn?" While the possibility might be remote, the inquiry's political context makes it important to address this pointed question. In response, Massey and Denton provided values for the model variables from a dataset of individual-level education, income, occupation, and residence data. Rather than rely on census tracts as the areal units, the dataset identified neighborhoods with a geographic clustering algorithm. Both the models based on census tracts and those based on the individual-level variables and geographic clustering algorithm indicated the same set of causal relationships, thus answering the pointed question

[8] For more on scientists' role-responsibilities, see Douglas (2003).

in the negative. This question raised the possibility that their conclusions were statistical artefacts, and justifying one's conclusions is a fundamental scientific role-responsibility. This undercutting question is also made relevant by the political context, since a politically motivated critic will probe precisely these potential weak spots. In a politically charged context like the study of residential segregation, it behooves a scientist to attend to a wider range of potential vulnerabilities.

Massey and Denton's (1985) follow-up to Massey and Mullan's (1984) original model shows how it is possible for answers to pointed questions to themselves be defeated. In our framework, this raises "further waves" of pointed questions. However, at some point, further waves become perverse. This is of a piece with the idea that in some contexts, inquirers are entitled to assert certain claims "by default" without having done evidential work in advance. Such default entitlement, can, however, be lost when an agent fails to adequately respond when asked an appropriate pointed question.[9] In science, default entitlement is often fixed by disciplinary role-responsibilities and highlighted by non-epistemic values.

Recall that epistemic priority as a genus is characterized by twin commitments to basic epistemic standards (EP1) and to values being filtered through non-basic epistemic standards (EP2). Inquisitive due diligence purports to apply to all correctly accepted hypotheses. Since absence of defeat is truth-conducive,[10] it thereby professes to be a basic epistemic standard. Additionally, it suggests that non-epistemic values only legitimately influence acceptance through the focal and pointed questions they make relevant. While this overlaps substantially with the risky account, there are important points of divergence. Specifically, inquisitive due diligence can diverge from internal consistency, predictive competence, and weak severity. Additionally, an epistemic standard can mitigate risk while failing to answer all relevant pointed questions, and vice versa. We'll now argue that these mismatches favor inquisitive due diligence.

17.4 Basic Epistemic Standards

As a basic epistemic standard, our account of inquisitive due diligence enjoys stouter defenses against anti-prioritarian arguments than our risk-infatuated brethren. The first of those arguments challenges the very existence of basic epistemic standards: "[e]pistemic priority implies evidence or epistemic standards can guide inquiry without being changed or reevaluated in the course of inquiry. Imposing such criteria stifles inquiry, and ignores its context" (Brown 2017, p. 72). Brown is surely correct that epistemic standards can evolve throughout the course of inquiry.

[9] See Williams (2015, pp. 263–264) for more on this "default and query" account of justification.

[10] More precisely, let h be *prima facie* confirmed by e, *i.e.*, $P(h|e) > P(h)$. Rebutting questions have as a potential answer some r such that $P(\neg h|e \,\&\, r) > P(\neg h|e)$. Undercutting questions have as a potential answer some u such that $P(h|e \,\&\, u) < P(h|e)$. Hence, defeaters always decrease the probability of h.

It is an interesting question whether *all* such standards are revisable in this way, or whether a core of basic epistemic standards serve as "fixed points," as Brown (ibid) pithily puts it.

As noted above, prioritarians can differ as to which epistemic standards are basic and whether basic epistemic standards apply to all or only most correctly accepted hypotheses. Douglas and Steel are mirror opposites in this regard. Douglas claims that relatively *thin* basic epistemic standards apply to *all* correctly accepted hypotheses. By contrast, Steel claims that a comparatively *thick* basic epistemic standard—weak severity—only applies to *most* correctly accepted hypotheses. We agree with Douglas about that the scope of basic epistemic standards should be universal (Sect. 17.4.1). Furthermore, inquisitive due diligence's thickness as an epistemic standard constructively crosscuts the standards that Douglas and Steel entertain (Sect. 17.4.2).

17.4.1 The Scope of Basic Standards

Like Douglas, we claim that an epistemic standard is basic only if it is required for all correctly accepted hypotheses. On the face of it, scientists generally have to kick the tires on a hypothesis before accepting it, and IDD provides relatively flexible constraints on what this tire-kicking involves, as it is ecumenical about defeaters (and hence pointed questions) and puts no *a priori* constraints on correct answers to pointed questions. Hence, inquisitive due diligence appears plausible as a universal epistemic constraint on acceptance.

However, recall that a basic epistemic standard must also be immutable throughout the course of inquiry. Here, we can dispel Brown's worries about basic epistemic standards' rigidity while holding fast to our position. On the one hand, we think that *inquisitive due diligence* is not subject to revision. In other words, every correctly accepted hypothesis must confront all of the relevant pointed questions, answer them correctly, and evade defeat in the process. On the other hand, the *set of relevant pointed questions* is not the same across different inquiries and need not stay constant over the course of a single inquiry.

To see how this approach accommodates Brown's concerns, we highlight two points. First, recall that pointed questions can target any statement in science: hypotheses, methodological claims, background assumptions, and, crucially, epistemic standards themselves. Consequently, it is consistent with our position that anything other than inquisitive due diligence can be revised throughout the course of inquiry.

Second, this one caveat is hardly a scientific straitjacket, for rather than *impeding* these kinds of transformative inquiries, as Brown asserts, inquisitive due diligence—a basic epistemic standard—*enables* epistemic standards to be revised in *principled ways*. To see why, recall that the measurements supporting Massey and Mullan's model were prone to bias, and this meant that standard significance tests badly underestimated error. Now, testing for statistical significance is a common epistemic standard, so Massey and colleagues were making a significant modification to the standards for hypothesis acceptance by declining to rely on them. Instead of significance tests, they showed that the results were unlikely to be statistical artefacts by replicating them on a different kind of dataset. Practicing inquisitive due diligence is the basic epistemic standard in this case, and the possibility of error is addressed by responding to relevant pointed questions.

Importantly, many capricious alterations to epistemic standards might preserve internal consistency and predictive competence, e.g., valorizing *p*-hacking.[11] Hence, inquisitive due diligence appears to provide a more informative answer to Brown's challenge to basic epistemic standards than Douglas's account.[12] Our differences with Steel on this point are even more striking, since the modest scope he assigns to basic epistemic standards actually concedes the key points that Brown raises. To see why, suppose we adopt Steel's definition of basic epistemic standards. Then, by his own lights, there is no epistemic standard that applies to all correctly accepted hypotheses. Yet, this is precisely what Brown is alleging. This would mean that on Steel's account of basic standards, there's no substantive disagreement between champions and critics of basic epistemic standards and the debate collapses into a merely verbal dispute. Thus, insofar as the epistemic priority debate reflects a genuine and not merely verbal disagreement, basic epistemic standards must be universal in scope.

Thus, we see that inquisitive due diligence satisfies the more demanding requirement that a basic epistemic standard should be universal, yet it does not shackle scientists with inflexible epistemic standards. Since this inflexibility was one of Brown's chief concerns, we take inquisitive due diligence to be a shining example of a basic epistemic standard. Moreover, in comparison to risky epistemic priority, inquisitive due diligence has richer resources for navigating how epistemic standards might be revised throughout the course of inquiry.

17.4.2 *Which Epistemic Standards Are Basic?*

Thus far, we've argued that basic standards should be universal constraints on all correctly accepted hypotheses. Furthermore, we've argued that inquisitive due diligence is a basic epistemic standard. However, because IDD purports to offer both necessary and sufficient conditions, we are committed to the stronger claim that it is the *only* basic epistemic standard. We do not pretend to establish that here. Instead, we will argue that risky prioritarians' candidates for basic epistemic

[11] See Steel and Whyte (2012, p. 169) for how Douglas's account is compatible with "'widely used tricks of the trade' for making a newly developed pharmaceutical look more effective or safe than it actually is." These "tricks" also figure in *p*-hacking.

[12] In later work, Douglas (2013) suggests that acceptance of scientific hypotheses should include more substantive epistemic desiderata (what she calls 'Group 4' or 'epistemically assuring' values). Our criticisms of Steel's variant of REP apply to Douglas's later view.

standards—consistency, predictive competency, and weak severity—fall short when they are contrasted with inquisitive due diligence. Our arguments aim to be both constructive and critical. The constructive part involves integrating these candidates into our inquisitive framework. The critical component involves showing that insofar as they contravene that framework, their status as basic epistemic standards is cast into doubt.

On the constructive side, we hope that it's clear that inconsistency, failed predictions, and failing severe tests can be defeaters that fuel pointed questions. However, as the example with Massey and Mullan's handling of autocorrelations illustrates, there are several cases in which pointed questions are irrelevant or can be safely "bracketed." When these irrelevant pointed questions concern consistency, predictive competence, or weak severity, our account permits correct acceptance in such cases, while Douglas's and Steel's views do not.

Enter the critical side of our argument, for scientific practice favors our view. First, as others have shown, acceptance in the face of inconsistencies and predictive failures is sometimes correct (Kuhn 1962; Feyerabend 1975; Frisch 2005; Bhakthavatsalam and Cartwright 2017; De Regt 2017). Thus, scientific practice favors our view over Douglas's. Similarly, there are always more severe tests that could be performed than are actually performed. For instance, on a conservative estimate, there were at least five (and arguably over 20) legitimate measures of segregation by the end of 1980s (Massey and Denton 1988). This suggests four additional severe tests for any model of the persistence of Black segregation considered at the time, yet segregation researchers are no different than most other scientists. Scientists often exercise their judgment to identify potential defeaters to their views, rather than performing a robustness check with every available measure of a quantity of interest. This favors our view over Steel's variant.

Importantly, our arguments resonate with Brown's (2017, p. 70) critique of internal consistency, predictive competency, and weak severity as basic epistemic standards. Our view only says that a rebutting question needs to be asked, but how it is answered cannot be settled on high. For example, a rebutting question fueled by an internal inconsistency *might* be so damning that it leads one to abandon *h*. It's also possible that the answer problematizes the prevailing reasons for accepting not-*h*, or that the answer highlights that the contradiction is only apparent and dissolves under more careful disambiguation (e.g., under conditions *A*, *h*; under conditions B, *not-h*). Or perhaps it is high time to adopt a paraconsistent logic. Which of these is the correct answer depends on the specifics of the science being discussed. As such, we agree with Brown that inquiry plays an important role in seeing where foolish consistency becomes the hobgoblin of small minds. Parallel points apply to predictive competency and weak severity. Those hobgoblins notwithstanding, epistemic priority is alive and well—so long as it embraces inquisitive due diligence.

17.5 Epistemic Filters: Risk or Relevant Questions?

Having shown that inquisitive due diligence outperforms REP(a) in Section 4, we'll now argue that inquisitive due diligence also outperforms REP(b)—hereafter: "the risk filter." To see why, let's reconstruct an argument that Brown pitches against the risk filter:

Brown's Argument.
(B1) It's sometimes correct to close inquiry before satisfying all epistemic standards as modified by values to suitably mitigate risk.
(B2) For all *h*, if it's correct to close inquiry into h, it's correct to accept *h*.
(¬REPb) ∴ For some hypotheses *h*, it's correct to accept *h*, but *h* does not satisfy all epistemic standards as modified by values to suitably mitigate risk.

Brown (2017, p. 68) defends B1 by observing that scientists frequently make "immediate decisions" about accepting hypotheses that can lessen their epistemic burdens:

> …the immediate decision will be made for *non-epistemic* reasons. The non-epistemic reasons that guide us to bring inquiry to a close are various: we have to act, to pass a law, to graduate, to publish or perish, to get the next grant, to get famous writing a controversial book, or to move on to a more interesting problem (ibid).

Although these examples of non-epistemic inquiry-closers can be used in risk calculations (indeed, damn near anything can!), it's crucial to Brown's argument that some are not. Otherwise, REP's proponents could insist that those values close inquiry by, e.g., lowering the standards to the point where they can be met. Moreover, prioritarians presumably think some of these values, e.g., grant-seeking and writing controversial books, should not be used in *suitably* mitigating risk. Consequently, B1 is fairly attributed to Brown.

While Brown is less explicit about B2, it's essential if he's going to circumvent the risk filter. Without B2, Brown's (2017, p. 68) claim that B1 "problematize[s] the general commitment to epistemic priority" does not follow. Moreover, Brown (p. 77 fn. 6) is clear that epistemic priority is a thesis about "the 'internal' aspects of scientific inquiry—characterization of data, statistical analysis, acceptance of hypotheses, theory choice." While we think that Brown's argument faces some steep hurdles (Sect. 17.5.1), we reject the risk filter on different grounds—all while brandishing our prioritarian badge (Sect. 17.5.2).

17.5.1 Problems with Brown's Argument

We reject Brown's argument primarily because it equivocates on what it means to "close inquiry." It's uncontroversial that inquiry is a collection of activities oriented toward the *settling* of a question—or in a Deweyan spirit—the *resolving* of a problem. This is why inquiry is always inquiry *into something*. Inquiry into a hypothesis, *h*, is inquiry into a question or problem, *Q*, for which *h* is a potential answer or solution—i.e., *Q* is a focal question for *h*. So, one sense of what it means to close inquiry refers to just those cases where we cease inquiring into *Q* because we've answered our focal question. Let's call an inquiry *settled* when inquiry into *Q* is correctly closed *because* we've correctly accepted an answer to *Q*.

In a *broader* sense, closing inquiry is just a matter of ceasing the activity of inquiring. Accepting an answer is but one of many reasons why we do this: grant funding has run out, publication deadlines are looming, new results make the question uninteresting. In general, when the costs associated with further pursuit of a question outweigh the prospective benefits of finding an answer, it's correct to close an inquiry without settling it. Consequently, whether an unsettled inquiry is correctly closed is a matter of practical, prudential, or otherwise non-epistemic considerations.

Which sense of "closing inquiry" does Brown's argument employ? Since B2 requires that *h* be correctly accepted, B2 is true when "close inquiry" means that the inquiry has been *settled*. However, as we've just seen, B2 is false when used in its more inclusive sense, since we often terminate inquiry without achieving its goal and thus without accepting a hypothesis, correctly or otherwise. Unfortunately, construing closure as settling spells trouble for B1. This premise is true just in case we can correctly accept *h* without satisfying all the epistemic standards as modified by values to suitably mitigate risk. But we can't assume this without begging the question and simply assuming that REF(b) is false. B1 *does* come out true if "close inquiry" is used in its broad sense, since we often cease inquiring into *Q* without satisfying any epistemic standards and may do so correctly when closure without settling realizes certain non-epistemic values. So, to avoid begging the question, B1 requires a broad reading of "close inquiry", but B2 requires the settling reading. The argument, therefore, equivocates.

17.5.2 A Different Filter

Despite our reservations with Brown's argument, we do not accept the risk filter either—but not because unfiltered non-epistemic values legitimately influence the acceptance of hypotheses. Rather, we maintain that considerations of risk only bear on correct acceptance insofar as they contribute to the relevance of pointed questions. IDD can thereby accommodate legitimate risk-based factors influencing hypothesis acceptance. Absent this constraint, risky prioritarianism implausibly entails that every risk-based consideration legitimately influences hypothesis acceptance.

Our argument's lynchpin is that answering all of the relevant pointed questions needn't satisfy all of the epistemic standards as modified by values to suitably mitigate risk. On typical accounts of inductive risk, filtration amounts to adjusting one's epistemic standards in light of foreseeable moral consequences of error. When the moral costs of error are high, the risk filter entails that epistemic standards should be altered accordingly. However, on this prescription, most research in residential segregation is guilty of incorrect hypothesis acceptance, for the moral stakes associated with research into residential segregation are undeniably high—not least because its findings contribute to local and national policy decisions. Naturally, some other areas of social science have lower moral stakes. So, according to the risk filter, there ought to be epistemic standards of differing stringency operative in segregation research when compared with many other social scientific research. However, that's not what we see. Rather, segregation studies tend to employ the same standards for hypothesis acceptance, e.g., *p*-values, as the rest of social science. It would be perverse to deem the acceptance of segregation research incorrect.[13] So, the risk filter ought to be rejected.

Additionally, even when segregation researchers consider the non-epistemic consequences of accepting a false hypothesis, they do not always respond by toughening their statistical tests. For example, Massey and Mullan (1984) explicitly mention the politics of segregation in both the introduction and conclusion of their essay, demonstrating a clear awareness of how their empirical results would undermine the *laissez-faire* political response to segregation. To risky epistemic prioritarians, this awareness ought to compel them to adopt more stringent epistemic standards. However, we find the exact opposite: when the character of the biases suggested that the standard significance tests were not strict enough to be useful, they chose not to publish significance levels at all. Thus, rather than taking the political consequences of their work as a reason to tighten their standards, they argued that the pointed question was irrelevant, fulfilling their inquisitive due diligence.

Importantly, however, a rejection of the risk filter does not mean considerations of risk never play a role in proper hypothesis acceptance; quite the opposite, and inquisitive due diligence explains why. When the moral costs of some remote error-possibilities are high, inquisitive due diligence requires scientists to ask "tougher" questions—e.g., pointed questions that target these possibilities. Such questions may concern (non-basic) epistemic standards—e.g. Does *h* meet some lower *p*-value?—but they need not. Often, they aim at uncovering or ruling out particular defeaters which, in less morally charged circumstances, might be deemed too unlikely to be worth considering. IDD can thus explain why segregation researchers cognizant of the moral risks associated with their work nonetheless observe the same non-basic epistemic standards as their colleagues. Rather than demanding modifications to their standards of hypothesis acceptance, IDD requires that these researchers pursue various pointed questions.

[13] The alternative—that other social scientists are being needlessly risk averse—would be just as perverse.

17.6 Epistemic Insufficiency

Thus far, we've shown that inquisitive due diligence is a more promising candidate when it comes to providing necessary conditions for correctly accepted hypotheses. But does it provide sufficient conditions for correct acceptance? Brown provides an interesting objection to our risky comrades' claim to sufficiency with his example of an antiracist psychologist:

> Imagine you are a psychologist doing research on factors that contribute to intelligence. You have gathered data carefully, run your analyses, and your data seem to support the claim that race is a factor in intelligence, independent of socioeconomic status and education level. This is not what you set out to discover, just a result of your analyses of your data. You believe strongly in empirical rigor and scientific integrity, and you have held yourself to high epistemic standards. You are also a strong believer in racial justice, and the idea that there might be innate racial differences in intelligence is anathema to you. This belief has led you to apply even stricter standards in this particular analysis. Further, you know that in our racist society, publicizing such results is likely to have a negative impact on the racial justice you hope to promote. (Brown 2017, p. 64)

Presumably, one satisfies "high epistemic standards" only if one satisfies *basic* epistemic standards. Furthermore, note that the psychologist's belief in racial justice has led them "to apply even stricter standards"—just as would be required to mitigate risk. This means that the antiracist psychologist has satisfied REP's conditions and ought to accept the results of her study. Brown (p. 65) contends that the psychologist should *not* accept and publish these results; hence risky epistemic prioritarianism fails to provide sufficient conditions for hypothesis acceptance.

Here, we agree with Brown: REP does not provide sufficient conditions for hypothesis acceptance. However, this isn't the death knell of epistemic prioritarianism. It only shows that we prioritarians shouldn't fetishize inductive risk. Confronted by Brown's example, we claim that the antiracist psychologist may still have not fully exercised her inquisitive due diligence even after satisfying all the epistemic standards modified by values to suitably mitigate risk. Brown (p. 74) seems congenial to this possibility:

> You might revisit your methods of data collection to see if they were biased. You might turn a critical eye to these concepts of "intelligence" and "race" that you are using in the study. You might publish with strong disclaimers about drawing inferences about those particular connections, given open questions [!] about the data and their interpretation.

These potential responses are readily translated into pointed questions. Are the methods of data collection biased? Are the constructs for intelligence and race valid? Are there alternative interpretations of the data that are equally plausible? Moreover, with Brown, we think that these can be relevant questions to ask regardless of whether some of their answers suitably mitigate risk. After all, addressing questions about bias in data collection are typically part of a scientists' role-responsibility, and values can make them relevant. Consequently, until these pointed questions are answered, it can be correct to withhold acceptance and thereby incorrect to accept, regardless of how the risk calculations shake out.

17.7 Conclusion

Inquiry anchors many critiques of epistemic priority. Brown (2013a, p. 75) nicely summarizes this "inquisitive ethos" when he aims to "treat value judgments, sets of evidence, and hypotheses and theories as parts of a linked network modified by the process of scientific inquiry." The result, he argues, is a more promising alternative to epistemic priority. In this essay, we have seen that two dividing lines have been entangled in the debate. The first division is between epistemic prioritarianism and its denial; the second, between those who live by the inquisitive ethos and those who do not. As we have argued, prioritarians who embrace inquisitive due diligence get the best of both worlds. They propose a basic epistemic standard that is not only worthy of its name but that can fruitfully illuminate how non-basic epistemic standards are transformed by the engines of questions and answers. They can use the risk filter when it sparks relevant pointed questions, but bypass it without equivocating on the closure of inquiry. They can raise pointed questions even when doing so runs afoul of expected utility theory. So, what's not to love about inquisitive due diligence? We hope that our interlocutors appreciate the relevance of this pointed question.

References

Banfield, E. 1974. *The Unheavenly City revisited*. Boston: Little.

Bhakthavatsalam, S., and N. Cartwright. 2017. What's so special about empirical adequacy? *European Journal for Philosophy of Science* 7 (3): 445–465.

Brown, M. 2013a. The source and status of values for socially responsible science. *Philosophical Studies* 163 (1): 67–76.

Brown, M. 2013b. Values in science beyond Underdetermination and inductive risk. *Philosophy of Science* 80 (5): 829–839.

Brown, M. 2017. Values in science: Against epistemic priority. In *Current controversies in values and science*, ed. K. Elliott and D. Steel, 64–78. London: Routledge.

De Regt, H. 2017. *Understanding scientific understanding*. New York: Oxford University Press.

Douglas, H. 2003. The moral responsibilities of scientists. *American Philosophical Quarterly* 40 (1): 59–68.

Douglas, H. 2009. *Science, policy, and the value-free ideal*. Pittsburgh: University of Pittsburgh Press.

Douglas, H. 2013. The value of cognitive values. *Philosophy of Science* 80 (5): 796–806.

Dutant, J., and C. Littlejohn. 2021. Defeaters as indicators of ignorance. In *Reasons, justification, and defeat*, ed. J. Brown and M. Simion, 223–246. Oxford: Oxford University Press.

Feyerabend, P. 1975. *Against method: Outline of an anarchistic theory of knowledge*. London: Verso Press.

Frisch, M. 2005. *Inconsistency, asymmetry, and non-locality: A philosophical investigation of classical electrodynamics*. New York: Oxford University Press.

Hart, H. L. A. 1968. *Punishment and Responsibility: Essays in the Philosophy of Law*. New York: Oxford University Press.

Hicks, D. 2014. A new direction for science and values. *Synthese* 191 (14): 3271–3295.

Kaivanto, K., and D. Steel. 2019. Adjusting inferential thresholds to reflect nonepistemic values. *Philosophy of Science* 86 (2): 255–285.

Khalifa, K., and J. Millson. 2020. Perspectives, questions, and epistemic value. In *Knowledge from a human point of view*, ed. A.-M. Crețu and M. Massimi, 87–106. Cham: Springer International Publishing.

Kuhn, T. 1962. *The structure of scientific revolutions*. Chicago: University of Chicago Press.

Longino, H. 1990. *Science as social knowledge: Values and objectivity in scientific inquiry*. Princeton: Princeton University Press.

Massey, D., and B. Bitterman. 1985. Explaining the paradox of Puerto Rican segregation. *Social Forces* 64 (2): 306–331.

Massey, D., and N. Denton. 1985. Spatial assimilation as a socioeconomic outcome. *American Sociological Review* 50 (1): 94–106.

Massey, D., and N. Denton. 1988. The dimensions of residential segregation. *Social Forces* 67 (2): 281–315.

Massey, D., and B. Mullan. 1984. Processes of Hispanic and black spatial assimilation. *American Journal of Sociology* 89 (4): 836–873.

Steel, D. 2010. Epistemic values and the argument from inductive risk. *Philosophy of Science* 77 (1): 14–34.

Steel, D. 2013. Acceptance, values, and inductive risk. *Philosophy of Science* 80 (5): 818–828.

Steel, D. 2017. Qualified epistemic priority: Comparing two approaches to values in science. In *Current controversies in values and evidence*, ed. K. Elliott and D. Steel, 49–63. London: Routledge.

Steel, D., and K. Whyte. 2012. Environmental justice, values, and scientific expertise. *Kennedy Institute of Ethics Journal* 22 (2): 163–182.

Sudduth, M. 2023. Defeaters in epistemology. Internet encyclopedia of philosophy. https://iep.utm.edu/defeaters-in-epistemology/. Accessed 25 July 2023.

Williams, M. 2015. What's so special about human knowledge? *Episteme* 12 (2): 249–268.

Part VI
Reply from Matthew J. Brown

Chapter 18
Values, Pluralism, and Pragmatism: A Career Mediospective

Matthew J. Brown

Abstract In this chapter, I reflect on my career to date and my current thinking on the core topics of this volume: values, pluralism, and pragmatism in science and philosophy of science. Since I am, I hope, merely in the middle rather than the end of my career, this is not a retrospective but a *mediospective*, if you will, though I will begin with retrospective reflections and end with prospective ones. Happily, there are many opportunities to reference and engage with the excellent contributions to this volume in the course of things, to discuss what I have learned from my generous colleagues as well as indicate the very few places where our views differ.

Keywords Values in science · The argument from inductive risk · The contingency argument · The value free ideal of science · Varieties of pluralism · Pragmatism in philosophy of science

Am I old? Is my career nearly at an end? It is a confusing thing, at my age, to be looking at the contributions to a volume in my honor. This collection and the 2023 conference it grew out of help mark a major turning point in my career, my departure from the University of Texas at Dallas—where I worked for 13 years as faculty in philosophy, humanities, and history of ideas as well as director of the Center for Values in Medicine, Science, and Technology (CVMST)—to take up the position of director of the Center for Dewey Studies and the Boydston Chair of American Philosophy at Southern Illinois University Carbondale. I helped hire Jonathan Tsou in 2022 as the Stone Distinguished Professorship of Humanities in Medicine and Science at UTD, only to turn around and take the position at SIU before we had a chance to work together. To my relief, Jon agreed to take up the directorship of the CVMST, and he absolutely blew my mind by organizing a conference in my honor in 2023, which eventually led to this volume.

I am particularly honored to have this volume put together by three "generations" of colleagues in the field: Carla Fehr, who I consider a mentor, Jonathan Tsou, my

M. J. Brown (✉)
Center for Dewey Studies, Southern Illinois University, Carbondale, IL, USA
e-mail: matt.brown@siu.edu

J. Y. Tsou et al. (eds.), *Values, Pluralism, and Pragmatism: Themes from the Work of Matthew J. Brown*, Boston Studies in the Philosophy and History of Science 347, https://doi.org/10.1007/978-3-031-92958-8_18

contemporary, and Jamie Shaw, who I consider something of a mentee, a very promising young colleague in the field. All three have been good friends to me, and I am gobsmacked by their generosity in organizing and contributing to this volume and the event that preceded it. Likewise, I am touched that 14 other colleagues from all stages have contributed, many of their essays deeply engaged with my own work.

My plan here is not to try to respond in detail to every chapter, which would be too great a task. Instead, I wish to take a moment to reflect on my career to date and my current thinking on the core topics of the volume: values, pluralism, and pragmatism in science and philosophy of science. Since I am, I hope, merely in the middle rather than the end of my career, this is not a retrospective but a *mediospective*, if you will, though I will begin with retrospective reflections and end with prospective ones. Happily, there are many opportunities to reference and engage with the excellent contributions to this volume in the course of things, to discuss what I have learned from my generous colleagues as well as indicate the very few places where our views differ.

18.1 How Did I Get Here? From Parkview Philosophy Club to the Dewey Center

I attended Parkview High School in Lilburn, Georgia from 1995–1999, and from my freshman year, I was part of the "Philosophy Club." Prior to college, students in the U.S. rarely encounter any philosophy; I had one English teacher who introduced us to Plato's Cave and Aristotle's theory of tragedy, but otherwise had little sense of what philosophy entailed. The Philosophy Club was conceived at least in part as an atheist/agnostic answer to the larger Christian after-school clubs in our Bible-Belt suburb. I was a curious, intellectual kid, and many of my friends were part of this club, so I happily joined. Maybe the older kids who founded the club knew what philosophy was, but I certainly did not; we happily discussed all manner of topics, from atheism to transfinite numbers, from libertarian politics to interpreting song lyrics. Towards the end of high school, my friend Benji Kmack recommended a book that would be my introduction to genuine (if pop) philosophy: Robert Pirsig's *Zen and the Art of Motorcycle Maintenance* (*ZAMM*)—though I wouldn't get around to reading it until my freshman year of college.

I started at the Georgia Institute of Technology in Fall of 1999 as a Physics and Computer Science double major. These connect with two of my enduring interests: I wanted to understand how everything works at a fundamental level, and I enjoyed tinkering with systems and making them do new and interesting things. What I did not enjoy was the software engineering focus of the CS program, which I dropped after three semesters (though I would work as a CS teaching assistant for much of my remaining time at Tech); I also did not do well with the "shut up and calculate" attitude that was a common response to "big questions" about what we were doing in advanced physics class. But Pirsig's book, combined with my first reading of

Kuhn's *Structure* in a freshman English class, started to open my mind to philosophy. I was sitting outside the classroom before Calculus reading *ZAMM*, when my classmate and friend Jenn Hou came up to me and suggested I take a philosophy class.

Professor Jon J. Johnston, who taught the class Ms. Hou recommended, was the best, most important teacher I have ever had. Every lecture was full of genuine passion, amusing anecdotes, and hilarious witticisms; the "socratic" discussions were just what my intellectually hungry young self craved. The other philosophers at Tech were all inspirations in their own right: Nancy Nersessian the historian and philosopher of science and cognitive ethnographer, Bryan Norton the pragmatist and environmental ethicist, Andy Ward the pragmatist philosopher of science (and later public health expert), and the epistemologist Chris McClellan; all had a major impact on me, and in many ways, my work steers closer to them than to Professor Johnston's own philosophical commitments. Jon was really an old-school, traditionalist philosopher, a professed monist, who particularly loved the history of philosophy—ancient Greek, early modern European, and East Asian particularly. He saw his mission as making the students at this soulless engineering school a little more humane and thoughtful, something that really resonated with me. When a group of us[1] wanted more than just classes and decided to form a club, "The Philosophical Society," Jon proposed our first reading: Feyerabend's provocative essay, "How to Defend Society Against Science." During this time, I also worked with David Ritz Finkelstein, from whom I learned about philosophy of physics (or "the physics of philosophy," as he preferred to say), and he helped me navigate my way to a degree from the School of Physics with several of the physics requirements substituted for philosophy courses and independent studies on theoretical physics (great for me, because I was a disaster in advanced lab classes).

When I decided to apply to graduate schools in philosophy rather than physics, Andy Ward suggested that I add UC San Diego to my list of schools, which is a lucky thing, as it was the only place I got in. I came to understand that Nancy Cartwright went to bat for me on the admissions committee, for which I am enduringly grateful. So in Fall 2003, I showed up in San Diego. UCSD was a great graduate program, in a beautiful place, with supportive fellow students and first-rate faculty who challenged me. I ended up in the shared office P.D. Magnus had just vacated, having finished and taken a job just before I came; lucky for me, he made several visits back to San Diego during my first years there. (I am hence particularly honored by his contribution here). Although I learned many new things at UCSD, one thing no one was talking about was pragmatism, which I had encountered from multiple sources as an undergrad. When I started a pragmatism reading group with some other grad students, Ryan Hickerson suggested we read Dewey's *Experience and Nature*, which began my long appreciation for Dewey; I had not liked Dewey's work much when first exposed to it. Although several of my interests were not

[1] The group included Jenn Hou, Jeremy Farris (now the executive director of the New Mexico State Ethics Commission), and Dana Habeeb (now faculty in Informatics at Indiana University Bloomington).

particularly appreciated by most of the faculty at UCSD—I received criticism of my supposed penchant for "heterodoxy for heterodoxy's sake" in more than one review—I was lucky that Nancy Cartwright, Paul Churchland, and Jerry Doppelt accepted and supported my pragmatist and Feyerabendian leanings and let me write a dissertation that ended up focusing on Dewey's philosophy of science and its contemporary relevance.

In Fall 2008, I was lucky to have Paul Churchland insist that I go out on the job market despite my half-finished dissertation—he saw the implications of the unfolding financial crisis in a way I was totally naive about. I was lucky in that cycle to get a job starting Fall 2009 at the University of Texas at Dallas, primarily to teach philosophy of technology—which I knew very little about! So I defended my dissertation and moved to Dallas. At the same time, Dennis Kratz, the Dean of Arts and Humanities there, received a large grant to reboot the Center for Values in Medicine, Science, and Technology, and over the summer he asked for input on what the Center should do. I was surprised to find when I arrived that Fall, that my email response had become a sort of vision document for the CVMST, and I quickly moved from a part of the small planning committee to being named director in 2011. This was an incredible boost to my career, and suited well my growing interest in the area of science and values, a natural place to focus my interests in pragmatism, pluralism, and philosophy of science. I worked at UTD for 13 years, and I consider my book, *Science and Moral Imagination (2020b)*, to be a culmination of this period of my career. Having spent the main part of the COVID-19 pandemic as Program Head for History and Philosophy, I was ready in 2022 for a change of scene, and as already mentioned, left UT Dallas to take up a new position at SIU, which brings me to the current moment.

18.2 Where Am I Now? An Opinionated Review of the Current Scene

Using the three key terms from the title of this volume, I will now combine a review of the current state of work in values in science, scientific pluralism, and pragmatist philosophy of science, with a stock-take of my own views in these areas and responses to the chapters in this volume as they become relevant in that narrative.

18.2.1 Values in Science

I have devoted much of my career to the broad topic of values in science, not only through my publications and presentations, but also having organized 12 conferences on that topic, as well as many lectures, workshops, panels, etc. So I'll begin by saying something about the current field as I see it and what I have tried to accomplish within it, engaging along the way with those chapters in the current volume that directly address the issues of key concern to me.

It is undeniable, in my view, that the most important work in the area of science and values in at least 30 years is Heather Douglas's articulation of the argument from inductive risk (2000). This clear, careful argument, which is almost certainly also deductively valid and sound (Havstad 2022; Brown and Stegenga 2023), brought new life and attention to this area within philosophy of science. It shows without a doubt that the value-free ideal of science (VFI) is untenable as a normative ideal. While science and values was still a pretty marginal topic at the turn of the century, Douglas and the philosophers of science she inspired have made it one of the hottest, most central topics in the field. At the same time, while improving in rigor and clarity on earlier arguments, sometimes I feel that the focus on inductive risk has unduly narrowed our attention. It is for this reason that I tried to capture the rigor and clarity of the argument for inductive risk while synthesizing it with various other arguments for values in science through what I have called "the contingency argument":

1. Scientific inquiry has many contingent moments, with reasonable alternative possible options.
2. Each contingent moment is a decision-point, that is, it is potentially an unforced choice.
3. The different options in these contingent moments may have implications and consequences for things that we care about, including ethical and social, as well as political, cognitive, and aesthetic values.
4. Value judgments should be made [when] making choices that may have ethical and social implications and consequences.
5. Thus, value judgments should be made to settle scientific contingencies, that is, scientists have a responsibility to make value judgments where there are scientific contingencies. (Brown 2020b, p. 63)

The contingencies referred to in the argument should include not only inductive risk, but other dimensions of epistemic risk (Biddle and Kukla 2017), including representational risk (Harvard and Winsberg 2022), arguments based on thick concepts and mixed claims (Alexandrova 2018), and various other moments where values must enter science. The contingency argument recognizes a broader set of reasons for rejecting the VFI.

Ultimately, I think this even this argument does not go far enough. It suggests that there may be moments in scientific inquiry where there are no contingencies or there are no foreseeable consequences or implications for how those contingencies are settled, and so no place for value judgments in those cases. Values are even more pervasive than the contingency argument suggests, because science is ultimately a practice, and every inference, decision, judgment, acceptance, and assertion is an *action*. Actions require practical reasons, also known as values, and what are typically referred to as "epistemic values" are not the right kinds of reason to justify actions. Every single move in science has a value at the back of it, justifying it, in principle if not as a matter of psychological fact.

My attacks on the VFI have assumed a clear target. Is Elliott right that there are a multiplicity of value-free ideals rather than a single value-free ideal (VFI)? Yes and no. Yes, in the sense that the VFI is a pretty schematic claim that has many broad terms to be filled in by more careful analysis, thus allowing for different versions to be created by different precisifications of the core components. It can be helpful here to introduce a distinction brought into philosophy by H.L.A. Hart and John Rawls, but made present to my mind by Michael Hardimon, between a *concept* and a *conception*. There is one core concept of the value-free ideal, but many conceptions. For the concept, I think one cannot do better than Douglas (2009), who is worth quoting at length:

> It does not hold that science is a completely value-free enterprise, acknowledging that social and ethical values help to direct the particular projects scientists undertake, and that scientists as humans cannot completely eliminate other value judgments. However, the value judgments internal to science, involving the evaluation and acceptance of scientific results at the heart of the research process, are to be as free as humanly possible of all social and ethical values. Those scientific judgments are to be driven by values wholly internal to the scientific community. Thus, the value-free ideal is more accurately the "internal scientific values only when performing scientific reasoning" ideal. (p. 45)

There are a number of free variables here, which are well-canvassed in Elliott's contribution. But whatever specification of the conception of value-freedom we entertain, it must fit the core concept above. We must be monists about the concept, but we might reasonably be pluralists about the conception. And I agree with Kevin that there may be substantive and interesting differences between the different conceptions; for some purposes, it may be relevant to treat the differences rather than the shared concept each is attempting to articulate. However, the kinds of arguments that I have made above, and many other classic arguments against the VFI, assume only the core concept and not a specific conception of the VFI, so this "multiplicity" gives no aid nor comfort to defenders of the VFI. The core concept of the VFI also rules out another form of pluralism that I have heard articulated recently in informal contexts: that the VFI is not a yes-no affair but a spectrum. This is incoherent; central to the concept of the VFI is its all-or-nothing character. One cannot be *free* or *neutral* to some degree; there is no middle term between partial and impartial, even if partiality can be more or less extreme in certain respects. The idea of a spectrum between the VFI and its denial should be abandoned.

Ultimately, I largely agree with the further move by Greg Lusk and Janet Kourany, that the VFI is dead and it is time to stop talking about it. I'm enough of a pluralist and a fallibilist to sound a note of caution, here. We should be open to hearing novel versions and defenses of the VFI, as they may contribute to our intellectual progress, right or wrong. Menon and Stegenga (2023; cf. Stegenga and Menon 2023) provide a good example: they show a form of "action guiding" VFI contra Lusk, though I think it doesn't work in the end (see Brown 2024, pp. 19–22). Still, for the most part, I think it is right to advocate that we move on, to dismiss rehearsals of the same tired arguments, and to set our sights on more fruitful projects. Lusk is right that it would be better to focus on the normative grounds behind the value-free ideal, and this opens up multiple levels of discussion. I think he is wrong to say that, "This suggestion is not a significant loss or gain for either side of the VFI debate" (this volume), for he recommends basically a full rejection of the VFI in favor of a series of value judgments to be made by philosophers of science and *by scientists*; this is the core point that opponents of the VFI have sought, despite their other differences.

The demise of the VFI requires us to have an alternative framework for the role that values are to play in science. I have proposed and defended a framework in my book that I call the *ideal of moral imagination* (IMI):

> Scientists should recognize contingencies in their work as unforced choices, discover morally[2] and epistemically salient aspects of the situation they are deciding, empathetically recognize and understand the legitimate stakeholders and their interests, imaginatively construct and explore possible options, and exercise fair and warranted value judgment in order to guide those decisions. (cf. Brown 2020b, p. 186)

This ideal lays out four considerations that should guide practical decision-making by scientists: the recognition and articulation of contingencies, the identification of stakeholders and their interests, the recognition of relevant values, and the multiplication of options in order to find one that better integrates the aims, values, and stakeholder interests at play.

My approach already fully embraces the normative, humanist, "value-full" ideal articulated by Kourany, only it gives that ideal a particularly pragmatist, pluralist spin. I think it appropriate for the IMI to emphasize that ethical and political principles, like methodological and epistemological standards, are not deliverances of Reason, a reflection of the Moral Law within us all, but products of often difficult social learning processes, and those processes can be improved (made more "intelligent," in Dewey's terms), through deliberate inquiry into what works best in the way of conduct (ethical or cognitive). This is not to deny that we've learned a lot in the process, and we have identified some pretty sturdy ethical and political principles that are rarely if ever up for grabs. It is also not purely procedural ideal in the Habermasian vein. To make this clearer: it seems to me that Janet's example of the food scientist actually fails to meet my ideal, because they fail to recognize certain contingencies or possible options for resolving their problem, because they're stuck in a corporate setting in which producing breakfast cereals for profit is their only option. It seems like the scientists fail to give due weight to health outcomes and they overvalue economic values and the interests of certain stakeholders, namely their shareholders and their profit motives, at the expense of others, consumers and their health. To me, the value judgments being made here sound far from fair and warranted in the light of all the relevant factors. Contra Kourany, I don't *eschew* "normative assessment," but rather I seek to provide a framework for it. I don't avoid moral principles, but I hope to provide some guidance for scientists and others in thinking about how to resolve conflicts in our moral principles and between our moral principles and desires, interests, practical needs, epistemic standards, and other normative considerations.

---

[2] Here I mean "morally" broadly, to include any so-called non-epistemic evaluations.

We really need metaethics to make sense of all this. In the book I suggest that bad metaethics (flat-footed noncognitivism) is the source of much bad thinking about values in science, and that an alternative "shallow" cognitivism should be adopted instead. Paul Franco takes up this suggestion in an excellent chapter that traces the intersecting histories of philosophy of science and metaethics. I'm grateful to Franco for fleshing out the rather sketchy claims I made in my prior work, showing in detail that the influence of noncognitivism in philosophy of science has influenced preference for the VFI and other dysfunctions.[3]

The metaethics I favor is, obviously, pragmatist and Deweyan. I take evaluative claims to be as cognitive as any other kind of claim, and value-judgments to be truth-evaluable in the same way that scientific claims are. An evaluation, like a scientific explanation, is a hypothesis concerning the solution of a specific kind of problem; it is amenable to a kind of empirical inquiry. A value judgment, like a scientific one, is true if it actually successfully resolves the problematic situation that led to the inquiry that produced that judgment. All judgments involve imagination and experimental testing; value judgments especially require means-ends reasoning that depends on cause-effect information. But this means-ends reasoning cannot be distinguished from *categorical* or *appraising* value judgments (contra Feigl, Hempel, and Nagel in Franco's narrative), because ends also have to be experimentally tested, evaluated by trying them out in practice. Hence why I read Jamie Shaw's chapter on Mill and Feyerabend with such interest, as Millian *experiments in living* seem to be a good place to think about the role of science *in* values, i.e., the empirical nature of value judgments. Though I find no substantive references to this idea of Mill's in Dewey, there is certainly significant convergence of views if not direct influence.

Fernández Pinto faults my approach for not taking into account the larger social organization of science, and particularly the significant role of industry funding and privatization in contemporary science. Can we make genuine progress on values in science by focusing on the midstream of scientific inquiry and the value-laden decisions therein without first addressing the social structures in which scientific research proceeds, and especially the incentive structures and funding regimes? I hope so, because scientists must make decisions, and they must do so in the existing context today. They can do better and worse at this decision making, and the IMI gives both forward-looking guidance and a standard of evaluation. In my view, the privatization of science does not alter the core normative responsibilities of scientists qua scientists, even if it makes them harder to meet.[4] It is thus not a problem for my view that commercialized science contravenes many of the core scientific values

[3] My only concern about Franco's narrative is how it squares with the (limited) acceptance of the argument from in inductive risk by Hempel and Nagel.

[4] Fernández Pinto calls me out for saying that science is largely publicly supported, because the majority of funding now comes from the private sector. But in a sense, the public still supports this research as well, as private industries fund science through money gathered by consumers, i.e., the public in their economic roles. The role of such funding is in practice, of course, quite different from government and non-profit funding for science in relevant respects.

I describe as well as the IMI. Demarcation of science from non- or pseudoscience is not my goal in *Science and Moral Imagination*, and so the fact that we continue to classify such science *as* science is no objection to my account, although I am sympathetic to the idea that some such violations disqualify some research *as* science (Bright and Heesen 2023). It is also not a problem for my or Dewey's pragmatist approach to ethics, which is never required to take existing practices for granted. This is not because we apply some epistemological or ethical framework from outer space, but because we can see that there are tensions in the practice and we can zoom out to the wider context. Dewey says "Democracy is Radical" and so is the ideal of moral imagination; consideration of all extant values and the interests of all legitimate stakeholders will strongly challenge business-as-usual in commercialized science. I will say, I would have written a very different book if my target had been institutional designers and high-level political figures who have the power to remake major scientific institutions. The main focus is action-guiding advice for practicing scientists and consumers of science. The scientist at the lab bench cannot make large-scale changes on their own, though they might be in a bad enough situation that their best option is to focus their career on advocating for systematic change.[5] Indeed, focusing on *science* at all, even at the large institutional scale, is probably as least as unhelpful as my focusing on the small-scale, because the real problems that Fernández Pinto may be problems of global neoliberalism and capitalism in general.

What about Khalifa, Millson, and Risjord (KMR) on inquisitive due diligence (IDD)? Their proposal may prove an excellent instantiation of Lusk's appeal to reframe the issues. I think their focus on inquiry and their engagement with socially relevant literature in social science is undertaken with great care and bears close consideration. In their piece, KMR address my arguments against a view I call "the epistemic priority thesis." Defenders of thesis assume the VFI is wrong, but that it is still right to worry about wishful thinking and to put a priority on epistemic considerations. Epistemic prioritarianism (EP) is thus an attempt to try to keep the best insights of the VFI in light of the irrefutability of the argument from inductive risk. Do KMR capture what is right in EP and evade my critiques of that position? While I much admire the clarity and care that they have brought to bear on the question, I have a number of concerns: first, the definition of epistemic standards as truth-conducive (and by implication, values as not truth-conducive) is a mistake. This is so because many epistemic standards are orthogonal to the truth as most understand it, and because many social and ethical values are themselves reliably truth-conducive, such as egalitarian or feminist values in the social sciences.[6] Second, echoing Feyerabend's criticism of Lakatos's "rationalism," IDD provides a rich

---

[5] Philip Kitcher makes a similar recommendation for scientists working in social-institutional conditions where anything like well-ordered science is unachievable in practice (Kitcher 2001, pp. 192–97).

[6] While one can avoid the former point by making an ad hoc distinction between properly epistemic values and cognitive values (what are sometimes called either superempirical virtues or supraempirical virtues) and the latter by saying that some values are also epistemic standards, this begins

vocabulary for the redescription of any violation of epistemic priority as a fulfillment of it, but provides no restrictions or guidance for the activities of scientists. If KMR find scientists implicitly or explicitly appealing to values in a seemingly non-prioritarian way, it seems to me that they can recast the argument as one about whether a question is focal or pointed, or whether a pointed question is relevant or irrelevant, etc. While they acknowledge that there is a point at which such machinations become perverse, no criterion of perversity is on offer.[7] Finally, IDD is clearly too demanding of a criterion of acceptance; to paraphrase Lakatos, every sufficiently interesting hypothesis grows in a permanent ocean of defeaters.[8]

A crucial element in the contemporary discussion of values in science is the move towards the political, whereas work like mine has tended to be understood as situated in a more ethical framework. One of the most important moves in this literature is the move away from "deficit model" or "public understanding of science" framings of *the disconnect problem* between experts and society (Brown and Havstad 2017). Using the work of Fleck, Victoria Wang provides a novel argument against such framings and suggests alternative resources for addressing the disconnect problem. Paul Howatt reminds us how important Dewey's work in this area is. While he rightly points out some tensions and ambiguities in Dewey's view of all this, the interpretation or elaboration of Dewey's view that he provides I think gives even better resources than Fleck for addressing these issues. In my own view, what Dewey (along with his inspiration and regular interlocutor, Jane Addams) was attempting to articulate was a form of democratic integration between communities and experts in order to solve public problems through interactive, iterative inquiry.

---

to muddy the waters considerably. (Testament to the rampant confusion in this area is that the very same things can be referred to as both super- and supra-.)

[7] Perhaps they take Potter Stewart's criterion to be the obvious way to sort the acceptable from the obscene. I think such criteria are appropriate for many everyday questions, but unhelpful in plumbing these deep epistemological waters.

[8] They thus misinterpret my approach to the issue when they accuse me of equivocation on what it means to "close inquiry." To close inquiry is to settle on a hypothesis as a solution to a problem, to make a judgment that we ought to act on that solution. This does not beg the question, but simply acknowledges that the cost of continuing inquiry has to be factored into the question of when inquiry should be closed in practice, as well as the evident fact that inquirers often use much less demanding standards than REP or IDD suggest. KMR also misinterpret the consequences of inductive risk for values in science and the REP view in a too-demanding direction. They say, "On typical accounts of inductive risk, filtration amounts to adjusting one's epistemic standards in light of foreseeable moral consequences of error. When the moral costs of error are high, the risk filter entails that epistemic standards should rise accordingly." This is incorrect. When the cost of *false positive* error are high, epistemic standards for acceptance should rise. When the cost of *false negative* error (including both wrongly rejecting and wrongly withholding judgment from or continuing to inquire about a hypothesis) are high, epistemic standards for acceptance should fall. Whether residential segregation research should have higher, lower, or equivalent standards to less pressing questions thus depends on content, context, and consequences of the particular questions and hypothesis under investigation. It seems reasonable to me that the case of Massey and Mullan as KMR describe it could reasonably be the case where the REP recommends lower standards rather than higher.

And Howatt is right when he interprets Dewey as providing strong roles for publics in both initiation and control of democratic inquiry.

The move away from framing the disconnect problem in terms of a deficit of understand has been accompanied by an equally important move towards a framing of *public (dis)trust in science* (Almassi 2012; Goldenberg 2016, 2021; John 2018; Melo-Martín and Intemann 2018; Oreskes 2019; Contessa 2023). LaRosa's chapter provides a crucial overview and application of these ideas. While I think this has been an importantly progressive problem-shift in the field, I think the focus on trust is itself problematic. One problem is accounts of trust in science treat the question in relatively wholesale terms (Oreskes 2019), which is inappropriate. I also think the framing of trust takes as a given the separation of science and society and the distinctiveness of the two. We don't only need to rethink the relationship between publics and experts in political contexts, as Howatt urges; we need to democratize science itself and eliminate the distinctions between scientific assessment and political decision-making—not because it will be better if science serves the public whim, but because science can only be trustworthy if it serves the public *interest*, and it can only serve the public interest by being part of democratic politics that engages in open inquiry (cf. Pamuk 2021). Just as LaRosa argues we need to emphasize patient-centered care in the face of an AI-driven return of paternalism in medicine, we need public-centered, genuine science to combat technocracy and authoritarianism in society in general.

One of the reasons that philosophers of science have been better-poised than most to address the ethical and social problems posed by "artificial intelligence" or "machine learning" is precisely the mix of sophisticated technical analysis and sensitivity to values that the field has cultivated over the last 20 years or so in the values in science literature. The problems of transparency, justice, fairness, explainability, and trustworthiness for machine learning and artificial intelligence that Fehr, Bollander, and LaRosa center are places where philosophers of science can bring crucial insights. I would like to suggest in this case that we go further in Bollander in not only applying social and political lenses to the problems of ethical AI/ML, but that we think of these issues in terms of political economy. It is manifest that the main outcomes of recent research in generative AI/ML serve not genuine human progress but the purposes of capital exploitation by threatening to eliminate forms of labor that have been difficult to deskill and automate and at which humans are uniquely good: writing and art. Whether this has been an intentional move or an accidental byproduct of research that is low stakes, easy to access, with low costs of failure, the consequences are grim. Rather than automate labor that is onerous, difficult, and unrewarding for humans, research in AI/ML ends up endangering forms of labor that actually bring people joy through creative expression, forms of labor already undervalued in our society. There are few[9] public goods served by generative image models (like DALL-E) and large language models (like GPT), and many harms. And because of the way that these technologies function in the logic of

[9] Some tedious tasks can also be automated, of course.

capitalism at work—deskill, exploit, accumulate—we can expect them to be pushed further in the future, unless we regulate them heavily.[10]

One of my major pushes has always been to see science and values as an interdisciplinary area of research, not just a subfield of philosophy of science; work on values in science should also bring in ethics and political philosophy, the historical, social, literary, and cultural studies of science, and collaborations with scientists and engineers. Philosophers of science have both a lot to teach and a lot to learn here. My success in this has been modest, and I would love to see the interdisciplinary aspect of research in values in science develop further.

18.2.2 Pluralism

Of the three core topics here, pluralism may be the most vexed, the one I have least fully articulated to my own satisfaction. I have engaged with forms of perspectivism and pragmatic pluralism throughout my career, as well as exploring the pluralistic worldview of Paul Feyerabend (Brown 2009, 2016, 2020a). Given the contemporary fashion for pluralism, I think it may be useful to articulate explicitly a distinction between types of pluralism that has proven useful to me, and to try to situate my own thinking and the thinking of others within this distinction.

The first distinction I want to draw is between ontic and epistemic pluralism. Ontic pluralism insists that there are multiple things or multiple truths about things where we might naively expect only one. Anti-reductionism and ideas of "the disorder of things" and "the dappled world" are forms of ontic pluralism. Feyerabend's "abundant world" is a kind of ontic pluralism (and not, as some hold, a form of neo-Kantianism). The kind of pluralism I have been tempted to defend in metaethics, according to which there are many kinds of incommensurable, irreducible values, judgments about which we should treat as cognitive, is a kind of ontic pluralism. Epistemic pluralism is a pluralism about representations or methods of knowing, or both, rather than pluralism about the things represented or known. Ontic pluralists may have reason to be certain kinds of epistemic pluralist, and certain epistemic pluralisms may naturally suggest certain kinds of ontic pluralism, but there is some distance between the two.

Epistemic pluralism has many forms that are not always as clearly distinguished as they should be. First, there is *relativism*. "Relativism" is of course as vexed a term as "pluralism," but I will stipulate a use of the term that captures the distinction I hope to make. Relativism (in this sense) is the idea that representations or methodological strategies are indexed to frameworks (traditions, cultures, paradigms, research programs, worldviews, belief-systems, etc.) that are generally self-contained, with clear boundaries. It is a mistake to think there is any friction between claims or approaches from within distinct frameworks. This form of relativism is

[10] My gratitude to my friends Nicholas Gans and Daniel Hooper for help with this paragraph.

occasionally found in Feyerabend's middle period work, though even then, he is never fully committed to such a picture.[11] The problem with relativism is lack of critical interaction, and so loss of the epistemic value of pluralism.

Another form is *complementarity*, which I think is the kind of pluralism seen in many versions of perspectivism. The "levels of analysis" pluralism criticized by Sandra Mitchell (2002) is another form. Unlike relativism, which resolves competition through indexing to frameworks of some kind, complementarity resolves it through partitioning our space of analysis in some way. Different kinds of explanations in biology might be appropriate at the level of development or at the level of evolution, but not both. Another familiar example from philosophy of biology is species-concept pluralism, where different definitions of "species" are appropriate to different kinds of organisms or different kinds of explanatory projects in biology. Yet another example is Bohr's theory of complementarity according to which different kinds of questions or experimental setups might call for different, apparently contradictory explanatory models that cannot be reconciled. The benefit of complementarity over relativism is that the plural things are seen as mutually necessary to a more complete understanding; not only should we tolerate them, but we should also actively accept and engage with them all as pieces of a fuller picture, however fragmented.

Another important form of pluralism is what I call *agonistic pluralism*, with a nod to Chantal Mouffe. This is the kind of pluralism I most strongly associate with Feyerabend and later Longino. This is also the kind of pluralism found in Kathleen Okruhlik's reading of Mill, who in her view advocates the promotion of cognitive diversity for the purposes of expanding criticism, which better serves the public interest. Okruhlik rightly challenges Feyerabend's reading of Mill's political philosophy, but to my mind, the key question here is whether (for Mill or Feyerabend or in the best view) the pluralism is permanent or transitory. I regret that I'll never be able to discuss this further with Kathleen. I take it this is also Shaw's version of Mill and Feyerabend in the ethical sphere. This pluralism must be agonistic in order to genuinely test alternative moral claims. But aspects of Shaw's argument seem in danger of falling into relativism, if we take the harm principle to give us not only the only grounds for holding ways of life to be impermissible (coercively), but rather the only grounds on which to *criticize* ways of life. And here again, the question of whether the pluralism is permanent or whether certain ways of life become discredited going forward is key. In my view, the existence of very strong criticisms is never in itself sufficient to rationally require giving up a criticized view, but agonistic pluralism should require some form of honesty about the disadvantages of the view thus identified, and at least an earnest attempt at uptake of criticism, whether

[11] On Tsou's claim that Feyerabend's "relativist" or "pluralism in society" middle period begins earlier than Ian Kidd and I claimed (Brown and Kidd 2016), I agree, but on his claim that such a view is already in place by 1970, I doubt. I would revise the start of the middle period to 1975 (with "How to Defend..." and the later chapters of *Against Method*), while acknowledging that "Experts in a Free Society" contains some precursors of the ideas that are genuinely developed in 1975. I will address this question further in my contribution to *Against Method at 50*.

or not the critic is satisfied that their concerns have been addressed. But this is a complex issue, and I am uncertain of how to resolve it.

I also think that Feyerabend's focus on agonistic pluralism explains a concern I have about Tsou's interpretation of Feyerabend's "pluralistic realism." Tsou reads Feyerabend's pluralistic realism as consistent with his relativism. But I would argue that the main purpose for Feyerabend's pluralistic realism is that it maximizes the grounds of testability and criticism, which requires *agonism* rather than *relativism*, as I understand it.

Finally, there is *integrative pluralism*, an idea pursued by Sandra Mitchell (2002). There is something paradoxical in Mitchell's combination of integration and pluralism, resolved because the pluralism and integration happen at different levels. At the level of theories and models, pluralism reigns, and as in complementarity, having more is better (within constraints of quality). But at the level of "actual, concrete phenomena" (Mitchell 2002), we want to integrate the information provided by multiple models to provide a richer causal story. Of course, this raises the question about whether integrative pluralism can retain the benefits of agonism over relativism and complementarity. But it also seems very relevant to the political space as well, where we surely want to encourage a diversity of values and value-systems, ways of life, experiments in living, etc., but where we also in a democracy need some ways of generating unity at the point of action that is responsive to the plurality of values and forms of life in our society. My teacher Bryan Norton (1991) is particularly good on this point. Norton's form of pluralism was one major source of inspiration for my moral imagination framework as well.

I think many pluralists in different areas of philosophy would benefit from carefully distinguishing these approaches to pluralism and being clear about which type applies to them. For example, I found Knox's approach to autism and neurodiversity both extremely interesting and largely satisfying on the specifics, as well as edifying as a detailed application of my own views. Still, I believe it would help to think through what sort of pluralism the "hermeneutic pluralism" framework embraces. At the beginning and end of the chapter, it seems like they are suggesting a kind of permanent agonistic pluralism; but I think my moral imagination framework is more a form of integrative pluralism, and their use of the framework bears that out, as some form of integration is needed to generate a specific protocol for the *Spectrum 10k* study. Dewey's pluralism has elements of complementarist, agonistic, and integrative pluralism, and I hope in the future to be able to sort out these elements and give a coherent picture of his complex pluralism.

18.2.3 Pragmatism

In my mind, pragmatism is the philosophical framework in which all of this comes together; I have already referenced it repeatedly above. Now is a good time to be a pragmatist. Philosophical pragmatism has been enjoying something of a renaissance, lately, particularly in philosophy of science.[12] The trend is growing but not entirely new: Hilary Putnam began his turn to pragmatism more than 40 years ago, Peter Godfrey-Smith and Philip Kitcher began to engage seriously with John Dewey's thought about 20 years ago, and a few years later, Cheryl Misak announced the arrival of the *New Pragmatists* (Misak 2007), among their number Ian Hacking (who disputed the label) and Arthur Fine. The last 10 years have seen an explosion of work in pragmatist philosophy of science (Barrotta 2018; Brown 2020b; Gronda 2020; Chang 2022; Andersen and Mitchell 2023; Gronda et al. forthcoming).

Lately I find myself wondering, what does *pragmatism* amount to in the philosophy of science? What makes a philosophy of science *pragmatist*? These are vexed questions, and we ought to acknowledge from the get-go that there is probably no entirely univocal of "pragmatism," even among card-carrying pragmatists. William James and Charles Peirce both understood pragmatism as a view about the clarification of the meaning of ideas and the significance of disputes: the meaning of an idea is the practical consequences for conduct or experience we take it to have, and there is a genuine dispute if and only if the different views have different consequences (otherwise there is a merely verbal difference). Both considered "truth" to be a key idea for clarification, though they articulated their accounts very differently. Dewey, by contrast, takes pragmatism to be the view that all judgments are practical judgments (in the of practical reasoning), that "all judgments of fact have reference to a determination of courses of action to be tried and to the discovery of means for their realization" (Dewey 1915, MW 8: p. 22). In my own work, I follow Dewey in thinking of pragmatism as primarily a thesis about inquiry and judgment as practical, and I take it to be a normative stance (Brown 2020b, pp. 11–12, 41–43, 238). In his recent book, Chang (2022) characterizes pragmatism in terms of knowledge and action: "Pragmatism regards knowledge as an outcome of humble on-the-ground inquiry, and locates it in actual intelligent activities we carry out in life" (p. 5), and "Pragmatist philosophers have clearly recognized the need to understand and assess knowledge in the context of action" (p. 19). Chang specifically distances himself from versions of pragmatism concerned primarily with the analysis of meaning, in favor of pragmatism as "a philosophy that helps us think better about how to do things" (p. 59), a sentiment I very much like.

David Danks offers a helpful definition: "a pragmatist approach to, or theory of, X is one that focuses on the practical functions and impacts of X, as well as the functional needs of users of X" (Andersen and Mitchell 2023, p. 160). Danks also provides a helpful distinction between "philosophy of pragmatist science"—where the scientists themselves are described as taking a pragmatist approach towards their subject-matter—and "pragmatist philosophy of science"—where the philosopher takes a pragmatist approach to science (pp. 161–162). This permits the possibilities of "realist philosophy of pragmatist science" (which Danks sees most

---

[12] The first few paragraphs of this section are adapted from parts of, "What Makes a Philosophy of Science *Pragmatist*? A review of H.K. Andersen and Sandra D. Mitchell (eds), *The Pragmatist Challenge: Pragmatist Metaphysics for Philosophy of Science* (2023)," *BJPS Review of Books* (forthcoming).

avowed contemporary pragmatist philosophers of science doing) and "pragmatist philosophy of realist science" (which Danks things characterizes much of the work on methodology despite not being typically called "pragmatist"), as well as "pragmatist philosophy of pragmatist science." Danks claims that there have been few examples of the latter (p. 177), but I would describe my own work (Brown 2020b) this way, and I think Chang (2022) would as well.

Most interesting in the new pragmatist philosophy of science, to someone who has been tracking these developments for a long time, is the new confidence in pragmatist ideas that had previously been considered disreputable. Earlier appropriations of pragmatism tended to accept broadly epistemological and metaphilosophical insights, while taking a critical approach to pragmatist metaphysical and semantic views on truth, meaning, and reality. Hasok Chang (2022), in contrast, rehabilitates recognizably pragmatist accounts of truth and reality, clarifying them by reference to ideas of "operational coherence" and "ontological pluralism" that he has defended previously. Andersen seeks to rehabilitate a Jamesian pragmatist theory of truth as *trueing*—by analogy to the process of truing a bicycle wheel (Andersen and Mitchell 2023, Ch 3). Truth becomes not a static or abstract property of propositions, but a process of bringing theories, models, maps, diagrams, etc. into alignment with the world in such a way as to permit us to successfully navigate it. This creative approach captures many of James's key claims about truth: that truth is temporally indexed, that truth is something that happens to an idea, that truths help us get on in the world. Mitchell provides a metaphysics of *affordances* in the of Gibson as the target of scientific realistic commitment (Andersen and Mitchell 2023, Ch 4). Although she does not mention Dewey, Mitchell's approach bears striking resemblance to Dewey's so-called "transactionalism," where he takes organism-environment complexes as the starting point of his philosophical analyses.

In my own work, what I have taken from pragmatism includes a process-oriented, contextualist theory of inquiry (Brown 2012, 2015), an approach to philosophy that emphasizes its social and human relevance as a measure of its adequacy (Brown 2010, 2021a), opposition to absolutism and invidious dualisms (Brown 2013, 2018, 2017), a deep kind of pluralism that avoids relativism (Brown 2009, 2020a), and a commitment to deep intertwining of fact and value, descriptive and normative (Brown 2020b, 2021b).

I worry that this latter point has not received enough attention in the contemporary resurgence of pragmatism. While issues of science and social values are mentioned several times by Chang (2022) and the authors in Andersen and Mitchell (2023), they never explicitly take up this theme. It is acknowledged only to be bracketed or left for future work. Earlier "revivals" of the pragmatist tradition have often erred by focusing exclusively on the epistemological and metaphysical, to the exclusion of the ethical and political. If we take Danks's definition of pragmatism above, the "practical functions and impacts of" science seem obviously to include ethical and political functions and impacts, and users of science seem desperately in need of analyses of these aspects of science in order to trust and use science. Some of the most valuable resources in pragmatism are in the area of value theory, and they can help us treat some of these pressing, socially relevant problems of science and values, the role of experts in democratic society, and public trust of scientists and scientific knowledge. My own work of course has focused on these topics, as has that of Barrotta (2018), and one hopes that these disparate pragmatisms in philosophy of science will become more engaged with each other. I am thus happy that related issues are taken up seriously by Magnus and Howatt in this volume, and I think it would be helpful to look earlier than Franco does as we think about science and values, particularly at the pragmatists who were doing a kind of metaethics while being grounded in philosophy of science.

Magnus raises a question about pragmatist metaethics and individual existential choices. Such choices involve major decisions that determine what kind of life one will lead or what sort of person one will become, for example, deciding whether to accept a job offer in a different city, or what sort of career to pursue. Following Putnam, Magnus worries that my view, like Dewey's, breaks down in the face of such choices. These are different from questions about which policy is best, which option is best supported by various values; concerning existential choices, stakeholder interests and perspectives seem irrelevant. So, Magnus says, the ideal of moral imagination is the wrong approach here; it is too demanding. Magnus proposes three possible answers, each of which comes with its own poison pill: acknowledge an exception to the IMI and try to draw a boundary, bite the bullet on the IMI and acknowledge its unintuitive consequences, and distinguish between the IMI as standard of rightness and as decision-procedure.

I am not sure the IMI is unable to address such cases without leading to unintuitive consequences. It seems to me that there are three characteristics of these sorts of existential choices. First, the relevance of other stakeholders is notably lower, but I reject the idea that it is nonexistent, for we all consist at least in part of a network of relationships to others. In Sartre's original example, the interests of the young man's mother, on the one hand, and of those who he might help by fighting, on the other, are certainly relevant to his decision. In choosing to take a new job, I had to think about my family, about the relationships I was leaving behind, with all the complicated obligations involved, and about the good I could do not just for myself but for others in my new position. This does not mean that these stakeholders outweigh other, more personal considerations, or make the decision for me. After all, it is I, not they, who must live the new life I choose. Second, the evidence one has to go on seems particularly scant in such cases. In what Magnus calls "doxastic efficacy" cases, there are a lot of things I cannot know until *after* I have committed to a decision; I do not know what it is like to choose to be this person rather than that person until after the fact. Third, existential choices seem to be undertaken in a state of (near) equipoise relative to all of the factors I *can* muster ahead of time. They are difficult decisions, and the evidence, values, and interests available to me may not be decisive. Even dramatic rehearsal of the different options may not settle my mind. I must choose; I must commit. The moral imagination framework offers only to help make decisions easier, more fair and warranted; it does not hold out the false promise of many moral theories to make them rationally decidable.

Thinking about these various ideas about what a pragmatist philosophy of science amounts to, brings me to Howard's argument that Quine is the natural inheritor of Dewey's pragmatist philosophy of science. In my misspent youth,[13] I thought it crucially important to distinguish Quine's philosophy of science from Dewey's, and to question even whether Quine should be called a "pragmatist." I no longer find this a particularly important question, as "pragmatism" has been, ever since William James introduced the word into the published record, an essentially contested concept. Two more interesting questions are whether Dewey influenced Quine, and whether there are interesting qualitative similarities between their views irrespective of historical influence. On the first point, as Quine himself said, "I am not much of a Dewey scholar."[14] Quine's references to Dewey are suggestive but remain brief and non-substantive, in one case a polite nod to the namesake of a lecture series, and another an invited paper for a 1975 conference on pragmatism.

Once the question of historical influence is off the table, the question of similarities remains relevant, as one could potentially say that Quine picked up the mantle of a Deweyan tradition in philosophy of science on that basis without asserting direct influence. Here, one can find many similarities and also many differences between the two, and how one weighs the similarities and differences will depend almost entirely on one's interests. (Perhaps this is true of any two sufficiently nuanced philosophers writing on the same topics?) For me, Dewey's theory of inquiry oriented towards scientific practice rather than abstract epistemology, his emphasis on the interplay of science, values, and society, his complex ecological and sociocultural psychology, his recognition of the inextricability of philosophy and politics, and his richer naturalistic metaphysics are all strong reasons to prefer Dewey's own work over Quine's pale shadow. For Don, Quine's naturalism, his attack on neo-positivism and traditional analytic philosophy, and his behaviorism are important continuities that outweigh the differences. This is a matter of preference and emphasis rather than a substantive disagreement over interpretation, as far as I can see.[15] I would personally rather start by going back to Dewey's rich and under-explored writings on science.

Lately I am trying to go back to Dewey not only for value theory but for his theory of democracy. I think we need to be thinking more democratically and our account of democracy needs to go deeper, and I see resources in Dewey and other pragmatists for precisely this. Again, Howatt is moving down a valuable line of inquiry in this respect.

---

[13] As a side note, it is strange to experience one's 15+ year old dissertation and pre-dissertation writings receiving critical attention from a leading light in one's field and something of a mentor!

[14] Hickman (2023), 1968.09.13 (16208): W. V. Quine to Pearl L. Weber.

[15] A similar point could be made about Howard's comment that Dewey's theory of science was displaced via a campaign by Ernest Nagel; while there are certainly important differences between Nagel and Dewey, a student of Dewey's philosophy of science cannot read Nagel's *Structure of Science* without constantly hearing Dewey's ideas and framings reappear. I am not inclined to call Nagel a Deweyan, either, but there is a *much* clearer case for both historical influence and qualitative similarities between Nagel and Dewey than with Quine. Also it is worth noting that Nagel almost certainly played some role in the invitation to Quine to be the first John Dewey lecturer at Columbia, the occasion for the aforementioned polite nod.

18.3 Where Do I Go from Here?

Here's where I think I'm headed next.

First, my new job at SIU has me leaning into my work on Dewey, pragmatism, and the history of philosophy. This is a welcome shift of emphasis, allowing me to pick up neglected projects and explore topics that bring me a lot of joy. Coinciding with this change, I have become more deeply involved with *HOPOS: The Journal of the International Society for the History of Philosophy of Science*, which I joined as an associated editor in 2023 and took over as editor-in-chief in 2025. I am excited to continue exploring the historical antecedents of our current predicaments as well as recovering neglected resources for both addressing current problems and overcoming them, making the field more vibrant and more socially relevant.

Second, this volume does *not* mark my retirement from philosophy of science in order to take up Dewey studies and HOPOS scholarship. It is truly a shift of emphasis. I continue to think about issues in science and values, and I am particularly excited at the moment to think further about the nature of expertise and the role of science in a democratic society. I also think we need more work on science and metaethics and to use insights from philosophy of science to improve value theory. I believe we need to be thinking more systematically about the nature of scientific practice and how to study it philosophically, and I believe I have more to say about the resources of pragmatism in this context. I want to use philosophy to fight both scientism and anti-science extremism. I am concerned about the way that science and contemporary philosophy have left us without a worldview for making sense of our lives, the inadequacy of materialism and standard scientific realism for filling that role, as well as the dangers of reactionary nostalgia for traditional, prescientific worldviews.

Also, I want to think more about democracy and inquiry from the political side. Philosophers of science who are inclined to think about the complexity of science, values, and politics tend to be concrete thinkers with little taste for so-called "ideal theory." But despite the excesses and problems of idealistic thinking in political philosophy, we *need* ideals, they do important guiding work in the critique of injustice and the urge to seek a better world. I want a better ideal of democracy in relation to science and inquiry. I also want to think about transitional forms, and about crises. We have faced crises: political, public health, and now, perhaps, climatological. More will likely come. Science and democracy have not always functioned well in these crises. How should they respond to crises? And what is the role of philosophers of science?

That's quite the bucket list of philosophical projects. But I am excited to get to work on at least some of them. I am happy, thanks to this volume, to know that I have many fellow-travelers to think about such topics with.

References

Alexandrova, A. 2018. Can the science of Well-being be objective? *The British Journal for the Philosophy of Science* 69 (2): 421–445.

Almassi, B. 2012. Climate change, epistemic trust, and expert trustworthiness. *Ethics and the Environment* 17 (2): 29–49.

Andersen, H., and S. Mitchell. 2023. *The pragmatist challenge: Pragmatist metaphysics for philosophy of science*. Oxford: Oxford University Press.

Barrotta, P. 2018. Scientists, democracy and society: A Community of Inquirers. In *Logic, argumentation, and reasoning*, vol. 16. Cham: Springer.

Biddle, J., and R. Kukla. 2017. The geography of epistemic risk. In *Exploring inductive risk: Case studies of values in science*, ed. K. Elliott and T. Richards, 215–238. Oxford: Oxford University Press.

Bright, L., and R. Heesen. 2023. To be scientific is to be communist. *Social Epistemology* 37 (3): 249–258.

Brown, M.J. 2009. Models and perspectives on stage: Remarks on Giere's scientific Perspectivism. *Studies in History and Philosophy of Science Part A* 40 (2): 213–220.

Brown, M..J. 2010. Genuine problems and the significance of science. *Contemporary Pragmatism* 7 (2): 131–153.

Brown, M.J. 2012. John Dewey's logic of science. *HOPOS: The Journal of the International Society for the History of Philosophy of Science* 2 (2): 258–306.

Brown, M.J. 2013. Values in science beyond underdetermination and inductive risk. *Philosophy of Science* 80 (5): 829–839.

Brown, M.J. 2015. The functional complexity of scientific evidence. *Metaphilosophy* 46 (1): 65–83.

Brown, M.J. 2016. The abundant world: Paul Feyerabend's metaphysics of science. *Reappraising Feyerabend, Special Issue of Studies in History and Philosophy of Science A* 57:142–154.

Brown, M.J. 2017. Against epistemic priority. In *Current controversies in values and science*, ed. K. Elliott and D. Steel, 64–78. New York: Routledge.

Brown, M.J. 2018. Is science really value free and objective? From objectivity to scientific integrity. In *What is scientific knowledge? An introduction to contemporary epistemology of science*. New York: Routledge.

Brown, M.J. 2020a. Pluralism and Perspectivism in the American pragmatist tradition. In *Knowledge from a human point of view*, Synthese library, studies in epistemology, logic, methodology, and philosophy of science, ed. A.-M. Creu and M. Massimi, vol. 416, 37–56. Springer.

Brown, M.J. 2020b. *Science and moral imagination: A new ideal for values in science*. Pittsburgh: University of Pittsburgh Press.

Brown, M.J. 2021a. Mary Parker Follett as integrative public philosopher. *Hypatia* 36 (2): 425–436.

Brown, M.J. 2021b. The descriptive, the normative, and the entanglement of values in science. In *Science, values, and democracy: The 2016 Descartes lectures by heather Douglas*, ed. T. Richards, 51–65. Consortium for Science, Policy & Outcomes, Arizona State University.

Brown, M. 2024. For values in science: Assessing recent arguments for the ideal of value-free science. *Synthese* 204:112.

Brown, M.J., and J. Havstad. 2017. The disconnect problem, scientific authority, and climate policy. *Perspectives on Science* 25 (1): 67–94.

Brown, M.J., and I. J. Kidd. 2016. Introduction: Reappraising Paul Feyerabend. *Reappraising Feyerabend, Special Issue of Studies in History and Philosophy of Science A* 57:1–8.

Brown, M.J., and J. Stegenga. 2023. The validity of the argument from inductive risk. *Canadian Journal of Philosophy* 53 (2): 187–190.

Chang, H. 2022. *Realism for realistic people: A new pragmatist philosophy of science*. Cambridge: Cambridge University Press.

Contessa, G. 2023. It takes a village to trust science: Towards a (thoroughly) social approach to public Trust in Science. *Erkenntnis* 88 (7): 2941–2966.

Dewey, J. 1915. The logic of judgments of practice. In *The middle works, 1899–1924*, ed. J. A. Boydston, vol. 8, 14–82. Carbondale: Southern Illinois University Press.
Douglas, H. 2000. Inductive risk and values in science. *Philosophy of Science* 67 (4): 559–579.
Douglas, H. 2009. *Science, policy, and the value-free ideal*. Pittsburgh: University of Pittsburgh Press.
Goldenberg, M. 2016. Public misunderstanding of science? Reframing the problem of vaccine hesitancy. *Perspectives on Science* 24 (5): 552–581.
Goldenberg, M. 2021. *Vaccine hesitancy: Public trust, expertise, and the war on science*. Pittsburgh: University of Pittsburgh Press.
Gronda, R. 2020. Dewey's philosophy of science. In *Synthese library: Studies in epistemology, logic, methodology, and philosophy of science*, ed. O. Bueno, B. Brogaard, A. Chakravartty, S. French, and C. D. Novaes, vol. 421. Cham: Springer.
Gronda, R., M. Janack, and G. Marchetti, eds. forthcoming. *Pragmatism and philosophy of science: Historical and contemporary perspectives. Routledge studies in the philosophy of science*. Routledge.
Harvard, S., and E. Winsberg. 2022. The epistemic risk in representation. *Kennedy Institute of Ethics Journal* 32 (1): 1–31.
Havstad, J. 2022. Sensational science, archaic Hominin genetics, and amplified inductive risk. *Canadian Journal of Philosophy* 52 (3): 295–320.
Hickman, L., ed. 2023. *The correspondence of John Dewey, 1858–2011, comprehensive edition*. Charlottesville: InteLex Corporation.
John, S. 2018. Epistemic trust and the ethics of science communication: Against transparency, openness, sincerity and honesty. *Social Epistemology* 32 (2): 75–87.
Kitcher, P. 2001. *Science, truth, and democracy*. New York: Oxford University Press.
Melo-Martín, I., and K. Intemann. 2018. *The fight against doubt: How to bridge the gap between scientists and the public*. Oxford: Oxford University Press.
Menon, T., and J. Stegenga. 2023. Sisyphean science: Why value freedom is worth pursuing. *European Journal for Philosophy of Science* 13 (4): 48.
Misak, C., ed. 2007. *New pragmatists*. Oxford: Oxford University Press.
Mitchell, S. 2002. Integrative Pluralism. *Biology and Philosophy* 17 (1): 55–70.
Norton, B. 1991. *Toward Unity among environmentalists*. New York: Oxford University Press.
Oreskes, N. 2019. *Why trust science?* Princeton: Princeton University Press.
Pamuk, Z. 2021. *Politics and expertise: How to use science in a democratic society*. Princeton: Princeton University Press.
Stegenga, J., and T. Menon. 2023. The difference-to-inference model for values in science. *Res Philosophica* 100 (4): 423–447.

Index

J. Y. Tsou et al. (eds.), *Values, Pluralism, and Pragmatism: Themes from the Work of Matthew J. Brown*, Boston Studies in the Philosophy and History of Science 347, https://doi.org/10.1007/978-3-031-92958-8

The manufacturer's authorised representative in the EU is Springer Nature Customer Service Centre GmbH, Europaplatz 3, 69115 Heidelberg, Germany. If you have any concerns regarding our products, please contact ProductSafety@springernature.com

Printed and bound by CPI Group (UK) Ltd, Croydon, CR0 4YY
01/10/2026
02225966-0014